Brock/Springer Series in Contemporary Bioscience

The Sulfate-Reducing Bacteria: Contemporary Perspectives

Brock/Springer Series in Contemporary Bioscience

Series Editor: *Thomas D. Brock*
University of Wisconsin-Madison

Tom Fenchel
ECOLOGY OF PROTOZOA:
The Biology of Free-living Phagotrophic Protists

Johanna Döbereiner and Fábio O. Pedrosa
NITROGEN-FIXING BACTERIA IN NONLEGUMINOUS
CROP PLANTS

Tsutomu Hattori
THE VIABLE COUNT: Quantitative and Environmental Aspects

Roman Saliwanchik
PROTECTING BIOTECHNOLOGY INVENTIONS:
A Guide for Scientists

Hans G. Schlegel and Botho Bowien (Editors)
AUTOTROPHIC BACTERIA

Barbara Javor
HYPERSALINE ENVIRONMENTS: Microbiology and Biogeochemistry

Ulrich Sommer (Editor)
PLANKTON ECOLOGY: Succession in Plankton Communities

Stephen R. Rayburn
THE FOUNDATIONS OF LABORATORY SAFETY:
A Guide for the Biomedical Laboratory

Gordon A. McFeters (Editor)
DRINKING WATER MICROBIOLOGY:
Progress and Recent Developments

Mary Helen Briscoe
A RESEARCHER'S GUIDE TO SCIENTIFIC AND
MEDICAL ILLUSTRATIONS

Max M. Tilzer and Colette Serruya (Editors)
LARGE LAKES: Ecological Structure and Function

Jürgen Overbeck and Ryszard J. Chróst (Editors)
AQUATIC MICROBIAL ECOLOGY:
Biochemical and Molecular Approaches

(Continued on page 290)

J.M. Odom Rivers Singleton, Jr.
Editors

The Sulfate-Reducing Bacteria: Contemporary Perspectives

Foreword by John R. Postgate

With 52 Figures

Springer-Verlag
New York Berlin Heidelberg London Paris
Tokyo Hong Kong Barcelona Budapest

J.M. Odom
Central Research and
Development Dept.
E.I. DuPont de Nemours and Co.
Wilmington, DE 19880
USA

Rivers Singleton, Jr.
School of Life and Health Sciences
University of Delaware
Newark, DE 19716
USA

Cover illustration: Sulfate-reducing bacteria are an integral part of the global sulfur cycle, which occurs on a small scale in sulfureta as illustrated on the cover. See p. 18. (Cover art by Margie L. Barrett, University of Delaware.)

Library of Congress Cataloging-in-Publication Data
The Sulfate-reducing bacteria: contemporary perspectives / [edited by]
J.M. Odom and Rivers Singleton, Jr.
p. cm. – (Brock/Springer series in contemporary bioscience)
Includes bibliographical references and index.
ISBN-13: 978-1-4613-9265-1 e-ISBN-13: 978-1-4613-9263-7
DOI: 10.1007/978-1-4613-9263-7
1. Sulphur bacteria. I. Odom, J.M. (J. Martin) II. Singleton, Rivers. III. Series.
[DNLM: 1. Bacteria, Anaerobic. 2. Environmental Pollution – prevention & control. 3. Oxidation-Reduction. 4. Sulfates. QW 75 S949]
QR92.S8S85 1992
589.9'6 – dc20
DNLM/DLC 92-2311

Printed on acid-free paper.

Production managed by Natalie Johnson; manufacturing supervised by Jacqui Ashri.
Typeset by Asco Trade Typesetting, Ltd., Hong Kong.

9 8 7 6 5 4 3 2 1

Dedication

The gifts of John Postgate, FRS, to the field of microbiology are legion and inspirational. Those of us who struggle to plow our narrow fields of concentration on small parcels of knowledge must stand in awe of a man who could write definitive monographs on diverse groups of organisms such as sulfate-reducing (Postgate, 1979 and 1984a) and nitrogen-fixing bacteria (Postgate, 1982a), a lay person's guide to microbiology (Postgate, 1992), and during the same career write an introductory book on jazz (Postgate, 1973).

Postgate's contributions to our understanding of the sulfate-reducing bacteria have been multifaceted. First, by developing methods to reliably and reproducibly manipulate these bacteria in the laboratory, he helped make them suitable subjects for rigorous microbial investigation. The explosion of new and diverse sulfate reducers discovered by Widdel, Pfennig, and others [see Widdel (1988) for a description of this diversity] is a consequence of Postgate's initial methodological work. For many years, his laboratory was one of the very few sources of these bacteria in pure culture.

Second, along with the classical work of Ishimoto, Koyama, and Nagai (1954), he demonstrated (Postgate, 1954, 1956) the presence of cytochromes in the obligative anaerobic sulfate reducers. [For a commentary on the development of this work see Postgate (1986).] His demonstration of cytochrome c_3 in *Desulfovibrio* opened vast new fields of different and exciting biochemical processes and proteins for study. This observation also required a revision of contemporary dogma and led to a profound rethinking of our notions of anaerobic metabolism. One example of this new way of looking at anaerobic physiology was Peck's demonstration that sulfate-reducing bacteria, like their aerobic counterparts, were capable of carrying out electron transport-coupled oxidative phosphorylation (Peck, 1960).

Finally, Postgate's many reviews of every aspect of these organisms, along with his classic monographs (Postgate, 1979 and 1984a), have shaped and helped the work of numerous investigators, including the contributors to this volume.

The Preface to the first edition of Postgate's *The Sulphate-Reducing Bacteria* closed with these words: "Microbiology is a science, but a touch of art and craft is always desirable, even essential, for progress to be made." Throughout his career, John Postgate has contributed greatly to the scientific understanding of sulfate reduction microbiology. But in the process of that science, he has illuminated the art and helped to define the craft of working in the microbial world. It is for these gifts that each of the contributors is pleased to dedicate this volume.

J.M. Odom
Rivers Singleton, Jr.

[References, see p. 211]

Foreword

John R. Postgate

The year 1993 is the centenary of the recognition of the sulfate-reducing bacteria as a distinctive group of microbes. The fact that Beijerinck, who did the recognizing, did not publish his seminal account of them (Beijerinck, 1895) for another two years merely reflects a somewhat more leisurely approach to publication in those days. Indeed, the whole tone of "*Spirillum desulfuricans* as the cause of sulfate reduction" carries pleasing echoes of a scientific world long vanished. Even the fact that his paper came out in three installments of the *Zentralblatt*, like a magazine serial, seems curious today. It carried no summary, and in it he referred to his "sulfide ferment" until, towards the end of installment 3, he could reveal that it was his new *Spirillum*: undoubtedly today's *Desulfovibrio*. He described his procedures for setting up enrichment cultures in meticulous detail and presented, with illustrations, two devices for separating aerobes and anaerobes, remarking that one was "quite spectacular" when used with a fluorescent pseudomonad. Neither device, it seems, yielded pure cultures of his sulfide ferment, and their bearing on the subject of his paper was distinctly peripheral. But he appeared to be almost as interested in the commensals in his cultures as in his sulfate reducers, and he dedicated considerable space to *Spirillum tenue*, which, he found, stimulated the sulfate reduction. In footnotes he digressed into such matters as a "beautiful demonstration on the isolation of nitrate-reducing bacteria" and the fact that he was really seeking a way of obtaining sulfate-free boiler water from the Dutch canals. At one stage he told the reader that he thought he had come upon a microbiological assay for sulfate, a possibility unhappily negated by "variations . . . for reasons not yet clear" (those with experience of sulfate reducers will sympathize).

His discourse (available in English translation by Doetsch, 1960) was an editor's nightmare by today's standards, inadequately referenced and discursive, but it was also clear and scientifically rigid, presenting many observations that have since proved to be fundamentally important in the study of these organisms. It is somewhat chastening to consider how many central themes of later research he adumbrated.

He remarked upon the pleomorphism of his new spirilla, which has been a perennial cause of alarm (or, conversely, overconfidence) among newcomers to the field. He noticed the inhibitory effects of H_2S on sulfate reduction, which, often forgotten, later led to problems with mass culture and ambiguities in growth yield experiments. He observed that acetate, butyrate, and formate were poor substrates for his ferment, thus signposting the convoluted path to the discovery of the slow-growing, acetate-utilizing genera of sulfate reducer in the 1980s; he also found that carbohydrates were poor substrates, which he attributed to the acidity generated by their anaerobic metabolism. He remarked, albeit obliquely, upon the inhibition of methanogenesis by sulfate reduction, a phenomenon now recognized to be central to anaerobic microbial ecology. He realized, rightly, that pure cultures would be very difficult to handle without contaminants to scavenge residual oxygen, and he speculated whether a chemical reductant (he suggested $FeSO_4$) might have to be used as "perhaps a last resort," thus presaging the use of reduced media for counting these bacteria (but he apparently foresaw no serious problem in obtaining pure cultures). He discussed the thought that some cycling of "nascent hydrogen" might be involved in sulfate reduction and dismissed it ("It is certain that with the sulfide ferment described here no hydrogen is produced")—well, he could not win them all! In his final paragraphs, Beijerinck asked what substrates other than sulfate they could reduce? How many species might there be? How might they be distributed in soils, the sea, and fresh water?

These observations and questions, some presented almost as asides, echoed down the twentieth century. Their resolution took many decades because, though he was quite right about the difficulty of handling pure cultures, he was over-optimistic about the readiness with which they would be obtained. Understanding of the chemical physiology and bacteriology of desulfovibrios progressed only slowly for a few decades, and there were false leads and confusions until the 1950s, when a phase of advance set in. As the decades progressed, the fascinating biochemistry and physiology of the desulfovibrios, with the desulfotomacula close behind, became revealed. In due course the discovery of many new genera, with a diversity of physiological and morphological types, widened our microbiological horizons; at the same time, systematic approaches to the ecology of sulfate reducers became possible, and the organisms became part of the mainstream of microbiological literature, regularly reviewed, and the subject of a book to themselves—which almost immediately had to be updated.

Now, in the early 1990s, the growth of knowledge of sulfate-reducing bacteria is still in its exponential phase. Molecular genetics has become feasible with these organisms, despite continuing problems with applying traditional genetical techniques to this group, and is opening new vistas. The taxonomic revolution engendered by r-RNA comparisons has been brought to bear upon them, revealing unexpected relationships. And more traditional biochemical and physiological studies have continued apace. Once more it is time for a book-length survey of the state of the art, and the present team, all active contributors to our newer knowledge, provides a valuable and necessary platform for even more spectacular advances as the twenty-first century approaches.

John R. Postgate

[References, see p. 211]

Contents

Contributors

Richard Devereux United States Environmental Protection Agency, Sabine Island, Gulf Breeze, FL 32561, USA

Theo A. Hansen Department of Microbiology, University of Groningen, 9751 NN Haren, The Netherlands

J.M. Odom Central Research and Development Department, E.I. DuPont de Nemours and Co., Wilmington, DE 19880, USA

Harry D. Peck, Jr. Department of Biochemistry, Life Science Building, University of Georgia, Athens, GA 30602, USA

John R. Postgate, FRS Professor Emeritus, University of Sussex, Brighton, BN1 9RQ, England

Rivers Singleton, Jr. School of Life & Health Sciences, University of Delaware, Newark, DE 19716, USA

David W. Smith School of Life & Health Sciences, University of Delaware, Newark, DE 19716, USA

David A. Stahl Departments of Veterinary Pathobiology, Microbiology, and Civil Engineering, University of Illinois, 2001 So. Lincoln Ave., Urbana, IL 61801, USA

Gerrit Voordouw Division of Biochemistry, Department of Biological Sciences, The University of Calgary, Calgary, Alberta, T2N 1N4, Canada

Judy D. Wall Department of Biochemistry, University of Missouri, Columbia, MO 65211, USA

Introduction

J.M. Odom and Rivers Singleton, Jr.

Sulfate-reducing bacteria are a distinctive and ubiquitous group of anaerobic prokaryotes. They are unified by a shared ability to carry out sulfate reduction as a principal component of their bioenergetic processes. All these organisms use sulfate or sulfur (and on occasion, other sulfur oxyanions) as a terminal electron acceptor to oxidize both organic and inorganic compounds. Despite this seeming physiological unity, work in recent years has demonstrated a tremendous morphological, ecological, nutritional, and metabolic diversity among this group of bacteria.

Sulfate reducers are found in diverse environments and are of great utilitarian and academic interest. Their practical importance arises from both economic and environmental concerns. Sulfate-reducing bacteria are of significant economic interest in many industrial sectors because of their role in contamination of petroleum products and anaerobic corrosion of steel. Postgate (1984a) referred to sulfate-reducing bacteria as "the penultimate stage of a grossly polluted environment"; they are thus also of great pragmatic interest because of their environmental impact.

Their academic importance arises from the unique ecological and evolutionary roles they play. Sulfate reducers are important organisms in many anaerobic environments in nature, and their physiological activities are of profound importance for many ecological communities. Furthermore, there is increasing biochemical and genetic evidence suggesting that these organisms began evolutionary divergence at an early time. This long evolutionary history has led to development in these organisms of a variety of unique proteins and biochemical processes with profound academic importance and fundamental insights.

For much of their history, as subjects of scientific study, these bacteria were difficult to grow and manipulate in the laboratory. Media were often poorly defined and difficult to reproduce. Techniques for maintenance of anaerobiosis were often primitive and difficult to achieve. In recent years, however, methods have been developed that facilitate growth and manipulation of sulfate reducers under defined and controlled conditions. These new methods have demonstrated a richness of metabolism and diversity of form not previously apparent in this group of organisms. The enhanced ability of laboratory manipulation of sulfate-reducing bacteria is leading to new advances in our understanding of their genetics, molecular biology, diversity, metabolism, and environmental and industrial importance.

For all of these reasons, interest in sulfate-reducing bacteria has increased greatly during the past decade. In 1979 and 1984, John Postgate published the first and second editions of *The Sulphate-Reducing Bacteria*. In the Preface to the second edition, he noted:

> There is an element of irony in the fact that, even as I was writing the first edition of this book, knowledge of the sulphate-reducing bacteria was undergoing a revolution.

Today, the revolution Postgate pointed to in 1984 was a mere spark that has assumed explosive dimensions, and the sulfate-reducing bacteria have become subjects of study from a variety of perspectives and disciplines. The literature seems so explosive since 1984 that it most likely is impossible for a single person to master. The present book is an attempt to describe some of the revolutionary aspects of our understanding of this important and unique group of prokaryotic organisms.

Publication of *The Sulfate-Reducing Bacteria: Contemporary Perspectives* was prompted by the explosive nature of this revolution and our realization that many of the traditional beliefs about these organisms are radically evolving. This evolution is due to two occurrences. First, the work of Widdel, Pfennig, and their colleagues, describing numerous new genera of sulfate reducers, demonstrated the taxonomic and ecological diversity of the sulfate reduction process. In his Foreword to this volume, Professor Postgate points out that Beijerinck's initial "observations and questions" about the sulfate reducers "echoed down the twentieth century." The new "observations and questions" of Widdel and Pfennig will undoubtedly echo down the remainder of the twentieth century and much of the twenty-first as well.

Second, though research on sulfate reduction has traditionally been a bastion of classical microbiology and enzymology, this has not deterred development of various molecular biology approaches by a few brave souls. This second evolutionary factor is just beginning to bear fruit and may ultimately resolve long-standing questions such as the nature of the intermediates in sulfate reduction, the physiological roles of various molecular forms of hydrogenase, the roles of many novel electron transfer proteins, and the place where these organisms fit into modern bacterial phylogeny and evolution.

We hope this book will have appeal for both novice and specialist. For both readers, we have two major goals. One is to demonstrate how our understanding of this important group of bacteria has been clarified by new advances of bioenergetics, genetics, and molecular biology. However, it is a truism of any truly exciting scientific investigation that as answers are discovered, new questions emerge. Thus, we hope the book also serves to illustrate new questions raised by those advances.

We close this preface with a few comments about terminology. The major growth product common for sulfate-reducing bacteria is bisulfide anion (HS^-). Consequently, there is a justified tendency among some workers to label them as *sulfidogenic bacteria* (Lupton et al., 1984b). There is merit in this tendency because it focuses attention on the common product of both sulfate and sulfur metabolism as new species of organisms are discovered which can reduce sulfur as well as sulfate. There is also a harmonious sense of nomenclature consistency with other organisms such as methanogenic or acetogenic bacteria. There are, however, difficulties with this label, as noted by LeGall and Fauque (1988). Many organisms produce bisulfide as a normal consequence of their metabolism, yet are not considered to be dissimilatory sulfate- or sulfur-reducing bacteria. It is this dissimilatory use of sulfate that distinguishes this group of bacteria from other members of the prokaryotic world. Thus, in this text we have chosen to adopt this view and refer to these organisms as *sulfate-reducing bacteria*.

Second, even during the period when only two genera of sulfate reducers were recognized, *Desulfovibrio* and *Desulfotomaculum*, both were universally abbreviated as *D*. In a complex text dealing with organisms in both genera, the reader was often in a state of bewilderment trying to decide which were the spore-formers (*Desulfotomaculum*) and which had cytochrome c_3 (*Desulfovibrio*). The diversity and taxonomy of sulfate-reducing bacteria has grown more complex over the years. Yet most new genera are named with the prefix *Desulfo-*. Thus, the standard method of abbreviating names of genera can lead to confusion when one has almost a dozen genera identified with the appellation *D*. To avoid this confusion, we have developed the following abbreviations for the genera of sulfate-reducing bacteria discussed in this volume:

Dba. = *Desulfobacter*
Dbt. = *Desulfobacterium*
Dbu. = *Desulfobulbus*
Dc. = *Desulfococcus*
Dmi. = *Desulfomicrobium*
Dmo. = *Desulfomonile*
Dmn. = *Desulfomonas*
Dn. = *Desulfonema*
Ds. = *Desulfosarcina*
Dtm. = *Desulfotomaculum*
Dv. = *Desulfovibrio*

J.M. Odom
Rivers Singleton, Jr.

[References, see p. 211]

1

The Sulfate-Reducing Bacteria: An Overview

Rivers Singleton, Jr.

1.1 Introduction

The intent of this chapter is to review the properties of the sulfate-reducing (occasionally referred to as sulfidogenic) bacteria in general terms to provide a broad framework for the remaining chapters of the book. The chapter is to provide a bridge between material familiar to the general microbiologist reader and the specialized material that makes up the remainder of the book. A major notion, which provides the general structure of the chapter, is that while these organisms seem to be unified by a common metabolic ability, they are in reality an extremely heterogeneous and diverse group of organisms.

The chapter has a two-fold objective. One is to survey the biochemical, physiological, taxonomic, and morphological complexity of these organisms. The intent here is not to preempt material that will be developed elsewhere in the book, but rather to illustrate broad patterns of sulfur and carbon metabolism that will be expanded and completed in other chapters. The goal here is a holistic one, to develop a picture of the organism and its place in nature. Bioenergetics, molecular biology and taxonomy, and genetics—which are subjects of later chapters—make little sense without an understanding of the fundamental physiology of the organism. A second objective of the chapter is to briefly explore the ecological, evolutionary, and economic role of sulfate-reducing bacteria. Again, the intent is not preemption of other book chapters but rather to suggest a general rationale for why these organisms are important objects of scientific investigation.

The chapter will open with a brief review of general patterns of metabolism and demonstrate that dissimilatory sulfate reduction is a manifestation of anaerobic respiration. Metabolic patterns can be defined by reduction reactions which are utilized to dispose of electrons generated via oxidation reactions. This section of the chapter will close with a summary of pathways that link carbon (and hydrogen) oxidation and reduction of sulfate in sulfate-reducing bacteria.

The second section of the chapter will focus on the explosion of taxonomic complexity of these organisms. Initially, these organisms were grouped into two distinct genera. However, since this original proposal by Campbell and Postgate (Campbell and Postgate, 1965; Postgate and Campbell, 1966), numerous genera with differing morphological and metabolic characteristics have been discovered.

Finally, the last section of the chapter will be given to consideration of why these organisms are important objects for study. The major point to be made here is that these organisms are of great significance for evolutionary, environmental, ecological, and industrial reasons.

1.2 Metabolic Considerations

General patterns of metabolism Sulfate-reducing bacteria are a unique and ubiquitous group of prokaryotic microorganisms, found in a variety of environmental niches. They are unified around the chemistry of the reaction (Postgate, 1984a; Thauer and Badziong, 1980):

$$4\ AH_2 + SO_4^{2-} + H^+ \rightarrow 4\ A + HS^- + 4\ H_2O \qquad (1.1)$$

These organisms can use a variety of electron donors (AH_2) and can couple oxidation of those compounds to reduction of sulfate (as well as elemental sulfur). This process is referred to as *dissimilatory* sulfate reduction, as opposed to *assimilatory* reduction. Dissimilatory processes are bioenergetic, whereas assimilatory processes reduce sulfur compounds for incorporation into biological compounds (e.g., cysteine or coenzyme A). Postgate (1984a) clearly illustrated the difference in these terms. Consider that a milligram of sulfur provides a growth yield of around 200 mg of cells for *Klebsiella aerogenes* as opposed to a yield of 0.5 to 1 mg of cells for members of the genus *Desulfovibrio*. This 200-fold difference in yield demonstrates the minor role of sulfur in *Klebsiella* metabolism and its major role in *Desulfovibrio* metabolism.

Dissimilatory sulfate reduction is a classical example of what is called "anaerobic respiration." A brief digression is necessary to clarify this term for general readers. Consider for a moment that *all* organisms have a similar basic problem in their metabolism. Energy necessary for growth is best obtained by oxidation of various reduced compounds in the environment. The metabolic problem for the organism is disposal of protons and electrons (often referred to as *reducing equivalents*) generated by those oxidations. Three general patterns of disposal have evolved over time (Figure 1.1).

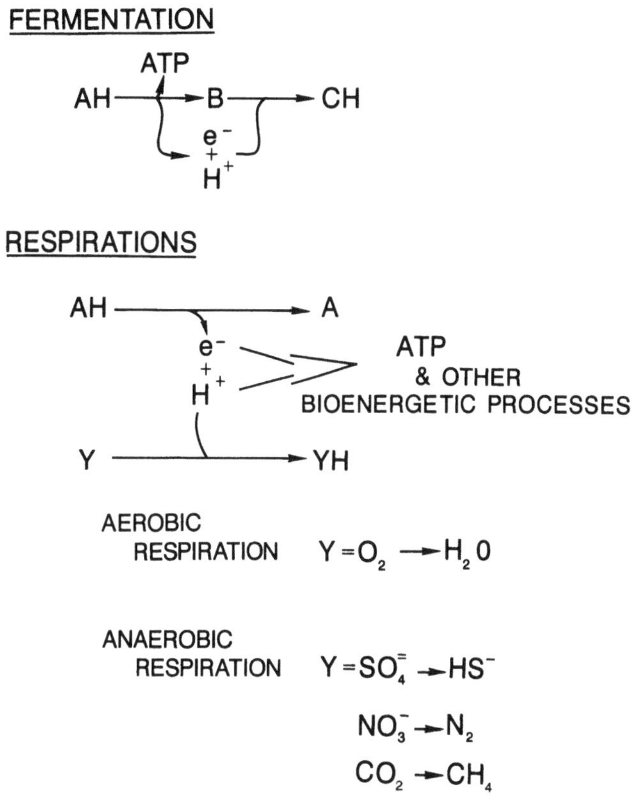

Figure 1.1 Metabolic patterns based on disposition of reducing equivalents (based on Brock and Madigan, 1991).

In *fermentation pathways,* electrons generated by oxidation of a compound are used to reduce a product generated during that oxidation process. Kelly (1987) noted that in a fermentation a single substrate provides both donor and acceptor molecules for reducing equivalents transferred in oxidation and reduction processes. For those individuals trained to calculate classical fermentation balances, he stated the case most elegantly: " . . . the oxidation states of the substrate and the sum of its products are equivalent." Traditionally, fermentation pathways involve three essential characteristics: an absence of exogenous electron acceptors, an absence of electron transport oxidative phosphorylation directly coupled to the oxidation process, and exclusive involvement of organic compounds. During the past decade, the later two points have been demonstrated to be questionable, as I note below.

The paradigm fermentation is conversion of glucose to lactic acid via the Embden-Meyerhof pathway. Glucose (AH) is initially oxidized to pyruvate (B). In these oxidation reactions, two moles of NAD serve as immediate electron acceptors and are reduced to NADH; four moles of ATP are synthesized by substrate level phosphorylation. Because the cellular concentration of electron-carrying cofactors (NAD) is not high, they must be continuously reoxidized in order for oxidative reactions to continue. Thus, NADH generated in these oxidations is used to reduce pyruvate to lactate (CH). In the overall physiology of the cell, the major source of ATP is generated via substrate level phosphorylation.

In *respiratory pathways,* electrons generated by oxidation of one compound are used to reduce a second compound that is independent of the initial oxidation pathway. Although ATP is often synthesized via substrate level phosphorylation, the amount produced is generally small in comparison with fermentation pathways and often is inadequate to support cellular growth. The major part of ATP synthesis in respiratory pathways occurs via electron transport-coupled oxidative phosphorylation. Cofactors (e.g., NAD, FAD, etc) reduced by oxidation of AH to A (Figure 1.1) are in turn reoxidized by terminal oxidases. (In most organisms terminal oxidases are imbedded in the membrane; however, as Peck demonstrates in Chapter 3, this may not be the case for some species of sulfate-reducing bacteria.) This oxidation results in transfer of protons from the inside to the outside of the cell and creates a proton and electrical potential gradient across the bacterial membrane. Dissipation of these gradients is used to drive membrane-bound ATP synthetases, as well as to carry out a variety of other microbial physiological functions, such as flagellar rotation for motility or transport of materials into or out of the cell.

Numerous electron acceptors can ultimately oxidize reduced cofactors. We recognize two broad patterns of respiratory pathways that depend upon the ultimate electron acceptor (Y in Figure 1.1). In *aerobic respiration,* electrons generated from reduced compounds are coupled to reduction of oxygen, whereas in *anaerobic respiration* electrons from oxidative reactions are used to reduce a variety of compounds, such as SO_4^-, NO_3^-, or CO_2.

An important caveat is essential in making these terminology distinctions; they ought not be invoked too rigidly, especially those distinctions regarding fermentations and respirations. The metabolic capabilities of any organism are not always seen with total clarity, and new processes should cause us to rethink these distinctions. Two examples serve to illustrate these murky terminology waters.

First, many microbial physiologists (I include myself in this category) would like to restrict respiratory processes to those which involve a transmembrane proton transport bioenergetic component for ATP synthesis. While it is true that in most fermentations (e.g., that of glucose discussed above) ATP is synthesized via substrate level phosphorylation, there is clear evidence that some classically "fermentative" organisms can generate proton gradients that may also be used to synthesize ATP. One example of this is transport of lactic acid across the cell membrane where it dissociates to lactate and H^+ (Konings, 1985; Morris, 1986), thereby establishing a transmembrane proton and charge gradient.

Second, in most common usage, the term *fermentation* usually refers to oxidation of organic compounds (Schlegel, 1986). This conventional usage has been challenged by the observation that some species of sulfate-reducing bacteria carry out disproportionation reactions involving both HSO_3^- and $HS_2O_4^-$ to SO_4^{2-} and HS^- (Bak and Cypionka, 1987; Jorgensen and Bak, 1991). A disproportionation reaction is defined as one in which "an element in a given oxidation state . . . reacts to form a higher and a lower oxidation state" (Porterfield, 1984); this definition has much in common with the definition of fermentation. Thus, as stated by Kelly (1987), this new dimension of sulfate reducer physiology is a clear demonstration of an *inorganic fermentation*.

Metabolic processes in sulfate-reducing bacteria: electron donors Traditional patterns of carbon and sulfur utilization by sulfate-reducing bacteria are summarized in schematic fashion in Figure 1.2. The intent of this figure is not to depict the metabolic capabilities of a single organism, but rather to present a skeleton of potential reactions observed in various organisms. Furthermore, some of the processes illustrated, e.g., direct coupling of ferredoxin to hydrogenase, have not been experimentally demonstrated. For metabolic details, the reader is referred to Chapters 2 and 3. A common laboratory growth compound for these organisms is sodium lactate, which is first oxidized to pyruvate, then to acetate and CO_2. The stoichiometry shown in Figure 1.2 demonstrates that two moles of lactate must be oxidized to provide the eight electrons necessary to reduce one mole of SO_4^{2-}.

As shown in Figure 1.2, other compounds can serve as electron donors for SO_4^{2-} reduction. In addition to serving as an electron donor for SO_4^{2-}-coupled growth, pyruvate can support growth of some organisms in the absence of SO_4^{2-}. The mechanism of sulfate-free growth on pyruvate is not clearly established, but several possibilities have been proposed (Postgate, 1984a; Sadana, 1954). Because acetate, CO_2, and H_2 are often seen as growth products under these conditions, the most likely possibility is a traditional phosphoroclastic reaction (see Chapter 3). A second possibility involves a fermentation in which one mole of acetate is reduced to ethanol, thereby consuming two reducing equivalents generated by pyruvate oxidation to acetate and CO_2. A third alternative mechanism for growth on pyruvate involves a "dismutation" of two molecules of pyruvate; one molecule is oxidized to acetate and CO_2, while the other is reduced to lactate.

Sulfate-reducing bacteria exhibit a variety of alternative metabolic processes, as illustrated in Figure 1.2. Some organisms can carry out a fumarate respiration, in which electrons from lactate or pyruvate oxidation are transferred to fumarate, to generate succinate (Barton et al., 1970). Other

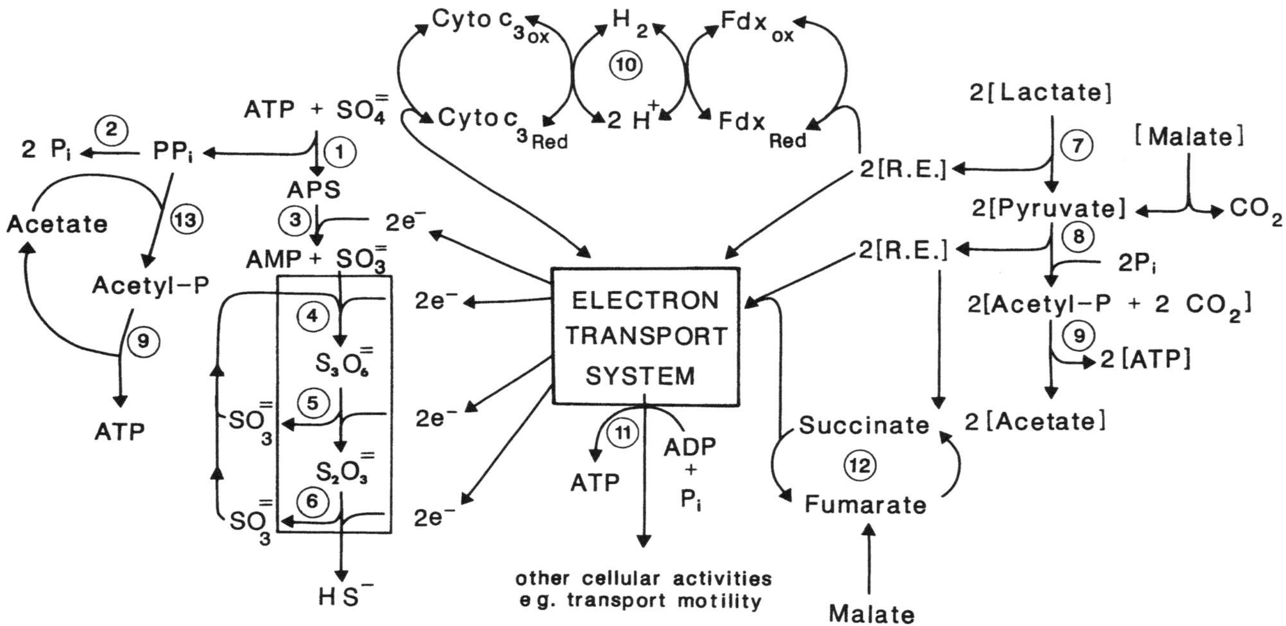

Figure 1.2 Schematic diagram of major metabolic patterns observed in many species of sulfate-reducing bacteria. Reactions are shown schematically to illustrate possible flow of electrons from carbon reactions to sulfur reactions and not to suggest mechanisms of electron transport. Also, some reactions, such as the direct coupling of ferredoxin to hydrogenase, have not been demonstrated to occur. See Chapters 2 and 3 for details of these reactions. The following enzymes or activities have been identified: (1) ATP sulfurylase; (2) pyrophosphatase; (3) APS reductase; (4) bisulfite reductase; (5) trithionate reductase; (6) thiosulfate reductase; (7) lactate dehydrogenase; (8) pyruvate dehydrogenase; (9) acetyl kinase; (10) hydrogenase; (11) ATPase; (12) fumarate reductase; and (13) acetate pyrophosphate phosphotransferase. (R.E. = reducing equivalents, see p. 2.)

strains can grow on malate + SO_4^{2-}, presumably oxidizing malate to pyruvate and CO_2 (Postgate, 1984a). Some strains can grow mixotrophically, utilizing H_2 and SO_4^{2-} as energy source and acetate as carbon source (Nethe-Jaenchen and Thauer, 1984). Some strains (e.g., *Desulfobacterium autotrophicum*) are capable of lithoautotrophic growth on H_2 and CO_2 (Brysch et al., 1987; Widdel, 1987; Chapter 2). Carbon fixation does not occur via the traditional ribulose bisphosphate carboxylase pathway but is fixed either via either a reductive citric acid cycle pathway or by direct acetogenesis by a carbon monoxide dehydrogenase pathway (see Chapter 2 for details).

Widdel (1988) has suggested that sulfate- and sulfur-reducing organisms can be organized into two metabolic groups. One group carries out an incomplete lactate oxidation to acetate and CO_2 shown in Figure 1.2. In Chapter 2 of this volume, Hansen refers to this group of organisms as "incomplete oxidizers." Although some members of this group can utilize low-molecular-weight alcohols, e.g., *n*-propanol, they rarely oxidize fatty acids. The second group grows more slowly; however, it is metabolically more diverse. Members of this group can completely oxidize a variety of compounds, including fatty acids and aromatic compounds, to CO_2.

It is now clear that the metabolic patterns shown in Figure 1.2 represent only a basic skeleton of metabolic capacities of sulfate-reducing bacteria. During the past decade and a half, the number of substrates that these organisms can use for energy and growth has increased substantially. In some regards, this new diversity of substrate utilization shuld not be surprising. Wake et al. (1977) completed a thermodynamic analysis of potential growth substrates for reactions coupled to sulfate reduction. Their calculations suggested that a wide variety of novel organic compounds might serve as growth agents for sulfate reducers. Details of the innovative and extensive metabolic capacities of these bacteria will be elaborated upon in Chapters 2 and 3.

Metabolic processes in sulfate-reducing bacteria: electron acceptors and bioenergetics Reducing equivalents generated by oxidations in sulfate reducers are in turn used to reduce SO_4^{2-} to S^{2-} via an electron transport system. While this system is treated as a "black box" in Figure 1.2, this is not meant to imply that little is known of the process. As noted above, in most "traditional" respiratory pathways proteins involved in terminal electron transfer processes are membrane associated. As a consequence of this spatial localization, protons are transferred from the interior to the exterior of the cell. Sulfate-reducing bacteria, as Peck discusses in Chapter 3, are unique in this regard, in that proteins involved in terminal electron transfer reactions frequently are soluble and cytoplasmic. Consequently, the "traditional" view of electron transport-coupled oxidative phosphorylation must be revised somewhat for these bacteria.

Reduction of SO_4^{2-} to HSO_3^- (reactions 1 and 3), by other oxidation half-reactions which are potentially possible, is thermodynamically unfavorable ($\Delta G_0' = +46$ kJ/mol, based on H_2 oxidation) [(Thebrath et al., 1989), based on data from Thauer et al., 1977)], and consequently the reaction requires activation by ATP. In many sulfate reducers, this reaction is "pulled to completion" by hydrolysis of PP_i to two moles of phosphate (reaction 2). It had been reported that in some species of the genus *Desulfotomaculum* energy in PP_i generated by SO_4^{2-} activation was salvaged by phosphorylation of acetate to form acetyl-P (reaction 13), which in turn phosphorylates ADP to generate ATP (reaction 9) (Liu et al., 1982). This hypothesis was attractive because it further extended notions of PP_i as an "energy carrier" (Baltscheffsky and Nyrén, 1987) to sulfate-reducing bacteria. The existence of this salvage pathway has been challenged, however, by Thebrath et al., (1989), who demonstrated that the requisite enzymes were either present in very low levels or absent in several species of the genus *Desulfotomaculum*.

The product of sulfate activation, adenylyl-SO_4^{2-} (APS), is then reduced to AMP and HSO_3^- by APS reductase (reaction 3), a non-heme iron flavoprotein ($\Delta G_0' = -68$ kJ/mol). The enzyme is located in the cell cytoplasm, and its natural electron carrier has not been identified (LeGall and Fauque, 1988).

There has been controversy regarding reduction of HSO_3^- to S^{2-} (Thauer and Badziong, 1980). One hypothesis held that the reaction is catalyzed in a single step by a single enzyme, bisulfite reductase (Peck and LeGall, 1982). An alternative hypothesis claimed that several enzymes (reaction 4, 5, and 6) and intermediates, such as those shown in the boxed area of Figure 1.2 (trithionate pathway), were involved in the reaction (Akagi, 1981). Present evidence supports the first hypothesis (Peck and LeGall, 1982). Although individual enzymes with activities shown in Figure 1.2 can be demonstrated, physiologically it is likely that a single enzyme catalyzes the overall reaction, with formation of a number of relatively stable intermediates. The green pigment desulfovirdin, which is of taxonomic significance for the genus *Desulfovibrio*, has bisulfite reductase activity and can catalyze conversion of HSO_3^- to S^{2-} in a single step (LeGall et al., 1979). The enzyme can also catalyze formation of trithionate and thiosulfate, depending upon assay conditions (see Chapter 3). These are major points of evidence supporting the first hypothesis.

Despite the evidence supporting the hypothesis for a single-step reduction of HSO_3^-, suggestions for a potential role of the trithionate pathway, or some mechanism resembling it, continue to be made. For example, in a study of proton translocation by a washed cell preparation of *Dv. desulfuricans* strain Essex 6, Fitz and Cypionka (1989) reported that proton translocation was stimulated by HSO_3^-, $HS_2O_3^-$, and $S_3O_6^{2-}$. Based on these observations, they suggested a hypothetical pathway for HSO_3^- reduction involving a trithionate pathway. As is undoubtedly true for many aspects of sulfate reducer physiology, the mechanistic question of HSO_3^-

reduction cannot be considered to be completely and unequivocally resolved. Indeed, it may well be the case that multiple pathways exist.

Regardless of the mechanism of HSO_3^- reduction, the overall process of sulfate reduction must ultimately be linked to electron transport-coupled phosphorylation for members of the genus *Desulfovibrio* (Peck, 1960, 1966). Two observations support this conclusion. As noted previously, sulfate activation is thermodynamically unfavorable and consumes both phosphoanhydride bonds of ATP; this is the energetic equivalent of two moles of ATP. Thus, the two moles of ATP obtained by substrate level phosphorylation in lactate oxidation to acetate are fully utilized by SO_4^{2-} activation. Furthermore, some strains of *Desulfovibrio* can grow with H_2 and SO_4^{2-} as a sole energy source (Sorokin, 1966a; Badziong and Thauer, 1978; Badziong et al., 1978). Both observations indicate that, for members of the genus *Desulfovibrio*, substrate level phosphorylation is inadequate to support growth and that without some type of electron transport-coupled phosphorylation the organism would be unable to grow. The demonstration of electron transport-coupled oxidative phosphorylation in sulfate-reducing bacteria (Peck, 1960, 1966) was the first indication that this process, once thought to be the exclusive domain of oxygen-respiring organisms, occurred in anaerobic bacteria as well.

Two mechanisms have been proposed for electron transport-coupled energy transfer in sulfate-reducing bacteria (LeGall and Fauque, 1988). Details of these mechanisms are the subject of Chapter 3. The intent here is to provide a brief sketch of the mechanisms.

One mechanism is based on studies of *Desulfovibrio vulgaris* (Marburg), which is capable of growth on H_2 and SO_4^{2-}. The mechanism proposes that electrons used to reduce SO_4^{2-} are generated by H_2 oxidation to protons in the periplasmic space (Badziong and Thauer, 1978, 1980; Thauer and Badziong, 1980; Thauer, 1989). Electrons are then transported across the cell membrane, via membrane-bound electron carrier proteins, and are used to reduce SO_4^{2-}. This mechanism involves traditional vectorial means of electron transfer and leads to an accumulation of protons, with a concomitant positive charge, on the outside of the cell membrane.

An alternative mechanism, based on studies of heterotrophically growing sulfate reducers, proposes (Peck, 1984; Peck and Odom, 1984; Steenkamp and Peck, 1981) that electrons from carbon oxidation are used to reduce protons to H_2 in the cell interior. Hydrogen is permeable and diffuses across the plasma membrane. Outside of the membrane, in the periplasmic space, H_2 is oxidized back to H^+ by means of a periplasmic H_2ase. Electrons are carried back across the membrane and used to reduce ferredoxin, which is in turn used to reduce HSO_3^-. The net effect of this is generation of a pH gradient across the membrane in a manner analogous to that which occurs in more conventional chemosmosis, presumably with similar bioenergetic consequences for the cell.

It is clear that bioenergetic processes in sulfate-reducing bacteria are complex and unique in comparison with many other bacteria. Further details of this complexity and uniqueness of electron transport processes will be discussed in Chapter 3.

1.3 Laboratory Manipulations of Sulfate-Reducing Bacteria

The product of sulfur metabolism is shown as bisulfide ion in the previous discussion; bisulfide is the true chemical species at neutral pH. However, most people who work with these organisms (along with their colleagues), are very much aware of the equilibrium that exists between bisulfide and its protonated and gaseous form, hydrogen sulfide. This metabolic product has led to a certain unpopular reputation for those researchers who grow these organisms in any quantity. The foul odors that often belch forth from the laboratories of those who work with these bacteria tend to make many workers feel like one of the cartoon characters shown in Figure 1.3. I leave it to the reader to assess with which of these characters he or she feels most affiliation.

At one time, sulfate-reducing bacteria were difficult to manipulate under laboratory conditions. Postgate (1984a) stated, "There is an element of art, as well as science in cultivating the sulphate-reducing bacteria." Several insights and innovations have helped to reduce the element of art in the manipulation of these organisms. One insight was the clear recognition that anaerobiosis is not the only requirement if one is to successfully grow these bacteria; a low reduction potential (E_h) is also essential (Postgate, 1984a; Pfennig et al., 1981). Most media are poised at an E_h value more negative than fully reduced resazurin, which has a midpoint reduction potential (at pH 7) of about $-.051$ V (Twigg, 1945; Clark, 1972).

Although most investigators (e.g., Pfennig et al., 1981) consider

Figure 1.3 The cartoonist Johnny Hart captured the essence of the way many colleagues react to those of us who work with the sulfate-reducing bacteria. (Used by permission of Johnny Hart and NAS, Inc.)

sulfate-reducing bacteria to be strict anaerobes, recent studies (Cypionka et al., 1985; Dilling and Cypionka, 1990) have raised questions about the obligate anaerobic nature of these organisms. These studies demonstrate that some sulfate reducers are not only oxygen tolerant, but actually have a capacity to reduce O_2 to H_2O. Although this oxygen tolerance may have implications for the bioenergetics and the ecological niches these bacteria can occupy [see Canfield and Des Marais (1991) and Chapters 3, 7, and 8], it is difficult to understand a potential role for O_2 in their general metabolism and physiology. The reduction potential of most redox proteins of sulfate-reducing bacteria is considerably more negative than the O_2/H_2O couple. For example, the reduction potential of the four hemes in *Dv. desulfuricans* tetraheme cytochrome c_3 ranges from −.125 V to −.319 V (Cammack et al., 1984; LeGall and Fauque, 1988). Other redox proteins in these organisms have similar reduction potentials. Consequently, any significant accumulation of O_2 within the cell would lead to rapid autoxidation of these proteins.

The study of sulfate-reducing bacteria has been promoted by development of media, both defined and undefined, that facilitate their growth and selection. Details for the preparation of these media have been described elsewhere (Pankhurst, 1971; Pfennig et al., 1981; Pfennig and Biebl, 1981; Postgate, 1984a; Widdel, 1980). Finally, manipulation and growth of these organisms has been greatly expedited by an increasing interest in anaerobic bacteria in general (Barnes and Mead, 1986; Zehnder, 1988) and in development of methods to handily manipulate microorganisms under anaerobic conditions (Aranki and Freter, 1972; Aranki et al., 1969; Balch and Wolfe, 1976; Hardie, 1986; Hungate, 1950; Hungate et al., 1964).

1.4 Taxonomic Considerations

The taxonomy of sulfate-reducing bacteria, which was once viewed as fairly simple, has become increasingly complex during the past decade. For many years, the organisms were thought to be unified solely about the ability to reduce sulfate. In 1965, Campbell and Postgate (Campbell and Postgate, 1965; Postgate and Campbell, 1966) proposed that sulfate reducers be organized into the two major genera, based on characteristics summarized in Table 1.1. These two genera probably still account for a significant number of organisms responsible for sulfate reduction in many natural systems. Since this tidy system was proposed, however, the pioneering work of Widdel (1980) demonstrated that sulfate reducers are much more diverse than previously recognized. At least 14 new genera, with different morphologies and metabolic capabilities, are now recognized (Pfennig and Biebl, 1976; Pfennig et al., 1981; Widdel, 1988; Devereux et al., 1989; Widdel and Hansen, 1991; and Chapter 6 of this

Table 1.1. Early taxonomic organization of sulfate-reducing bacteria according to scheme of Postgate and Campbell[a]

Desulfovibrio	*Desulfotomaculum*
Non-sporeformers	Sporeformers
Small vibrios to long spirilla	Sausage-shaped rods reminiscent of *Clostridium*
G + C content high	G + C content low
Pigments:	Pigments:
Desulfoviridin = green pigment, red fluorescence in base	Cytochrome *b*
Cytochrome c_3	

[a]Campbell and Postgate, 1965; Postgate and Campbell, 1966.

volume). One need only compare the simplicity of Table 1.1 with the complexity of forms and metabolic capabilities of the genera listed in Table 1.2 to grasp the explosive revolution that has occurred in the taxonomy of the sulfate-reducing bacteria [see also Table 6.3 as well as Table 1 in Widdel and Hansen (1991)].

Antigenic studies support further the notion that sulfate-reducing bacteria are highly diversified. Aketagawa et al. (1985b) prepared antisera against sulfite reductase, hydrogenase, and whole-cell antigens from three strains of *Dv. vulgaris* and compared the ability of these antisera to cross-react (as measured by immunoprecipitation) with corresponding antigens from other strains of *Dv. vulgaris* and other members of the genus *Desulfovibrio*. The study demonstrated a complex series of immunological relationships between these antigens with cross-reactions occurring between antigens from divergent genera of *Desulfovibrio*. Furthermore, there was no consistent pattern of cross-reaction between antisera against specific proteins.

Antibodies to surface antigens of various *Desulfovibrio* species (Abdollahi and Nedwell, 1980; Aketagawa et al., 1985b; Singleton et al., 1985; Smith, 1982) suggest that these structures are very diverse. On the basis of ELISA techniques, Singleton et al. (1985) demonstrated that the pattern of antibody cross-reactions for surface antigens is quite different from cross-reactions for soluble cytochrome c_3. Using immunoprecipitation and ELISA techniques, Singleton et al. (1982, 1984) demonstrated a variety of cross-reactions among antibodies to cytochrome c_3 from three species of *Desulfovibrio* and heterologous cytochromes c_3. The diversity of cross-reactions was such as to suggest that cytochromes c_3 from several strains within a given species were markedly different from each other. This diversity is also reflected in amino acid sequences of cytochromes c_3 from various strains of desulfovibrio (Meyer and Kamen, 1982).

As the data in Table 1.2 demonstrate, it is now clear that the once

Table 1.2. Characteristics of some contemporary sulfate-reducing bacterial genera [based on data from Widdel and Hansen (1991)]

Genus	Morphology	Cell wall/membrane[a]	Organic compound oxidation[b]	Desulfoviridin	Spore formation	Temperature effects
Desulfovibrio	Vibrio	$Gram^-$, Euba	Incomplete	+	No	Mesophilic
Desulfotomaculum	Rod	$Gram^+$, Euba	Both	−	Yes	Mesophilic and moderately thermophilic strains
Desulfomicrobium	Oval/rod	$Gram^-$, Euba	Incomplete	−	No	Mesophilic
Desulfobulbus	Oval	$Gram^-$, Euba	Incomplete	−	No	Mesophilic
Desulfobacter	Oval/vibrio	$Gram^-$, Euba	Complete	−	No	Mesophilic
Desulfobacterium	Oval	$Gram^-$, Euba	Complete	−	No	Mesophilic
Desulfococcus	Sphere	$Gram^-$, Euba	Complete	+	No	Mesophilic
Desulfosarcina	Oval (aggregrates)	$Gram^-$, Euba	Complete	−	No	Mesophilic
Desulfomonile	Rod	$Gram^-$, Euba	Complete	+	No	Mesophilic
Desulfonema	Multicellular filaments	$Gram^-$, Euba	Complete	±	No	Mesophilic
Desulfobotulus	Vibrio	$Gram^-$, Euba	Incomplete	−	No	Mesophilic
Desulfoarculus	Vibrio	$Gram^-$, Euba	Complete	−	No	Mesophilic
Thermodesulfobacterium	Rod	$Gram^-$, Euba	Incomplete	−	No	Thermophilic
Archaeoglobus	Sphere	Archaebacteria	Complete	−	No	Thermophilic

[a] Gram −/+, Euba = typical eubacterial Gram-negative or -positive cell wall and membrane.
[b] Incomplete = organic compounds oxidized to acetate; Complete = organic compounds oxidized to CO_2.

apparently simple taxonomical relationships of sulfate reducers of Table 1.1 are now extremely intricate. Indeed, the organization at the genera level shown in Table 1.2 may not reflect the true diversity of sulfate-reducing bacteria. For example, the variety found within organisms labeled *Desulfovibrio* is so great that Devereux and his colleagues (1990) proposed a new taxonomic family, *Desulfovibrionaceae,* to describe the phylogenetic depth and cohesiveness of these organisms. This complex diversity of sulfate-reducing bacteria is further illustrated by data in Table 6.3. The ecological and economic ramifications of this newly recognized diversity of these bacteria are just now being realized and will be discussed in greater detail in Chapters 7 and 8.

1.5 Importance of Sulfate-Reducing Bacteria

Despite the odoriferous physiology of these organisms, they are important to study for both practical and academic reasons. Their practical importance stems from the significant economic impact they have in many industrial and ecological areas (Hamilton, 1983; Postgate, 1982b; 1984a; Odom, 1990). Their academic importance is related to their unique environmental and evolutionary roles in nature. In this section, I will only briefly review the environmental and economic importance of these organisms; greater detail will be discussed in Chapters 7 and 8.

Economic importance Because they are ubiquitous to oil-bearing shales and strata (Dockins et al., 1980), sulfate-reducing bacteria play an important economic role in many aspects of oil technology. They are responsible for extensive corrosion of drilling and pumping machinery and storage tanks. Furthermore, because they contaminate resulting crude oil, the organisms cause release of hydrogen sulfide into petroleum products, thereby increasing the sulfur level of fuels. They are also important in secondary oil recovery processes, where bacterial growth in injection waters can plug machinery used in these processes. It has also been suggested that these organisms may play a role in biogenesis of oil hydrocarbons (Beerstecher, 1954; Davis, 1967). For all of these reasons, these bacteria are of vital importance in petroleum-producing and processing-industries.

A major economic consequence resulting from the metabolic action of sulfate-reducing bacteria is anaerobic corrosion of underground buried ferrous metals, such as tanks and pipelines. David White (Center for Environmental Biotechnology, University of Tennessee and Oak Ridge National Laboratory) has referred to bacterial corrosion as an "industrial venereal disease—it's expensive, everybody has it, and nobody wants to talk about it." While I do not believe the last point is quite accurate, the first two points are a germane assessment of the problem. For example, a

1954 estimate cites losses of buried pipelines to biological corrosion to be between 0.5 and 2.0 billion dollars a year (Greathouse and Wessel, 1954). This figure can only have increased in the intervening three decades. As noted above, this is a serious problem in many aspects of the oil technologies, where oil drilling and processing equipment are subject to extensive corrosion by the activities of these organisms.

Postgate (1984a) has pointed out that anaerobic corrosion of iron by sulfate-reducing bacteria is unique and has three main characteristics: (a) it is restricted to anaerobic environments; (b) corrosion tends to occur in localized areas or pits, rather than being evenly distributed over the iron surface; and (c) the corrosion is graphitized, i.e., the metallic iron is entirely removed, leaving the carbon skeleton behind. The role of sulfate-reducing and sulfur-oxidizing bacteria in corrosion processes has been reviewed by Cragnolino and Tuovinen (1984).

Despite this unique character, there is significant debate concerning the role of sulfate-reducing bacteria in corrosion of ferrous surfaces (Crombie et al., 1980; Iverson, 1974; also see Chapter 8). The controversy basically centers around two alternative, and perhaps complementary, hypotheses. (1) One hypothesis implicates members of the genus *Desulfovibrio* in a direct role in the corrosion process. This hypothesis claims that hydrogenase (H_2ase) present in these bacteria leads to cathodic depolarization of metal surface. The H_2ase strips molecular hydrogen from the metal, thereby leading to oxidation of exposed ferrous surface by sulfide anion produced by the bacteria. (2) The alternative hypothesis denies direct involvement of *Desulfovibrio* H_2ase in the process and claims that the corrosion process is initiated directly by sulfide produced as a product of bacterial metabolism. Because of the difficulty in separating out these various effects, the role and relative importance of these two hypotheses has not been resolved. One apparent way of testing the first hypothesis might be to develop specific inhibitors of H_2ase—inhibitors that would block action of that enzyme.

Sulfate-reducing bacteria play a vital role in a variety of other important economic areas. In many industrial processes, sulfidogenic microorganisms—by producing sulfide—cause precipitation of iron present in process water, thereby causing blackening and discoloration of products. In the paper industries, for example, sulfate-reducing bacteria can contaminate processing water, leading to sulfide production, which causes blackening of paper (Starkey, 1961). The organisms are responsible for a number of food spoilage factors, particularly thermophilic strains of *Desulfotomaculum*, which produces heat-resistant spores (Donelly and Busta, 1980; Lin and Lin, 1970).

Environmental importance These organisms are also of great ecological importance, which is in part related to their economic importance. Postgate (1984a) noted that sulfate-reducing bacteria can be the penultimate stage of a grossly polluted environment. He describes a progression that begins with aerobic respiring and faculatively aerobic, fermenting organisms. These organisms deplete oxygen from the ecosystem and produce a number of partially oxidized fermentation products. As O_2 is depleted from the system, the population shifts to become dominated by obligately anaerobic organisms. If sulfate is present during this period of increasing anaerobiosis, sulfate-reducing organisms become predominant. Sulfide produced by these organisms contributes to and accelerates the environment's anaerobic nature by rapidly making the E_h of the system negative. Finally, as the complexity of growth substrates is diminished or sulfate depleted, methanogenic organisms complete the process with formation of CO_2 and CH_4.

A pivotal ecological role of sulfate-reducing bacteria has also been clearly demonstrated in nonpolluted ecosystems. For example, in the marsh ecological system, more than half of the total decomposition is accounted for by sulfate reduction (Howarth and Hobbie, 1982). Furthermore, there appears to be an active interaction between sulfate-reducing bacteria, which utilize organic compounds from root exudates during the active growing season, and plants in the marsh (Hines et al., 1989). Numerous studies have shown that sulfate-reducing bacteria are either the final or penultimate organism in carbon mobilization (Dicker and Smith, 1985a, 1985b; Howarth, 1979, 1984; Howarth and Marino, 1984; Jorgensen, 1977a). These studies suggest a system with the following structure (see Figure 7.1). Carbon dioxide is fixed by the major plant of the community, e.g., *Spartina*. Vegetative biomass is degraded by mechanical action of a number of eucaryotic organisms. Much of this material is oxidized to carbon dioxide and water by aerobic respiring microorganisms. A significant portion, however, is oxidized to organic acids and alcohols by fermentative heterotrophs. Sulfate- and sulfur-reducing, bacteria can utilize these products by anaerobic respiration coupled to sulfate, brought in on tidal cycles, and produce acetate, carbon dioxide, and sulfide. Some sulfate reducers can couple acetate oxidation to CO_2 with sulfate reduction. Acetate can also be utilized by methanogenic bacteria to produce CH_4 and CO_2.

Based on the pioneering work of Bryant et al. (1977), Peck (1984) noted that *Desulfovibrio* species can play an ecological role under conditions of both high and low sulfate concentration. When sulfate concentration is high, complex fermentation products are oxidized to CO_2 and water. Under conditions of low sulfate concentration, sulfate-reducing bacteria oxidize lactate using protons as electron acceptors and generate H_2. Growth occurs, because H_2 levels are kept low by interspecies hydrogen-transfer to the methanogens. This process can be represented chemically as follows:

Desulfovibrio:

$$2\ \text{lactate} + 4\ H_2O \rightarrow 2\ \text{acetate} + 2\ CO_2 + 4H_2 \qquad (1.2)$$

Hydrogen-utilizing methanogens:

$$CO_2 + 4\ H_2 \rightarrow CH_4 + 2\ H_2O \qquad (1.3)$$

$$\text{Net: } 2\ \text{lactate} + 4\ H_2O \rightarrow \text{acetate} + CO_2 + CH_4 + 2\ H_2O \qquad (1.4)$$

Environmental consequences of sulfate-reducing bacterial growth can be devastating. They have been responsible for massive fish kills (Butlin, 1949). Postgate (1982b; 1984a) has noted a variety of environmental consequences of sulfate reducer growth, including killing of sewer workers by hydrogen sulfide, development of "poisonous dawn fogs," and killing of rice crops by sulfate-reducing bacteria depleting oxygen in paddies.

A final ecological note has to do with the sulfur cycle, which is a major biogeological cycle in nature. In the cycle, sulfate-reducing bacteria reduce SO_4^{2-} to S^{2-} or S^0. These compounds are in turn reoxidized by sulfur-oxidizing organisms, such as members of the genus *Thiobacillus*. Sulfate reducers themselves may play a more complex role in the sulfur cycle than previously realized. The discoveries (Bak and Cypionka, 1987; Bak and Pfennig, 1987) that these bacteria can carry out disproportionation reactions of thiosulfate and HSO_3^-, leading to production of SO_4^{2-} and HS^-, adds new complexity to their potential environmental role. Postgate (1984a) and Pfennig and Widdel (1981, 1982) have summarized the extensive contribution of sulfate-reducing bacteria to this cycle.

Although the sulfur cycle is a major natural biogeochemical process on a global scale, it can also be seen at local levels in sulfureta (see Figure 1.4). Sulfate ion from the surrounding land mass is leached into and diffuses to the bottom of these small aquatic systems, where it is reduced to S^{2-} by anaerobic sulfate-reducing bacteria. Sulfide ion then diffuses upward through the water column to the surface layers, where it is oxidized by aerobic organisms to either elemental sulfur (S^0) or SO_4^{2-}. The latter diffuses to the bottom, anaerobic region where the cycle continues. Because of its low solubility in water, however, S^0 precipitates from solution and forms a yellow crust on the water surface and near the edges of these systems (Postgate 1984a).

It is clear that sulfate-reducing bacteria play a diversified and important role in a multitude of environmental systems. To fully comprehend this environmental role will require a more thorough understanding of their biochemical and physiological diversity, an understanding that is slowly emerging. As noted by Devereux and Voordouw in this volume (Chapters 5 and 6), the tools of molecular biology will play an important role in facilitating this understanding.

Evolutionary importance There appears to be clear geochemical evidence that the process of biological sulfate reduction is evolutionarily

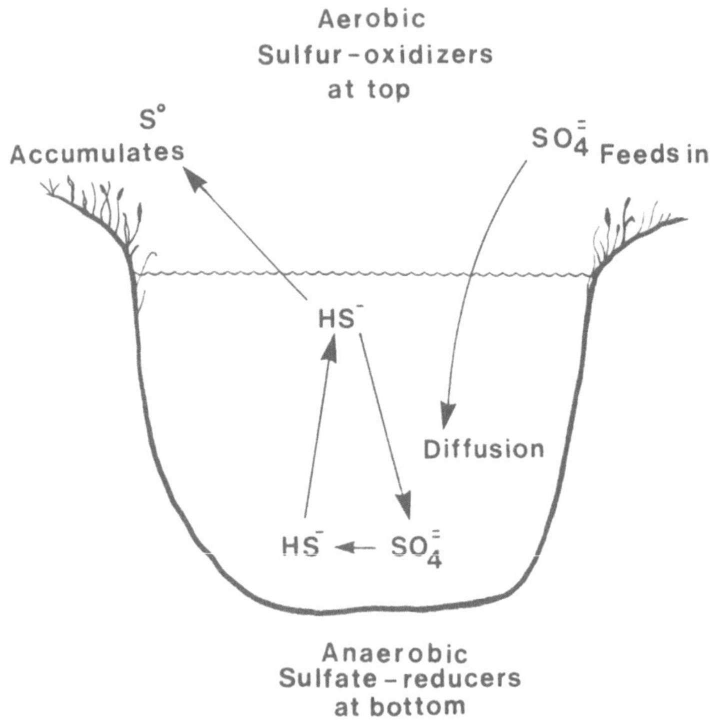

Figure 1.4 Schematic diagram illustrating the chemical processes of a sulfuretum (based on Postgate, 1984a).

ancient (see Chapters 3 and 6). This conclusion grows from the observation that bacterial reduction of sulfate shows a kinetic isotope effect in which $^{32}SO_4^{2-}$ is used more rapidly than the heavier isotope $^{34}SO_4^{2-}$ (Kaplan and Rittenberg, 1962, 1964; Widdel, 1988). Consequently, geological sulfur deposits that are enriched in the lighter isotope, relative to the heavier isotope are believed to be biological in origin. Evidence of this nature suggests that many of the world's sulfur reserves were biologically formed and that this action occurred prior to significant accumulation of atmospheric oxygen (Postgate, 1984a). This line of reasoning clearly suggests that the process of sulfate reduction was of bioenergetic significance before the process of oxygen reduction. Present arguments suggest that development of dissimilatory sulfate reduction occurred between 2.8 and 2.5×10^9 years ago and was simultaneous with oxygenic photosynthesis by ancestral cyanobacteria but prior to aerobic respiration (Schidlowski, 1979; Schopf et al., 1983; Widdel, 1988).

In his classic book on sulfate-reducing bacteria, Postgate (1984a) stated, "The evolutionary relationships within the sulphate-reducing bacteria present a fascinating problem." As will be seen in Chapters 3, 5, and 6, the tools of molecular biology have helped to illuminate this "fascinating problem" but have by no means provided a complete and satisfactory answer.

Such evolutionary insights will be valuable, for they may provide us with clues regarding the evolution of bioenergetic mechanisms in general. It is clear that sulfate reduction is coupled to ATP generation in a manner that is informally analogous to that of oxygen reduction in aerobic organisms. Furthermore, sulfate-reducing bacteria have undergone a long and protracted period of evolutionary history (Devereux et al., 1989; and Chapters 3 and 6). Within this group of bacteria, unified by a capability for dissimilatory sulfate reduction, we see all of the metabolic diversity—ranging from strictly fermentative pathways to truly respiratory (i.e., electron transport-coupled, energy-conserving) pathways—found in the rest of the microbial world.

Because of the metabolic diversity found in this group of microorganisms, sulfate reducers may represent stages in a process of bioenergetic evolutionary development. Members of the genus *Desulfotomaculum* may represent a transition from "strictly fermenting" organisms, such as the clostridia, which they resemble in that they possess a Gram-positive type of cell envelope (Nazina and Pivovarova, 1979; Norqvist and Roffey, 1985; Sleytr et al., 1969) and are spore formers. Furthermore, based on 16S rRNA sequence similarities, various *Desulfotomaculum* species are phylogenetically more closely related to organisms with typical gram positive cell envelopes, such as *Bacillus subtilis*, than they are to other genera of sulfate-reducing bacteria (Devereux et al., 1989). In the genus *Desulfovibrio* we see organisms with a more typical Gram-negative cell wall. While organisms in both genera appear to couple sulfate reduction to a chemosmotic mechanism of ATP synthesis, in at least the genus *Desulfovibrio* this process may be evolutionarily unique. Thus, information about the process of sulfate reduction can perhaps aid us in understanding the development and operation of energy-transfer systems in general.

1.6 Summary

This chapter began with the notion that sulfate-reducing bacteria were unified by a shared ability to utilize sulfate reduction as an important component of their bioenergetic processes. Indeed, this was the understanding during the early scientific history of these unique microorganisms. If it is not presently apparent, then by the end of this volume it should be absolutely clear that these organisms are extremely diverse from metabolic, morphological, antigenic, as well as ecological, perspectives. This diversity will have profound implications for our understanding of the environmental and economic role that these microorganisms play in the natural world. When this diversity of form and function is coupled with the ancient evolutionary age of these bacteria, their study promises a wealth of knowledge about basic biochemical and physiological processes.

[References, see p. 211]

Some of the personal work discussed in this chapter was supported by various grants from the National Science Foundation. Of greater importance, however, were the contributions of various colleagues and students, especially Marjory Hurley Middleton, Jean Denis, and Robert Ketcham. I especially thank Leon Campbell for his ongoing support, collegiality, and friendship.

2

Carbon Metabolism of Sulfate-Reducing Bacteria

Theo A. Hansen

2.1 Introduction

In the nearly 100 years that sulfate-reducing bacteria have been studied in pure, or supposedly pure, cultures, our ideas about their carbon metabolism have gone through periods of sometimes slow but occasionally drastic changes. In the past 20 years it has become clear that (1) far more compounds can be oxidized by sulfate reducers than previously thought; (2) organic compounds can be oxidized completely, beyond the level of acetate, to CO_2; and (3) some sulfate-reducing bacteria have the potential for fully autotrophic growth. Furthermore, the awareness has grown that a detailed knowledge of their carbon metabolism is required for a proper understanding of the bioenergetic aspects of sulfate reduction and of their role in nature.

2.2 Energy Substrates of Sulfate-Reducing Bacteria

Most of the compounds that have been shown to be utilized as energy sources by sulfate-reducing bacteria are summarized in Table 2.1. Some of these compounds are used by many strains, whereas others have been identified as suitable energy sources for only a single pure culture. Sulfate-reducing bacteria do not degrade natural biopolymers such as starch, glycogen, and proteins or lipids and thus depend on the activities of other organisms for providing them with fermentation and degradation products. The majority of compounds listed in Table 2.1 have been recognized as suitable energy substrates only during the past 15 years. Before 1977 only a limited number of compounds were considered to play such a role for sulfate reducers; these included hydrogen, formate, lactate, pyru-

Table 2.1 Compounds that can be used as substrates for dissimilatory sulfate reduction (see Widdel, 1988; Hansen, 1988; Schnell et al., 1989; Zellner et al., 1990; DeWeerd et al., 1990; Trinkerl et al., 1990; Aeckersberg et al., 1991)

Class of Compound	Specific Compounds Utilized
Inorganic	Hydrogen, carbon monoxide
Hydrocarbons	Straight alkanes (C_{12} to C_{20})
Monocarboxylic acids (aliphatic)	Formate, acetate, propionate, butyrate, higher fatty acids up to C_{20}, isobutyrate, 2-methylbutyrate, 3-methylbutyrate, 3-methylvalerate, pyruvate, lactate
Dicarboxylic acids	Succinate, fumarate, malate, oxalate, maleinate, glutarate pimelate
Alcohols	Methanol, ethanol, propanol-1, butanol-1, pentanol-1, isobutanol, propanol-2, butanol-2, ethylene glycol (mono-, di-, tri- and tetra-), 1-,2-propanediol, 1,3-propanediol, glycerol
Amino acids	Glycine, serine, alanine, cysteine, cystine, threonine, valine, leucine, isoleucine, aspartate, glutamate, phenylalanine
Sugars	Fructose, glucose, mannose, xylose, rhamnose
Aromatic compounds	Benzoate, 2-, 3-, and 4-hydroxybenzoate, phenol, p-cresol, catechol, resorcinol, hydroquinone, phloroglucinol, pyrogallol, 2-,4-dihydroxybenzoate, 3-,4-dihydroxybenzoate (protocatechuate), 2-aminobenzoate (anthranilate), 4-aminobenzoate, aminobenzene (aniline), hippurate, nicotinic acid, indole, quinoline, phenylacetate, 4-hydroxyphenylacetate, indolylacetate, 3-phenylpropionate,[a] trimethoxybenzoate,[a] *m*-anisate,[a] *p*-anisate,[a] vanillate,[a] 3-methoxysalicylate,[a] vanillin,[a] syringaldehyde,[a] *p*-anisaldehyde,[a] 2- and 3-hydroxybenzaldehyde[a]
Miscellaneous	Choline, betaine, oxamate, acetone, dihydroxyacetone, furfural, cyclohexanone, cyclohexanol, cyclohexanecarboxylate and a few other alicyclic compounds

[a] Utilization without destruction of the aromatic structure.

vate, some dicarboxylic acids such as malate and fumarate, a few primary alcohols (such as ethanol), choline, glycerol, and alanine. In a similar fashion, only a few genera of sulfate-reducing bacteria, with similar physiology, were recognized to exist. These bacteria are now sometimes designated as the classical sulfate-reducing bacteria.

There may be two reasons for the late discovery of the true metabolic diversity of sulfate-reducing bacteria as an ecophysiological group. First, most of the recently studied organisms grow more slowly and only under more carefully controlled conditions than classical sulfate-reducing bacteria. Second, unsuccessful attempts of several investigators to repeat the isolation of the acetate-oxidizing, sulfate-reducing bacteria described by Baars (1930) had led to the notion that sulfate-reducing bacteria do not oxidize fatty acids such as acetate or propionate; furthermore, oxidation of organic multicarbon substrates was believed always to be incomplete and not beyond the level of acetate [for older literature in this field see Postgate (1984a)]. This "dogma", of restricted metabolic and physiological diversity, persisted for several years until Widdel and Pfennig (1977) isolated *Desulfotomaculum acetoxidans*, which oxidized acetate to CO_2. Since then several new species and genera have been described, and a renewed interest has arisen in all aspects of the ecology, physiology, and biochemistry of sulfate-reducing bacteria.

It is perhaps not surprising that, in general, it has proved to be easier to enrich for and isolate strains with novel metabolic features from marine and estuarine than from freshwater environments. Sulfate concentration in freshwater is usually low, whereas in seawater SO_4^{2-} concentration is approximately 28 mM; consequently, sediments and stagnant water of freshwater environments easily become sulfate limited, and specialization in the degradation of relatively few substrates by the group of sulfate-reducing bacteria as a whole may be a logical consequence. It should, however, be kept in mind that several sulfate-reducing bacteria have the ability to grow fermentatively in the absence of sulfate.

Table 2.1 illustrates the enormous diversity of energy substrates for sulfate-reducing bacteria. Of particular interest is the recent isolation of a sulfate-reducing bacterium that grows on alkanes with a C_{12} to C_{18} chain; these hydrocarbons were oxidized to CO_2 (Aeckersberg et al., 1991). This observation implies the end of another dogma, namely, the impossibility of anaerobic attack on saturated aliphatic hydrocarbons. Indirect evidence for a possible anaerobic oxidation of methane by sulfate-reducing bacteria (see, e.g., Alperin and Reeburgh, 1985) has not led to isolation of methane-utilizing, sulfate-reducing bacteria.

2.3 Degradation of Various Organic Compounds in the Presence of Sulfate

Complete or incomplete oxidation The metabolic capabilities of sulfate-reducing bacteria seem to fall naturally into two broad categories—those organisms capable of totally oxidizing organic compounds to CO_2 and those organisms incapable of such complete oxidation processes (Widdel, 1988). In the discussion that follows, the phrase "completely oxidizing sulfate-reducing bacteria" will refer to the first category of organisms.

Many sulfate-reducing bacteria cannot completely oxidize to CO_2 substrates that contain at least one carbon-carbon bond and that are not more oxidized than acetate. This metabolic limitation usually reflects the absence of a biochemical pathway for oxidation of acetyl-CoA to CO_2. This, however, does not mean that a particular substrate is necessarily degraded to the level of acetate; a *Desulfovibrio* strain simply may not oxidize propanol beyond the propionate level because it lacks enzymes necessary to degrade propionate. Completely oxidizing sulfate reducers, on the other hand, do not always oxidize substrates completely to CO_2. Utilization of easily degradable organic compounds such as butyrate, lactate, or ethanol by such sulfate-reducing bacteria often leads to an excretion of acetate; thereafter acetate oxidation may proceed very slowly (cf. Brysch et al., 1987).

Two-carbon compounds that are more oxidized than acetate can be oxidized to CO_2 by sulfate-reducing bacteria that cannot oxidize acetyl-CoA to CO_2; this was shown to be the case in experiments with a *Desulfovibrio* that degrades oxamate and oxalate (Postgate, 1963). Similarly, glycine oxidation by two marine *Desulfovibrio* strains yielded mainly CO_2 and only 0.1 to 0.3 acetate per glycine oxidized. Most likely degradation of glycine is initiated by an oxidation to NH_4^+, CO_2, and methylenetetrahydrofolate (THF); methylene-THF is then either oxidized to CO_2 (and THF) or, in a reaction with glycine, converted to serine, which is oxidized to acetate via pyruvate (Stams et al., 1985). Thus, glycine degradation can be described as a process somewhere in between reactions 2.1 and 2.2:

$$\text{glycine} + H_2O + 0.75\ SO_4^{2-} \rightarrow 2\ HCO_3^- + 0.25\ H^+ + NH_4^+ + 0.75\ HS^- \qquad (2.1)$$

$$\text{glycine} + H_2O + 0.25\ SO_4^{2-} \rightarrow 0.5\ \text{acetate} + HCO_3^- + 0.25\ H^+ + NH_4^+ + 0.25\ HS^- \qquad (2.2)$$

A surprising result in the study of carbon metabolism of anaerobic bacteria during the past decade has been the discovery that sulfate-reducing bacteria can completely oxidize the acetyl group of acetyl-CoA to CO_2 by two entirely different mechanisms. *Desulfobacter* strains employ the citric acid cycle (Brandis-Heep et al., 1983; Gebhardt et al., 1983; Schauder et al., 1986). This pathway is modified from that found in most aerobes by (1) presence of an α-ketoglutarate dehydrogenase that is ferredoxin- as opposed to NAD-dependent; (2) a membrane-bound, NAD-independent malate dehydrogenase; and (3) an ATP-citrate lyase instead of a citrate synthase (Möller et al., 1987) (see Figure 2.1). Acetate activation proceeds via transfer of the CoA group from succinyl-CoA. ATP-citrate lyase enables the organism to obtain an ATP by substrate level phosphorylation by oxidation of acetate to 2 CO_2; this may be a

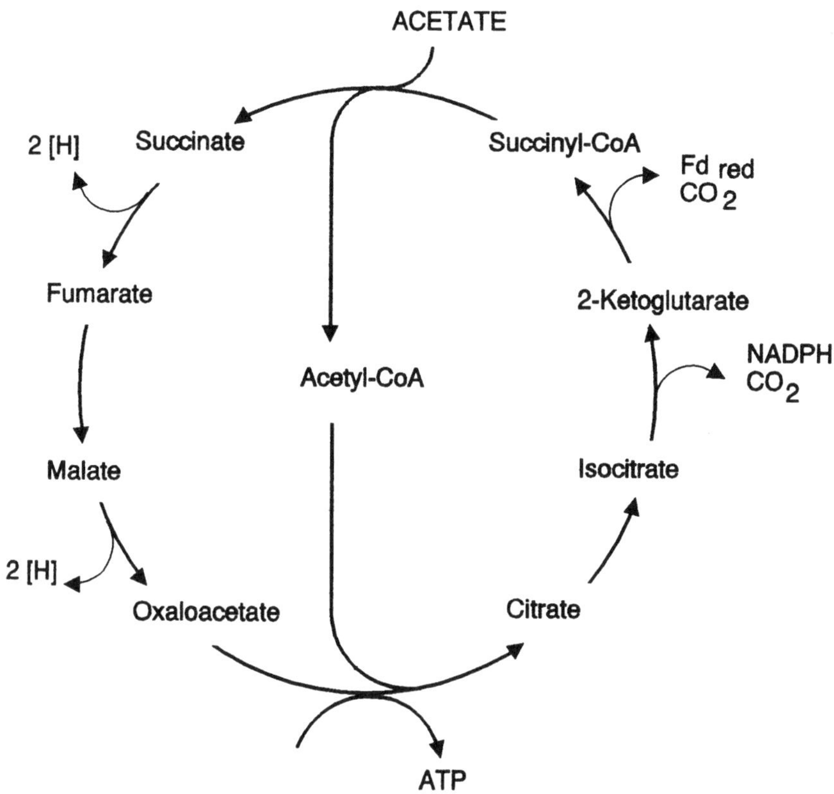

Figure 2.1 Pathway of acetate oxidation in *Desulfobacter postgatei*. For detailed information see Möller et al. (1987). Abbreviation: Fd_{red}, reduced ferredoxin

bioenergetic necessity because of expenditure of two ATPs in sulfate activation and because the ATP yield by electron transport-associated phosphorylation most probably does not exceed two ATPs (Möller et al., 1987). The NADPH generated by isocitrate dehydrogenase is reoxidized by a membrane-associated NADPH dehydrogenase, and reduced ferredoxin is reoxidized by NADP (Möller-Zinkhan and Thauer, 1988).

Acetyl-CoA oxidation in other genera of completely oxidizing sulfate-reducing bacteria is not cyclic but involves cleavage of the two-carbon unit into a methyl and a carbon monoxide moiety, each of which is oxidized via independent pathways (Fig. 2.2). In sulfate-reducing bacteria with this so-called oxidative carbon monoxide dehydrogenase pathway, high activities of carbon monoxide dehydrogenase were demonstrated and, characteristically, α-ketoglutarate dehydrogenase was absent (Schauder et al., 1986; Möller-Zinkhan and Thauer, 1990).

In all organisms, the carbon monoxide moiety is oxidized by a single enzyme, carbon monoxide dehydrogenase. During oxidation of the

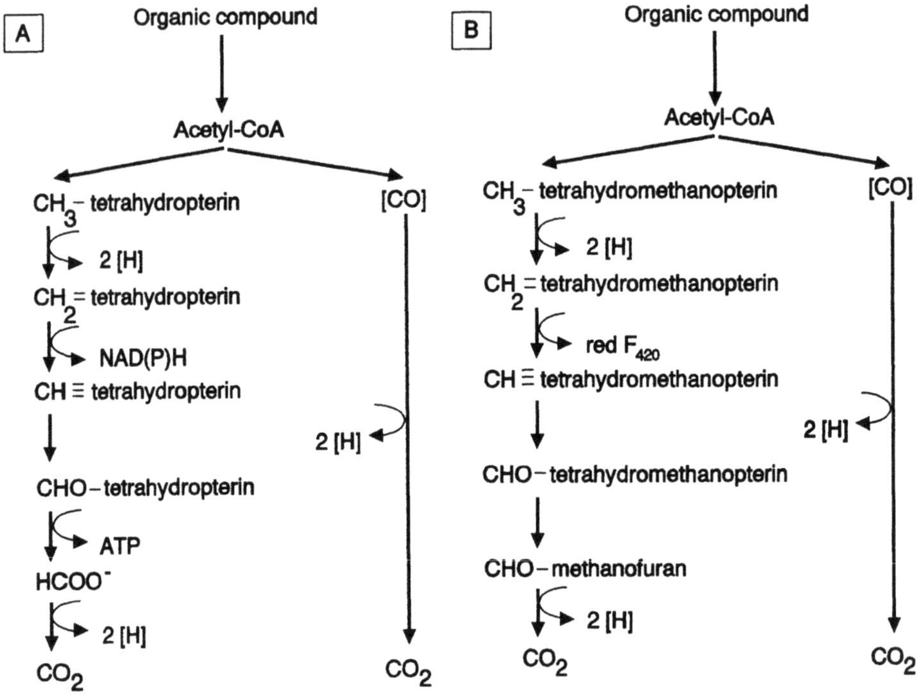

Figure 2.2 Oxidative carbon monoxide dehydrogenase pathway for the oxidation of acetyl groups; A. in sulfate-reducing eubacteria such as *Desulfotomaculum* and *Desulfobacterium*, B. in the sulfate-reducing archaebacterium *Archaeoglobus fulgidus*. Abbreviation: red Fd, reduced ferredoxin

methyl moiety, however, different methyl-group carriers are involved, depending on the organism. For example, the methyl group is oxidized to CO_2 via C_1-intermediates bound to tetrahydrofolates in the case of *Desulfotomaculum acetoxidans* (Spormann and Thauer, 1988, 1989), to tetrahydropterins containing four L-glutamates as in *Desulfobacterium autotrophicum* (Schauder et al., 1989), or even to tetrahydromethanopterin, as was found for the sulfate-reducing archaebacterium *Archaeoglobus fulgidus* (Möller-Zinkhan et al., 1989; Möller-Zinkhan and Thauer, 1990). In eubacteria the last steps in oxidation to CO_2 proceed via formate and a formate dehydrogenase, but in *Archaeoglobus* the C_1-unit is transferred to methanofuran and dehydrogenated to CO_2.

Acetate oxidation via the oxidative carbon monoxide dehydrogenase pathway in eubacteria does not lead to ATP by substrate level phosphorylation; this disadvantage, in comparison with the citric acid cycle (see above), may be outweighed by the fact that most reducing equivalents are generated at a very favorable reduction potential and none at the poor level of the succinate/fumarate potential, as in the citric acid cycle (Thauer, 1988).

A second possible advantage of the oxidative carbon monoxide dehydrogenase pathway may lie in the easy access to other C_1-substrates at the methyl level. Heijthuijsen and Hansen (1989) isolated a sulfate-reducing bacterium that oxidized betaine (*N,N,N*-trimethylglycine) to CO_2 and dimethylglycine. This organism turned out to be a *Desulfobacterium* strain that was similar to the type strain, *Desulfobacterium autotrophicum*, but was unable to grow as an autotroph; *Dbt. autotrophicum* was also found to oxidize betaine to dimethylglycine. There are other examples of utilization of C_1-substrates at the methyl level by strains with the oxidative carbon monoxide dehydrogenase pathway: methanol utilization by *Dbt. catecholicum* (Szewzyk and Pfennig, 1987); trimethoxybenzoate degradation to 3,4,5-trihydroxybenzoate by some spore-forming strains (Klemps et al., 1985); and metabolism of the methyl group of some methoxylated aromatics, including vanillate, by *Desulfomonile tiedjei*, an organism in which the carbon monoxide dehydrogenase route is assumed to be functioning (Mohn and Tiedje, 1990; DeWeerd et al., 1990).

Utilization of lactate Lactate is still one of the most widely used substrates for cultivating sulfate-reducing bacteria in biochemical studies. Several species or genera described in the past 15 years, however, are unable to grow on lactate, e.g., completely oxidizing *Desulfotomaculum* species and *Desulfobacter* (Widdel, 1988). Oxidation of L- and D-lactate to pyruvate is catalyzed by lactate dehydrogenases that are NAD(P) independent and are mainly membrane bound (Czechowski and Rossmore, 1980; Ogata et al., 1981; Stams and Hansen, 1982; Peck and LeGall, 1982; Pankhania et al., 1988; Möller-Zinkhan et al., 1989). None of the enzymes has been purified to homogeneity, and the natural electron acceptors are not known with certainty. Most likely electrons are accepted at the cytochrome *b*/menaquinone level. Pyruvate dehydrogenase is coupled to ferredoxins or flavodoxin as electron acceptors. See Fig. 2.3 for the pathway of lactate oxidation to acetate.

Oxidation of primary alcohols, secondary alcohols, diols, and glycerol Ethanol is used as an energy source by many *Desulfovibrio* species and by several representatives from other genera such as *Desulfobulbus, Desulfococcus, Desulfosarcina, Desulfobacterium, Desulfotomaculum*, and also by some *Desulfobacter* strains (Widdel, 1988). Oxidation of ethanol to acetate by certain *Desulfovibrio* species was reported not to be associated with growth, for instance in the case of *Dv. gigas* (Esnault et al., 1988), and allowed only poor growth of *Dv. vulgaris* (Widdel, 1988). In the author's experience both species do grow on ethanol. The molar growth yield of *Dv. gigas* on ethanol was about 3.6 g dry weight per mol of ethanol (Kremer et al., 1988a).

For growth of *Dv. vulgaris* Hildenborough on ethanol, a direct transfer from a lactate to an ethanol medium has to be avoided. This action

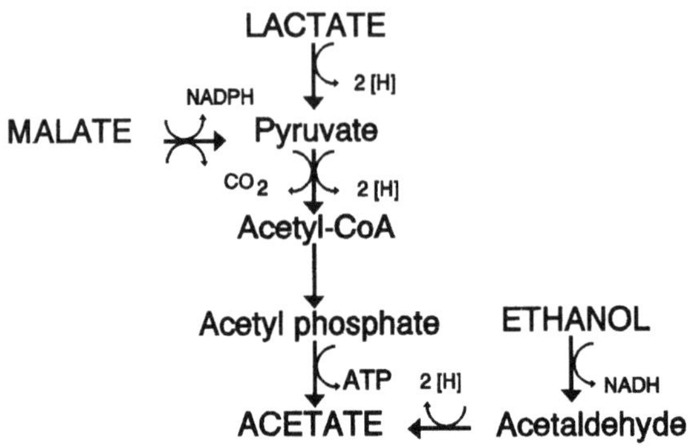

Figure 2.3 Pathways for the oxidation of lactate, ethanol and malate in *Desulfovibrio* species.

leads to production of an extremely malodorous compound or compounds, most probably as a result of a reaction of acetaldehyde with sulfide, and to cessation of growth (cf. Widdel, 1988). An intermediate transfer to a medium with lactate + ethanol avoids this problem.

The rod-shaped, desulfoviridin-negative strains [which should now be accommodated in the genus *Desulfomicrobium* (Devereux et al., 1990)], such as *Dv. desulfuricans* Norway (=*Dv. baculatus* DSM 1741), *Dv. baculatus* × (DSM 1743), and the type strain of the latter species (strain VKM), do not utilize ethanol. However, rod-shaped, desulfoviridin-negative *Desulfovibrio* strains that utilize ethanol do exist, e.g., *Desulfovibrio* HL21, maintained as *Dv. baculatus* DSM 2555 (Kremer et al., 1988a).

In experiments with *Dv. gigas*, *Dv. baculatus* HL21, *Dv. carbinolicus*, and *Desulfovibrio* 20020 (DSM 3099; most likely a *Dv. salexigens* strain), ethanol was shown to be oxidized by an NAD-dependent alcohol dehydrogenase (Fig. 2.3). The level of enzyme activity was strongly elevated in ethanol-grown cells as compared with lactate-grown cells (Kremer et al., 1988a). Acetaldehyde dehydrogenase was detected and assayed with oxidized benzylviologen. In crude extracts of *Dv. gigas* the enzyme was strongly stimulated by K^+. Even after extensive dialysis of the extract, aldehyde dehydrogenase remained CoA- and phosphate independent, which indicates that the reaction product is not acetyl-CoA or acetyl phosphate. Whether the aldehyde dehydrogenase is identical to the molybdenum [Fe-S] protein purified by Moura et al. (1976) and shown to have some aldehyde dehydrogenase activity (Turner et al., 1987) remains to be established.

Utilization of longer chain primary alcohols by incompletely oxidizing sulfate-reducing bacteria yields the corresponding acid and usually requires inclusion of a suitable carbon source in the medium (e.g., acetate + CO_2). Most likely these alcohols are utilized because of nonspecificity of ethanol and acetaldehyde dehydrogenases. Alcohol dehydrogenase in extracts of ethanol-grown *Dv. gigas* was active with ethanol (100%), propanol (72%), and butanol (36%), but not with isopropanol. The latter observation is in accordance with the organism's inability to utilize isopropanol (cf. Esnault et al., 1988). Similarly, aldehyde dehydrogenase activity towards propanal and butanal was demonstrated.

Activity of *Dv. gigas* alcohol dehydrogenase towards methanol, a substrate that is not utilized by the organism, was extremely low. Contrary to the species description in the ninth edition of Bergey's Manual (Postgate, 1984b), methanol-supported growth of *Desulfovibrio* strains is rather rare; if it occurs, it is very slow. Strains that have been shown to grow on methanol are *Dv. carbinolicus* EDK82 (Nanninga and Gottschal, 1987), *Dv. salexigens* DSM 2638 (Zellner et al., 1989), and *Dv. alcoholovorans* (Qatibi et al., 1991b). Kremer et al. (1988a) ascribed slow growth of *Dv. carbinolicus* on methanol to the presence of very high levels of NAD-dependent alcohol dehydrogenase with a slightly better affinity for methanol than the *Dv. gigas* enzyme. Methanol is also utilized by some *Desulfotomaculum* and *Desulfobacterium* strains (Klemps et al., 1985; Szewzyk and Pfennig, 1987).

Only a few strains of sulfate-reducing bacteria, such as *Desulfococcus multivorans* (Widdel, 1988) and *Dv. salexigens* DSM 2638 (Zellner et al., 1989a), appear to utilize secondary alcohols such as 2-propanol and 2-butanol.

Diols, such as ethanediol and propanediols, have not routinely been included in substrate tests with sulfate-reducing bacteria. Consequently, it cannot be said how widespread the ability is to utilize these compounds among these organisms. Ethanediol oxidation was reported for *Desulfovibrio* DG2 (Dwyer and Tiedje, 1986), *Dv. carbinolicus* (Kremer et al., 1988a), *Dv. vulgaris* strain Marburg (Tanaka, 1990), *Dv. alcoholovorans*, and *Dv. fructosovorans* (Qatibi et al., 1991b). The first organism also utilized oligomers of ethyleneglycol (up to the tetramer). Ethanediol is oxidized to acetate or, in the case of *Dv. vulgaris* strain Marburg, to acetate and glycolate.

Oxidation of 1,3-propanediol has been shown to occur in the same species and in addition for *Desulfovibrio* strain OttPdl (Oppenberg and Schink, 1990). Products of 1,3-propanediol oxidation were acetate + CO_2 (in the case of *Desulfovibrio* OttPdl and *Dv. alcoholovorans*) and 3-hydroxypyruvate in cultures of *Dv. carbinolicus* (Nanninga and Gottschal, 1987) and *Dv. fructosovorans* (Qatibi et al., 1991b), and, in the case of *Dv. vulgaris*, a mixture of 3-hydroxypropionate and acetate (Tanaka, 1990). A hypothetical scheme by Oppenberg and Schink (1990) suggested 3-hydroxypropanal, 3-hydroxypropionate, malonylsemialdehyde, and acetaldehyde as important intermediates in oxidation of 1,3-propanediol to acetate.

Utilization of 1,2-propanediol was reported for *Desulfovibrio* DG2 (Dwyer and Tiedje, 1986) and *Dv. alcoholovorans* SPSN and the rather similar strain *Desulfovibrio* DFG (Qatibi, 1990). Strain SPSN formed a mixture of acetate (+CO_2) and smaller amounts of propionate as products, whereas DFG formed exclusively propionate. Propionate production can be envisaged as the result of a dehydration of 1,2-propanediol to propanal, followed by an oxidation to propionate; one can speculate that the pathway from diol to acetate starts with an oxidation to lactaldehyde in a reaction catalyzed by an alcohol dehydrogenase.

Glycerol utilization among *Desulfovibrio* strains is not rare. It was demonstrated for (1) certain strains that require 2.5% NaCl in the growth media (most of them are of marine origin), such as *Desulfovibrio* El Agheila Z [its classification as a *desulfuricans* strain was shown to be unjustified (see Devereux et al., 1990)], *Dv. desulfuricans* Canet, *Desulfovibrio* DSM 3099, *Desulfovibrio* DSM 3100, and *Dv. salexigens* DSM 2638 (Baker et al., 1962; Stams et al., 1985; Zellner et al., 1989a); (2) strains that grow in freshwater media and media with low NaCl concentrations such as *Dv. vulgaris* Hildenborough (NCIB 8303), which grows slowly (Kremer and Hansen, 1987), *Desulfovibrio* OttPdl (Oppenberg and Schink, 1990), *Dv. carbinolicus* (Nanninga and Gottschal, 1987), *Dv. fructosovorans* (Ollivier et al., 1988), and *Dv. alcoholovorans* (Qatibi et al., 1991b). With the exception of *Dv. carbinolicus*, which forms 3-hydroxypropionate as sole product, all strains studied oxidize glycerol to acetate and CO_2. The pathway of glycerol dissimilation was shown to involve a glycerol kinase, a membrane-bound glycerol-3-phosphate dehydrogenase, and enzymes of the glycolytic route to pyruvate. Glycerol kinase also used dihydroxyacetone as a substrate, which explains the good growth of glycerol-utilizing strains on this substrate (Kremer and Hansen, 1987). Most likely glycerol oxidation to 3-hydroxypropionate is initiated by a dehydration to 3-hydroxypropanal.

Utilization of formate, acetate, and longer straight and branched alkanoic acids Formate can be oxidized by almost all *Desulfovibrio* species, by some *Desulfotomaculum* species, and by *Desulfococcus, Desulfobacterium,* and *Desulfonema* (Widdel, 1988). In *Desulfovibrio* formate dehydrogenase is a periplasmic protein. Purified cytochrome c_{553} could function as electron acceptor in experiments with partially purified formate dehydrogenase from *Dv. vulgaris* Miyazaki (Yagi, 1979); the dehydrogenase from *Dv. gigas* is thought to use cytochrome c_3 as electron acceptor (Riederer-Henderson and Peck, 1986).

Acetate and acetyl group oxidation may proceed either via the citric acid cycle or the oxidative carbon monoxide dehydrogenase pathway, as discussed above.

Propionate is a characteristic growth substrate for *Desulfobulbus* and is oxidized to acetate and CO_2. Several of the completely oxidizing sulfate-reducing bacteria also utilize propionate but do so more slowly. *Desulfobulbus* oxidizes propionate via a pathway involving a transcarboxylation of propionyl-CoA to methylmalonyl-CoA with oxaloacetate as donor (Stams et al., 1984; Kremer and Hansen, 1988) (see Fig. 2.4). Succinate is a typical

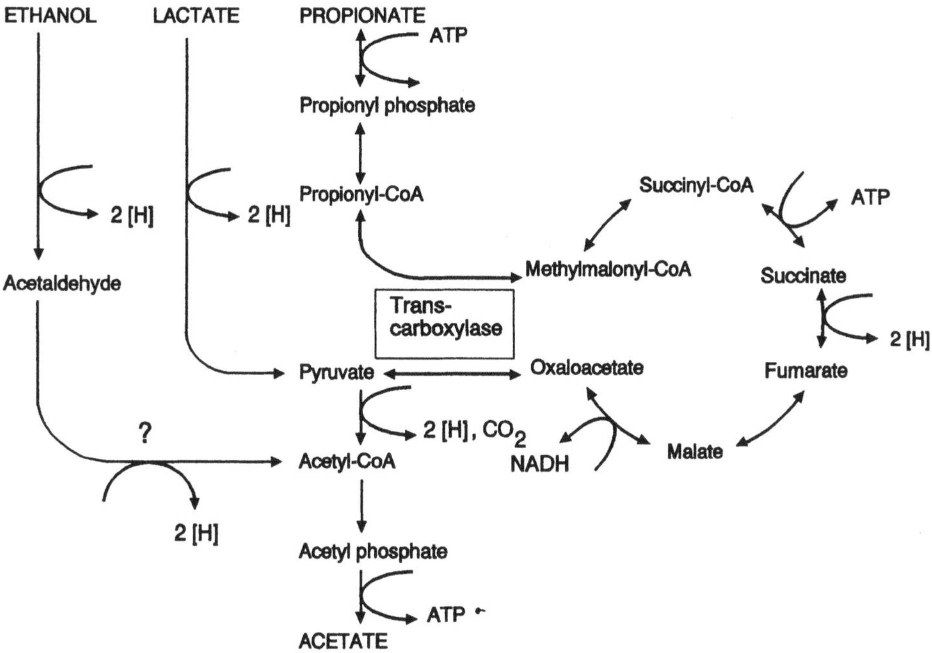

Figure 2.4 Pathways for ethanol, lactate, and propionate oxidation and propionate formation in *Desulfobulbus propionicus*.

C_4-intermediate. Succinate dehydrogenase from *Desulfobulbus elongatus* consists of three subunits and contains one cytochrome *b* and a flavin and eight non-heme irons (Samain et al., 1987).

Isobutyrate utilization has been reported only for certain completely oxidizing sulfate-reducing bacteria which generally do so via the oxidative carbon monoxide dehydrogenase pathway (Widdel, 1988). *Desulfococcus multivorans* degrades isobutyrate (and also propionate) via a succinate pathway. Isobutyrate is converted to methylmalonyl-CoA via isobutyryl-CoA, metacrylyl-CoA, 3-hydroxyisobutyryl-CoA, 3-hydroxyisobutyrate, methylmalonate semialdehyde, and propionyl-CoA. Propionyl-CoA is ultimately oxidized to acetyl-CoA and CO_2 via a pathway similar to that found in *Desulfobulbus* (Stieb and Schink, 1989).

Utilization of sugars Only few strains of sulfate-reducing bacteria have been shown to use sugars as energy substrates. *Desulfotomaculum nigrificans* DSM 574 (Klemps et al., 1985) and a second thermophilic spore-forming species, *Dtm. geothermicum* (Daumas et al., 1988), were reported to oxidize fructose. An earlier claim that the former species could utilize glucose was shown to have arisen from heat sterilization, which promoted conversion of glucose to fructose.

Only few representative *Desulfovibrio* species have been shown to utilize sugars. In this respect it is of interest that Joubert and Britz (1987) isolated four similar strains of rod-shaped, motile, Gram-negative bacteria that degraded several carbohydrates. Significant sulfate reduction was shown only in experiments with lactate as a substrate. Because of unsatisfactory carbon and redox balances in their experiments, however, the data should be interpreted with caution. *Dv. fructosovorans* grows on fructose and at a reasonable rate (μ_{max} 0.06 h^{-1}), albeit far more slowly than most fermenting bacteria; fructose is oxidized to acetate (+CO_2) (Ollivier et al., 1988). Recently, slow growth of *Dv. salexigens* DSM 2638 on glucose was described (Zellner et al., 1989a).

Several carbohydrates are utilized by a sulfate reducer that was isolated from the hindgut of a termite; this organism, which was described as *Dv. termitidis* (Trinkerl et al., 1990), also utilizes pyruvate and lactate. It remains intriguing why so few sulfate-reducing bacteria utilize carbohydrates as substrates. As described in section 2.5, both *Desulfovibrio* and *Desulfobulbus* strains accumulate and degrade polyglucose as a storage polymer; the pathway for breakdown of glucose phosphates to acetate is therefore present.

Utilization of aromatic compounds The range of aromatic compounds that can be degraded by sulfate-reducing bacteria is impressive. Some sulfate reducers have been shown to utilize certain aromatic compounds without destruction of the aromatic ring. Some strains of *Desulfotomaculum* convert trimethoxybenzoate to trihydroxybenzoate (Klemps et al., 1985). Three *Desulfovibrio* strains (*Dv. vulgaris* Marburg, *Dv. simplex* strain XVI, and *Desulfovibrio* strain MP47) oxidize certain substituted benzaldehydes (e.g., vanillin (4-OH-3-methoxybenzaldehyde), syringaldehyde (4-OH-3,5-dimethoxybenzaldehyde), *p*-anisaldehyde (4-methoxybenzaldehyde), and 2- and 3-OH-benzaldehyde) to the respective benzoic acids (Zellner et al., 1990). Reductive dehalogenation of 3-chlorobenzoate and a few other halobenzoates to benzoate (DeWeerd et al., 1990) by *Desulfomonile tiedjei*, a facultative sulfate reducer, and oxidation of 3-phenylpropionate to benzoate by *Desulfococcus niacini* (Imhoff-Stuckle and Pfennig, 1983) are other examples of utilization of aromatic compounds without ring structure degradation. Under sulfate-reducing conditions, however, *Desulfomonile tiedjei* does oxidize benzoate.

Degradation of aromatic compounds by sulfate-reducing bacteria which also breaks the aromatic ring structure has been reported for several compounds (see Table 2.1). The main natural sources of aromatic compounds are aromatic amino acids and lignin and its precursors and degradation products. Another source of aromatic compounds is xenobiotics, that is, synthetic compounds not normally present in the environment (Schnell et al., 1989), such as aniline (aminobenzene). Aniline is degraded via 4-aminobenzoate, 4-aminobenzoyl-CoA, and benzoyl-CoA (Schnell and Schink, 1991).

Utilization of dicarboxylic acids Malate and fumarate are used mainly by several *Desulfovibrio* species and also by certain *Desulfobacterium* strains (Widdel, 1988). Kremer et al. (1989) confirmed that *Dv. vulgaris* does not grow on these C_4-substrates. They also reported that *Dv. gigas* grows well on dicarboxylic acids, an observation not in agreement with data of Widdel (1988) and Postgate (1984b). In four *Desulfovibrio* strains studied, the degradation pathway involved an NADP-dependent malic enzyme as key enzyme in conversion of malate to pyruvate (Fig. 2.3). The *Dv. gigas* enzyme required Mn^{2+} or Mg^{2+} and was strongly stimulated by K^+. Succinate was also utilized by the strains; growth yields were very poor, a not unexpected finding in view of the unfavorable reduction potential of the succinate/fumarate couple.

Utilization of amino acids Although some strains have the metabolic capability to oxidize amino acids, in general, good growth of sulfate-reducing bacteria on amino acids is not very common. Some examples of amino acid oxidation include cysteine degradation by certain *Desulfovibrio* strains (Senez and Leroux-Gilleron, 1954; Forsberg, 1980) and utilization of several amino acids, including cysteine, by two marine *Desulfovibrio* strains (Stams et al., 1985). *Desulfotomaculum ruminis, Dtm. nigrificans,* and *Dtm. thermoacetoxidans* grow on alanine (Klemps et al., 1985; Min and Zinder, 1990). Cells of the marine *Desulfovibrio* strains and *Dtm. ruminis* contained high levels of NAD-dependent alanine dehydrogenase when grown on alanine (Stams and Hansen, 1986).

Miscellaneous compounds Acetone is a fermentation product of certain strains of *Clostridium*. Platen et al. (1990) reported that *Desulfococcus biacutus* grew slowly on acetone, which was degraded via acetoacetyl-CoA and acetyl-CoA to CO_2. The nature of the initial endergonic carboxylation reaction has not yet been elucidated. The ability of *Desulfovibrio furfuralis* to oxidize furfural, a product formed by dehydration of pentoses, was studied in detail by Folkerts et al. (1989). During the initial stages of growth on furfural, furfurylalcohol and 2-furoic acid accumulated; eventually approximately two acetates were formed per furfural oxidized. A pathway was proposed with succinic semialdehyde as a key intermediate.

Of special interest is the finding that *Desulfobacterium autotrophicum*, a complete oxidizer using the oxidative carbon monoxide dehydrogenase pathway, can reductively dechlorinate certain chlorinated compounds; low concentrations of tetrachloromethane were reduced to tri- and dichloromethane, and 1,1,1-trichloroethane was reduced to 1,1-dichloroethane (Egli et al., 1987). Similarly, *Desulfomonile tiedjei* reductively dechlorinates perchloroethene to trichloroethene (Fathepure et al., 1987). The transformations of chlorinated methanes by *Desulfobacterium* are thought to be catalyzed by metalloporphyrins such as corrinoids (Egli et al., 1990).

2.4 Fermentative Capacities of Sulfate-Reducing Bacteria

Sulfate-reducing bacteria in freshwater sediments and other similar anaerobic habitats (e.g., anaerobic waste water purification plants) often encounter sulfate limitation; even in the deeper layers of marine sediments sulfate limitation is not unusual. It is not unexpected, therefore, that several sulfate reducers do not entirely rely on sulfate reduction for their energy generation. Some strains reduce nitrite, and exceptionally even nitrate, to ammonium and thus can carry out another respiratory process; other strains can carry out a variety of fermentation processes.

As noted in Chapter 1, fermentation is a process in which a compound is degraded without the use of an exogenous electron acceptor. Fermentative capacities among sulfate-reducing bacteria vary but are often present. Interestingly, some *Desulfovibrio* strains even ferment the inorganic compounds sulfite or thiosulfate to sulfate and sulfide (Bak and Pfennig, 1987); this property was also described for *Desulfomonile tiedjei*, an organism that was originally isolated as an anaerobe that reductively dechlorinates 3-chlorobenzoate (Mohn and Tiedje, 1990).

Fermentations at low hydrogen partial pressure Of particular interest is the ability of some *Desulfovibrio* strains to ferment lactate and ethanol according to equations (2.3) and (2.4), but only if the H_2 partial pressure is kept low by a hydrogenotrophic methanogen:

$$\text{lactate} + 2\,H_2O \rightarrow \text{acetate} + HCO_3^- + H^+ + 2\,H_2 \qquad (2.3)$$

$$\text{ethanol} + H_2O \rightarrow \text{acetate} + H^+ + 2\,H_2 \qquad (2.4)$$

These processes were discovered by Bryant et al. (1977) after enrichments on ethanol in low-sulfate media inoculated with anaerobic sewage sludge led to isolation of a syntrophic association between a *Desulfovibrio vulgaris* strain and a methanogen. Similar syntrophic utilization of lactate and ethanol was demonstrated for *Dv. desulfuricans* (Bryant et al., 1977), *Dv. gigas* (Kremer et al., 1988a), and *Desulfotomaculum nigrificans* (Klemps et al., 1985). In a whey biomethanation system studied by Chartrain and Zeikus (1986), *Dv. vulgaris* was shown to play a major role in syntrophic lactate and ethanol degradation.

The first steps of reactions 2.3 and 2.4 (lactate or ethanol to pyruvate or acetaldehyde and hydrogen) are inhibited by H_2 accumulation. Even at low hydrogen concentrations, production of hydrogen from lactate is an energy-dependent process that is thought to be driven by the electrochemical proton potential (Thauer, 1989); such an energy requirement is not likely in the case of ethanol (Hansen and Kremer, 1990).

Recently, syntrophic utilization of 1,3-propanediol by *Desulfovibrio* strains (Oppenberg and Schink, 1990; Qatibi, 1990), 1,2-propanediol, and glycerol was reported (Qatibi et al., 1991a). In the absence of sulfate, *Dv. alcoholovorans* degrades 1,2-propanediol into propionate and H_2 in coculture with *Methanospirillum hungatei*. Degradation of 1,3-propanediol by the same species yielded acetate and CO_2 with considerable transient accumulation of 3-hydroxypropionate; glycerol was fermented to acetate, CO_2, and 3 H_2 (Qatibi et al., 1991a).

The reactions discussed in this section demonstrate that even in the absence of sulfate, bacteria belonging to the genus *Desulfovibrio* can actively ferment a variety of substrates with concomitant release of reducing equivalents as H_2, but only at the low hydrogen partial pressures that prevail in most anaerobic environments. In these processes protons, instead of sulfate, function as electron acceptors, and no reduced products other than H_2 are formed. It is important to stress that there is no net consumption of protons. Thus, protons do not function as typical exogenous electron acceptors, and consequently such processes are consistent with our conventional definition of fermentation.

Transfer of reducing equivalents from fermenting organisms to methanogens does not only proceed via formation of hydrogen. Fermenting organisms may also reduce CO_2 to formate, and formate may be utilized by methanogens. In certain systems interspecies formate transfer has been shown to be of major importance (Thiele and Zeikus, 1988).

Fermentations not dependent on a low hydrogen partial pressure Fermentations that are not dependent on a low hydrogen partial pressure are possible with malate and fumarate. This fermentation is not surprising since many *Desulfovibrio* strains have the metabolic capability of oxidizing dicarboxylic acids, including succinate to acetate (see Section 2.3). Transfer of the reducing equivalents (liberated in the oxidation of fumarate and malate to acetate) to fumarate instead of APS and bisulfite via an electron transport chain to the succinate dehydrogenase/fumarate reductase would not be difficult. *Desulfobacterium autotrophicum* and similar organisms also grow by fermentation of malate and fumarate (Brysch et al., 1987).

Substrates that are fermented in pure culture by only a few species are choline, glycerol, fructose, and serine. Choline (*N,N,N*-trimethylethanolamine) supports fermentative growth of *Dv. desulfuricans* and *Dv. vulgaris* subsp. oxamicus. Products are ethanol, acetate, and trimethylamine, with acetaldehyde as an important intermediate (Hayward, 1960). Energy conservation was proposed to occur via acetyl-CoA and acetyl phosphate. Glycerol is dismutated to 1,3-propanediol and 3-hydroxypropionate by *Dv. carbinolicus* (Nanninga and Gottschal, 1987) and

Dv. fructosovorans (Ollivier et al., 1988). The latter species ferments fructose to acetate, succinate, and ethanol. Fructose is also fermented by *Desulfotomaculum nigrificans* and *Dtm. geothermicum*, but the products are unknown (Klemps et al., 1985; Daumas et al., 1988). Fermentation of serine (with acetate, formate, and succinate as products) was reported for some marine *Desulfovibrio* strains (Stams et al., 1985).

Some genera of sulfate-reducing bacteria can carry out a propionic fermentation. This process was studied in detail in *Desulfobulbus propionicus*. The fermentation of lactate proceeds with the same propionate/acetate stoichiometry as in an ideal *Propionibacterium* fermentation:

$$3 \text{ lactate} \rightarrow 2 \text{ propionate} + 1 \text{ acetate} + 1\ CO_2 \qquad (2.5)$$

Of special interest was the finding that *Desulfobulbus propionicus* (Laanbroek et al., 1982), but not *Desulfobulbus elongatus* (Samain et al., 1984) also ferments some alcohols:

$$3 \text{ ethanol} + 2\ HCO_3^- \rightarrow 2 \text{ propionate} + 1 \text{ acetate} + H^+ \qquad (2.6)$$

$$3 \text{ propanol-1} + 2 \text{ acetate} + 2\ HCO_3^- \rightarrow 5 \text{ propionate} + H^+ \qquad (2.7)$$

The pathway of propionate formation was elucidated by measuring enzyme levels and by studying the fate of C-2-labeled ethanol (Stams et al., 1984). The reaction sequence from pyruvate to propionate is essentially the same as in *Propionibacterium* (Figure 2.4) with succinate and methylmalonyl-CoA as key intermediates. Unlike the *Propionibacterium* pathway, however, succinate is not converted to succinyl-CoA in a transferase reaction with propionyl-CoA, but rather via a succinate thiokinase; propionate is formed via propionyl phosphate. Levels of key enzymes of this pathway such as transcarboxylase were comparable. In the presence of sulfate the reaction sequence is used in the reverse direction to oxidize propionate to acetate (Kremer and Hansen, 1988; see above); the thermodynamics of the reactions involved allow such a reversiblity.

Some sulfate-reducing bacteria have the ability to carry out a slow homoacetogenic fermentation. *Dtm. orientis* could be cultivated in sulfate-free media on the following substrates: lactate, formate, ethanol + CO_2, methanol + CO_2, and trimethoxybenzoate + CO_2 (Klemps et al., 1985). Trimethoxybenzoate was demethylated to yield trihydroxybenzoate. Although *Dtm. orientis* does ferment $H_2 + CO_2$ to acetate, in particular when it is grown under sulfate limitation, this process does not lead to growth (Cypionka and Pfennig, 1986).

2.5 Storage Polymers

Organic storage polymers are synthesized by most prokaryotes. Their accumulation is usually a reflection that bioenergetic processes exceed cellular requirements for synthesis of structural materials such as proteins.

Storage polymer accumulation is, therefore, most pronounced when some component other than carbon and energy source is limiting growth, e.g., the nitrogen source. For example, in *Desulfovibrio vulgaris* Hildenborough and *Dv. baculatus* HL21, accumulation of intracellular polyglucose was brought about by limitation of Fe^{2+} or NH_4^+ in growth medium. In *Dv. gigas*, even in normal media, the presence of large amounts of polyglucose was observed (Stams et al., 1983). Early exponential phase cells and cells from lactate-limited continuous culture did not show this massive accumulation of polyglucose (Kremer et al., 1988b). Breakdown of polyglucose by *Dv. baculatus* HL21, in the presence of sulfate and hydrogen, yielded acetate (and CO_2); small amounts of ethanol and succinate were observed as additional products in the absence of sulfate. *Desulfobulbus propionicus* also uses polyglucose as a storage polymer. Bacteria belonging to the genera *Desulfococcus, Desulfosarcina,* and *Desulfonema* were reported to contain sudanophilic granules which are usually taken as an indication for the presence of poly-β-hydroxybutyric acid (Widdel, 1980).

2.6 Some Aspects of Carbon Assimilation

Carbon nutrition Most sulfate-reducing bacteria have simple growth requirements. They grow in mineral media with an organic or inorganic energy source and a suitable carbon source or combination of carbon sources. In many cases the organic energy source can also be used as the carbon source. Growth factors are required by some species or strains. Improvement of media and growth conditions often led to a disappearance of requirements for additions of yeast extract or other complex nutrients. Thus, Klemps et al. (1985) could cultivate most of their *Desulfotomaculum* strains in simple vitamin-supplemented media after solving the problem of sulfide toxicity; only *Dtm. nigrificans* still required 0.02% yeast extract for good growth.

Some organic compounds have been shown to be reasonable carbon sources for sulfate-reducing bacteria but cannot be used as energy source. A *Desulfovibrio* strain studied by Sorokin (1966b) used acetate, mannitol, alanine, and asparagine as single carbon sources (in the presence of some CO_2) but did not degrade these substrates as energy sources. In a bicarbonate-buffered medium *Dv. baculatus* HL21 grew with hydrogen as energy source and L-alanine, L-serine, or L-aspartate as carbon and nitrogen source (Stams et al., 1986). None of these amino acids was used as an energy source, but dense cell supensions very slowly oxidized alanine to acetate. Reasonable NAD-dependent alanine dehydrogenase activities were induced by growth with alanine as carbon and nitrogen source, but alanine dehydrogenase activities were higher in strains that use the substrate as an energy source (Stams and Hansen, 1986). Because this strain oxidizes ethanol in an NAD-dependent reaction (see Section 2.4), the inability to grow on alanine most likely arises from a very slow degradation rate (either because of a low transport capacity across the cytoplasmic membrane or slow conversion to pyruvate) and not by problems with linking the oxidation of NADH with APS and bisulfite reduction, as suggested by Stams et al. (1986).

C_2 assimilation involving $C_2 + C_1$ Sulfate-reducing bacteria that incompletely oxidize substrates such as lactate to acetate, e.g., *Desulfovibrio* and *Desulfobulbus* species, can use acetate + CO_2 as their carbon sources in the presence of an energy source such as H_2 or formate. This was shown by Sorokin (1966a,b), Badziong et al. (1978), and Brandis and Thauer (1981). The key reaction in acetate assimilation is reductive carboxylation of acetyl-CoA to pyruvate as shown by the fate of labeled acetate (Badziong et al., 1979):

$$\text{acetyl-CoA} + CO_2 + 2\,[H] \rightarrow \text{pyruvate} + \text{CoA} \qquad (2.8)$$

Autotrophic growth Enrichment and isolation of truly autotrophic sulfate-reducing bacteria has proved to be difficult. Use of a mineral medium with sulfate as electron acceptor and H_2/CO_2 as a gas mixture consistently led to enrichment of a coculture of an autotrophic *Acetobacterium*, which produced acetate from hydrogen and carbon dioxide, and a *Desulfovibrio*, which utilized hydrogen as an energy source and acetate + CO_2 as carbon source, as described in the preceding section (Brysch et al., 1987).

By using direct dilutions of mud samples in agar media, Brysch et al. (1987) succeeded in isolating pure cultures of facultatively autotrophic sulfate-reducing bacteria. From marine mud, strains were isolated that grew autotrophically on hydrogen with doubling times of 16 to 20 h; a characteristic representative of these bacteria was described as *Desulfobacterium autotrophicum*. Some of the sulfate-reducing bacteria enriched and isolated on certain organic substrates, namely *Desulfococcus niacini, Desulfonema limicola, Desulfosarcina variabilis, Desulfotomaculum orientis, Dtm. geothermicum, Desulfobacter hydrogenophilus, Desulfobacterium vacuolatum*, and *Desulfomonile tiedjei*, have also been shown to grow autotrophically (Brysch et al., 1987; Schauder et al., 1987; Klemps et al., 1985; Daumas et al., 1988; DeWeerd et al., 1990). Because growth of all autotrophic sulfate reducers studied thus far is slower than that of homoacetogens, such as *Acetobacterium woodii*, and far slower than that of *Desulfovibrio* strains which grow on H_2/CO_2 + acetate, it is obvious that enrichments on H_2/CO_2 lead to the cocultures described above. *Desulfovibrio baarsii* [a species that on the basis of phylogenetic data should be removed from the genus *Desulfovibrio*; see Devereux et al. (1989)] cannot grow on $H_2 + CO_2$ but does grow on formate + CO_2 and therefore is able to obtain all of its cell carbon from C_1-compounds (Jansen et al., 1984).

Biosynthesis of all cell components from C_1-compounds has been

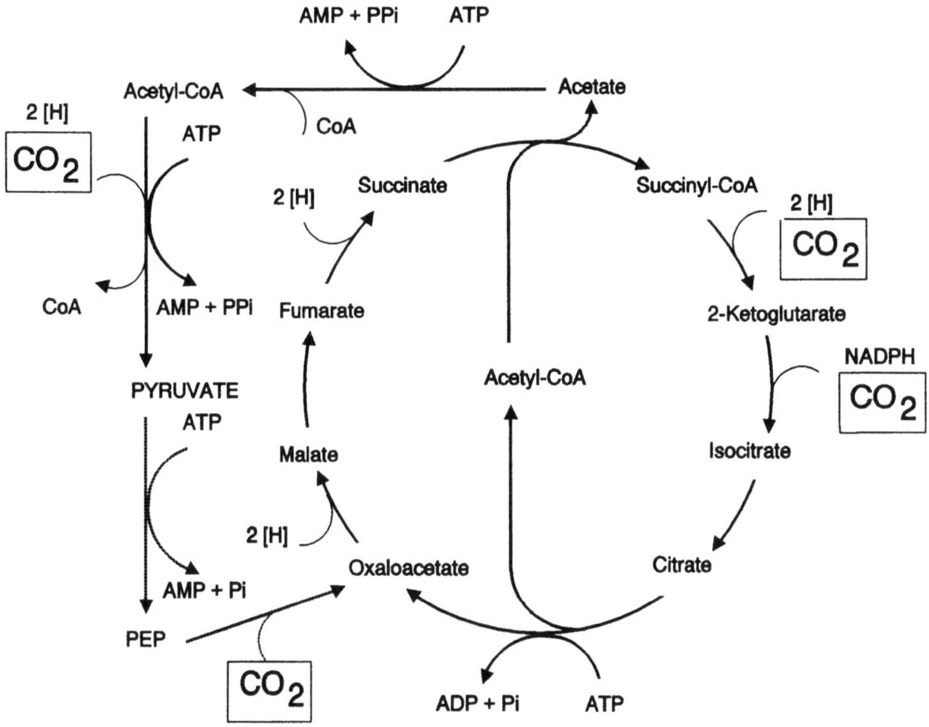

Figure 2.5 Reductive tricarboxylic acid cycle for CO_2 assimilation in *Desulfobacter hydrogenophilus*. Dotted lines indicate the anaplerotic sequence from pyruvate to the citric acid cycle.

shown to involve pathways other than the well-known ribulose bisphosphate pathway that functions in almost all aerobic autotrophs. Labeling studies and enzyme measurements have demonstrated two different mechanisms of C_1-fixation in sulfate-reducing bacteria. *Dba. hydrogenophilus* uses a reductive citric acid cycle. Key enzymes of this autotrophic pathway are an ATP-citrate lyase and two reductive carboxylases (for synthesis of α-ketoglutarate and pyruvate) (see Figure 2.5 for details). *Dba. hydrogenogenophilus,* like other *Desulfobacter* species, oxidizes acetyl-CoA via an oxidative citric cycle when grown on acetate. The ATP-citrate lyase is thought to function in both oxidative and reductive cycles in *Dba. hydrogenophilus* (Schauder et al., 1987).

An entirely different pathway was found in *Desulfobacterium autotrophicum* (Schauder et al., 1989; Länge et al., 1989). The enzyme, carbon monoxide dehydrogenase, which functions as an acetyl-CoA synthesizing enzyme, plays a key role in autotrophic growth of this organism. This route, in which a C_2-compound is synthesized from 2 C_1-compounds in a noncyclic mechanism, is essentially the same as the pathway of acetogenesis of anaerobic bacteria with a homoacetate fermentation and a reversal of the oxidative carbon monoxide dehydrogenase pathway shown earlier (Figure 2.2). Other sulfate-reducing bacteria in which the reductive carbon monoxide dehydrogenase pathway is thought to function in the synthesis of cell material from C_1-compounds include *Desulfovibrio baarsii* and *Desulfotomaculum* (Jansen et al., 1984; Schauder et al., 1986).

2.7 Summary and Concluding Remarks

Sulfate-reducing bacteria can metabolize a wide variety of organic compounds as well as a few inorganic compounds. The knowledge that has been acquired over the past decades has shed new light on the metabolic potential of these bacteria and on the anaerobic degradation of organic compounds in nature. The study of the breakdown of organic substrates by sulfate-reducing bacteria has led to the discovery of novel pathways for oxidation of acetyl groups to CO_2 (the variants of the oxidative carbon monoxide dehydrogenase pathway).

Biochemical details of the utilization of several substrates remain to be elucidated. What has become clear from the available data, however, is that during degradation of a particular organic substrate in most cases reducing equivalents are liberated at different redox levels and transferred to different redox carriers. These carriers can be ferredoxins, flavodoxins, NADH and NADPH, or membrane-associated components such as certain cytochromes. This complex and diversified electron carrier system obviously complicates bioenergetic studies of the process of sulfate reduction; this is especially true since reduction of sulfate to sulfide is also a rather complex process in comparison with the reduction of oxygen to water.

[References, see p. 211]

3

Bioenergetic Strategies of the Sulfate-Reducing Bacteria

Harry D. Peck, Jr.

3.1 Introduction

It has become evident that the basic chemiosmotic mechanisms that govern electron transfer-coupled phosphorylation, including photophosphorylation and oxidative phosphorylation, must have evolved under anoxic conditions existent in the early environment of earth. In this anaerobic environment there does not appear to have been a single major energy source, and it is possible that a diversity of bioenergetic systems evolved around specific redox couples, such as $H_2/S°$, H_2/CO_2, CO/SO_4^{2-}, etc. Thermodynamically, these reactions do not provide for biosynthesis of large amounts of ATP, and growth of most anaerobes is generally viewed as being severely limited by ATP availability. With the advent of photosynthetic systems and subsequent generation of an oxygen atmosphere, evolution became focused upon higher-yielding, ATP-generating systems involving light and oxygen, resulting in loss of the rich diversity of anaerobic bioenergetic systems from the metabolic potential of most photosynthetic and aerobic prokaryotes and eukaryotes. In this regard, it should be noted that a number of redox enzymes, such as anaerobic carbon monoxide dehydrogenase (Ragsdale et al., 1983) and some of the "iron only" hydrogenases (Chen and Mortenson, 1974), involved in anaerobic metabolism are highly oxygen sensitive, and these proteins may not have been able to survive the change to oxygen-based bioenergetics regardless of how slowly this change occurred.

Thus, oxidative phosphorylation and photophosphorylation became the single most important energy-conserving reactions on our planet, and other mechanisms of energy conservation were largely lost in eukaryotes and strictly aerobic prokaryotes. Nevertheless, one may speculate that a diversity of energy-conserving mechanisms has persisted in strictly anaerobic bacteria that have never adapted to aerobic conditions. Study of these different and unique bioenergetic systems should provide new and different insights into the evolution and mechanisms of energy coupling in biological systems. Three distinct issues must be considered in this study.

Evolutionary considerations of bioenergetic mechanisms First, a perennial operational problem concerned the identification of existing, strictly anaerobic microorganisms whose bioenergetic systems may be profitably investigated as examples of bioenergetic diversity. The discovery and description of the Archaebacteria as the third form of life (Balch et al., 1979) led to the tentative conclusion that these microorganisms were primitive bacteria related to the first forms of life on earth. Initially, development of phylogenetic trees for prokaryotes, with techniques of molecular biology, established that prokaryotes were organized in at least two super kingdoms, the Eubacteria and the Archaebacteria (Woese, 1987). Recently, however, the evolutionary tree for prokaryotes has been rooted, with the result that Eubacteria are now regarded as ancestral to Archaebacteria (Iwabe et al., 1989). This development removed a number of problems associated with the proposed ancestral position of Archaebacteria and indicates that bioenergetic diversity may be profitably reinvestigated in the Eubacteria.

A second aspect of this problem has been to match a vision of the early surface environment of the earth with various physiological types of bacteria. Initially, a bare surface environment was proposed to be populated with bacterial chemoautotrophs oxidizing sulfur and ammonia and fixing carbon dioxide as a sole carbon source. Subsequently, our concepts of early earth embraced an anaerobic and reduced surface laced with pools of nutrient-rich "organic soup," which was populated with fermentative anaerobes (Abelson, 1966). This speculation was provocative and challenging but was of little scientific substance, except that it emphasized the essential nature of a match between constant energy sources, prior to photosynthesis in the early environment, with physiological capabilities of various microorganisms surviving today in similar unique environments.

Over the past decade extremely thermophilic anaerobes, largely Archaebacteria, capable of growth above 90°C on inorganic substrates such as H_2, CO_2, and sulfur have been characterized. Recognition of these new physiological types of microorganisms plus the description of the biota of the deep ocean thermal vents have encouraged a view of the early earth environment, which was anaerobic with temperatures up to and above 100°C with a diverse array of bioenergetic substrates continuously provided by geologic activity (Corliss et al., 1981). Among these substrates, H_2 would be of central importance as an electron donor for electron transfer-coupled phosphorylation. Reduction of various inorganic species coupled to H_2 oxidation would allow for support of different physiological types of metabolism such as respiratory sulfate reduction, sulfite reduction (Postgate, 1951), respiratory sulfur reduction (Stetter and Gaag, 1983), methanogenesis (Bryant, 1979), respiratory nitrate reduction (ammonia forming; Wolin et al., 1961), and acetogenesis (Wieringa, 1940). Thus, it might be anticipated that the diversity of substrates originating from geological activities would encourage evolutionary development of different embellishments of a basic chemiosmotic system in order to extend and maximize mechanisms for biological energy production.

A third approach has been to examine the fossil and geochemical record for evidence of microbial activities (Schopf and Walter, 1982). Because of their small size and the possibility of artifacts, the direct fossil record has not been particularly revealing with regard to the evolution of bacteria or establishing their existence within the various geological time frames of the earth. In addition, from inspection of putative microbial microfossils, it is not possible to unequivocally deduce the physiological type of microbe. However, a number of geological features and chemical alterations have been interpreted to indicate the existence of certain physiological types of bacteria at specific times within the geological record. A most important conclusion to emerge from this approach was the realization that the curious structures, termed stromatolites, found in the geological record (beginning 3.6 billion years ago up to recent times) resulted from microbial activities (Schopf and Walter, 1982).

In all respects, fossil stromatolites appear to be identical to modern stromatolites, which represent microbial mats composed primarily of cyanobacteria, phototrophic bacteria, and fermentative bacteria. Their rock-like structures are due to the accretion of inorganic materials by cyanobacteria, the primary producers in this complex microbial association. The existence of cyanobacterial microfossils, in association with these ancient stromatolites, which are similar to modern cyanobacteria supports the concept that ancient and modern stromatolites result from metabolic activities of similar physiological types of bacteria. These arguments are compelling for the early existence of bacteria. If one accepts this general conclusion, it is evident that cyanobacteria, which are biologically highly developed and catalyze oxygenic photosynthesis, were in existence around 3.6 billion years ago—a time perceived as the beginning of the fossil record. Thus, we must conclude that a substantial portion of metabolic evolution occurred prior to development of the fossil record.

This conclusion is further supported by changes in the isotope ratios of carbon and sulfur observed in geological formations. Microbial activities have also resulted in small changes in the isotopic composition of certain geological deposits containing carbon and sulfur which have been employed as indicators of CO_2 fixation and respiratory sulfate reduction in various geological strata. One must keep in mind that changes in these isotope ratios reflect major environmental biological activities involving large pools of these compounds and not localized events. Interpretation of changes in stable carbon isotopes is complicated by the existence of several different pathways for CO_2 fixation in plants and prokaryotes, but perturbations of the ratios of carbon isotopes can be detected at 3.5 billion years ago, well before there was a record of plant fossils (Schidlowski et al., 1983). Consequently, one can infer that this change in the composition of carbon isotopes arose from microbial activities; however, this information reveals nothing about the physiological type of bacteria responsible for this fractionation of carbon isotopes.

Sulfur is composed of two stable isotopes, ^{32}S and ^{34}S, and their ratios vary depending on the source of the sulfur. The lighter isotope, ^{32}S, has been clearly shown to be preferentially metabolized during respiratory sulfate reduction; this discrimination leads to changes in sulfur isotope ratios comparable to those found in nature. It is also recognized that respiratory sulfate reduction, but not assimilatory sulfate reduction, is the only biological process that will produce kinetic isotope effects of the magnitude found in the geological record. The importance of large geological sulfur reservoirs such as metal sulfides, ocean sulfates, and elemental sulfur plus the importance of an oxygen atmosphere for the generation of sulfate, i.e., an oxygen-linked versus an anaerobic sulfur cycle, are essential concepts required to understand sulfur isotope data.

These data (Chambers and Trudinger, 1979) suggest that respiratory sulfate reduction began to significantly and extensively influence the isotopic composition of sulfur reservoirs about 2.8 billion years ago and reached its modern development about 2.0 billion years ago; this time period probably corresponds to the formation of the modern oxidizing environment and an aerobic sulfur cycle (Kaplan and Rittenberg, 1962, 1964). Prior to 2.8 billion years ago, alterations in sulfur isotope ratios occurred but were less frequent and extensive. This evidence, then, is less supportive of the presence of respiratory sulfate reduction, but it has been proposed to indicate limited sulfur cycling by a strictly anaerobic sulfur cycle. The isotope data provide significant evidence that microbial activity involving sulfur occurred at least 2.8 billion years ago and that respiratory sulfate reduction can be traced to this time frame and probably earlier, as suggested by the microbial ecology of stromatolites.

These brief considerations suggest that a variety of bioenergetic mechanisms were evolved by anaerobic eubacteria in response to a collection of inorganic substrates such as CO, CO_2, H_2, SO_4^{2-}, S° and S^{2-}. Many of these substrates were continuously provided by geological activities and H_2 possibly generated from pyrite formation (Drobner et al., 1990). In a later geological time, substrates such as H_2, CO_2, and possibly CO were provided by the fermentation of complex biological materials generated by photosynthesis. Most of these bioenergetic systems probably evolved before 3.6 billion years ago but, with the formation of an extensive and stable oxygen atmosphere, this diversity of bioenergetic pathways was supplanted in most organisms by a common cytochrome electron transfer system coupling ATP formation to reduction of oxygen to water. Nevertheless, the basic anoxic environments, rich in inorganic substrates, have continued to exist in one form or another up to the present time and may have supported bioenergetic and physiological derivatives of this earlier time period.

Sulfate reduction and bioenergetic mechanisms The sulfate-reducing bacteria were the first strictly anaerobic, non-photosynthetic prokaryotes in which the presence of electron transfer-coupled phosphorylation was demonstrated (Peck, 1960, 1966). Subsequently, oxidative phosphorylation was shown to be an essential feature of the anaerobic metabolism of methanogenic bacteria (Vogels et al., 1988), acetogenic bacteria (Ljungdahl, 1986), sulfur-reducing bacteria (Fauque et al., 1980), and nitrate-respiring (ammonia-forming) bacteria (Steenkamp and Peck, 1981). Elucidation of mechanisms responsible for coupling of sulfate reduction to energy generation is an important general biological problem, and significant progress towards understanding these mechanisms has been forthcoming.

It has become evident that there is not just one but rather multiple mechanisms for coupling oxidation of various substrates to reduction of sulfate and other oxyanions of sulfur to electron transfer phosphorylation. Clearly, sulfate-reducing bacteria are defined by their physiological capability of utilizing sulfate as their terminal electron acceptor, but they constitute a large, heterogeneous group of microorganisms that have evolved in response to a multiplicity of electron donors and acceptors (Widdel, 1988). One might speculate that initial evolutionary pressures were towards utilization of geologically produced substrates, and subsequent evolution involved utilization of organic substrates generated by degradation of anoxygenic and oxygenic photosynthetic products.

3.2 General Bioenergetic Considerations

Basic chemiosmotic theory requires a membrane that is impermeable to protons, as well as many other cations, and that allows the development of chemical gradients that can be utilized to drive reversible ATPases for ATP synthesis. In this discussion, the term *scalar protons* refers to protons generated directly by oxidation of various substrates, and the term *vectorial protons* refers to protons transported across the cytoplasmic membrane as a result of electron transfer processes. Respiratory enzymes may be positioned relative to the cytoplasmic membrane such that chemical gradients are generated by scalar, as well as vectorial, reactions on the external surface of the cytoplasmic membrane (the periplasm in Gram-negative bacteria) by electron transfer across the membrane. Examples of such reactions are the Q cycle or specific transmembrane redox enzymes that translocate protons or other cations during catalysis (e.g., cytochrome oxidase). Thus, oxidation of H_2 by hydrogenase in the periplasm can generate a scalar proton gradient. The same reaction in cytoplasm would result in acidification, and protons at some point must be translocated across the membrane with expenditure of energy. Additionally, proton utilization in the cytoplasm to produce H_2 or methane can contribute to formation of a pH gradient by decreasing the concentration of protons in the cytoplasm relative to the periplasm.

In classic Gram-negative bacteria, such as *Escherichia coli*, there are few if any redox proteins among the periplasmic proteins. In sulfate-reducing bacteria, phototrophs, and bacteria catalyzing inorganic oxidations and reductions, however, there appear to be comparatively large numbers of periplasmic redox enzymes that play major roles in mechanisms of specific bioenergetic systems centered on different electron acceptors and donors (Hooper and DiSpirito, 1985). Nevertheless, a central feature of aerobic oxidation of reduced sulfur and nitrogen compounds is a classical cytochrome electron transfer chain with unique periplasmic redox proteins that feed electrons from oxidation of these inorganic substrates into the electron transfer chains (Kelly, 1989). The cellular location of these enzymes and electron transfer proteins in *Desulfovibrio vulgaris* is shown in Figure 3.1. In each instance, it appears that specific cellular localization of a redox enzyme can be correlated with optimization of the potential to generate an electrochemical gradient, i.e., charge separation. Thus, transfer of two electrons to a Q-cycle from a *cytoplasmically localized hydrogenase* will result in net translocation of only *two protons*, i.e., two produced in the cytoplasm by oxidation of H_2 and four translocated by the Q-cycle. However, transfer of two electrons from a *periplasmically localized hydrogenase* will result in the production of *six protons*, i.e., two from the periplasmic oxidation of H_2 and four from the Q-cycle. Bioenergetically, it is more efficient for hydrogen to be oxidized on the external surface of the cytoplasmic membrane; however, the possibility of protonophores specific for translocation of protons from cytoplasmically oriented hydrogenase has not been investigated.

Another important general concept concerns the stoichiometry of proton translocation. The chemical hypothesis of electron transfer-coupled phosphorylation assumed that there existed a constant stoichiometry of one phosphate esterified per two electrons transferred. This idea became transformed to a requirement that pairs of protons were generated during electron transfer to drive proton-translocating ATP synthases. It is now believed that three protons are required for synthesis of a molecule of ATP from ADP plus phosphate, and Thauer and Morris (1984) proposed that protons could be generated singly rather than two or three simultaneously. Thus, reactions that generate a single proton by scalar reactions or vectorial proton translocation might have bioenergetic significance in anaerobic bacteria, and one could think in terms of fractional amounts of ATP generated by specific reactions.

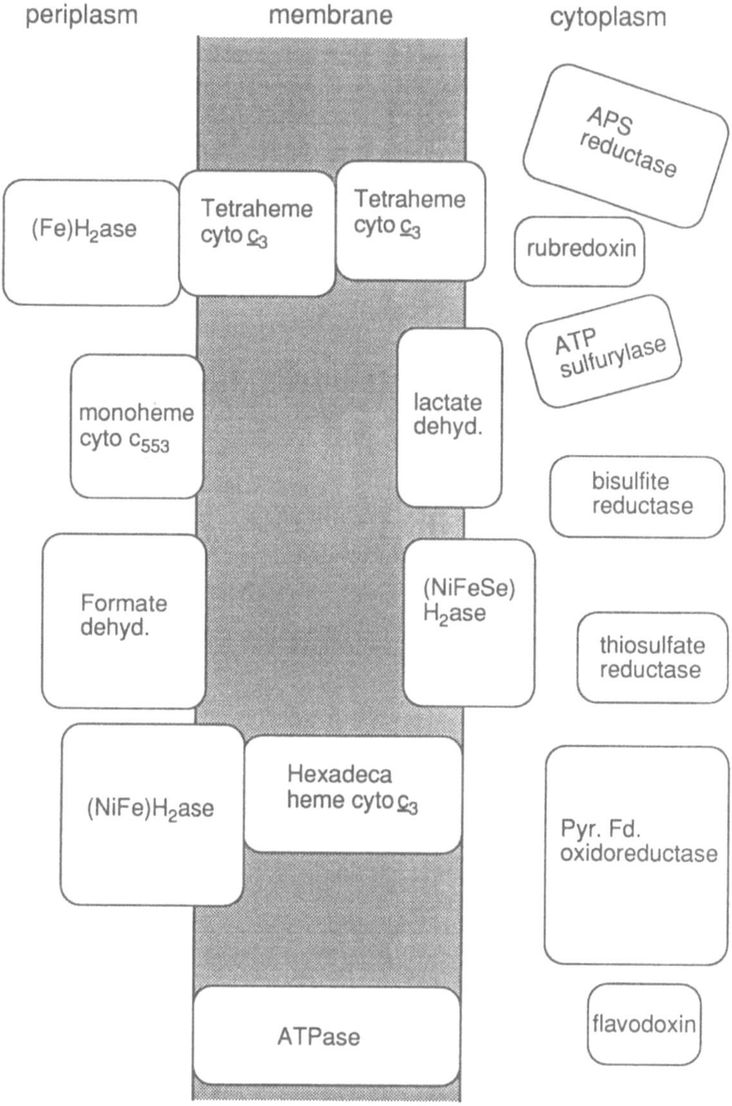

Figure 3.1 Cellular localization of electron transfer proteins and reductases in *Dv. vulgaris*.

3.3 Unique Biochemical and Physiological Aspects of the Sulfate-Reducing Bacteria

Sulfate-reducing bacteria belonging to the genus *Desulfovibrio*, and probably other genera of these bacteria, possess a number of attributes that have a general impact on our basic concepts concerning their physiology and bioenergetics. These attributes include: requirement for ATP to reduce sulfate (Peck, 1959; Ishimoto, 1959); cytoplasmic localization of enzymes involved in the pathway of respiratory sulfate reduction (Peck and Le Gall, 1982; Kremer et al., 1988b); an abundance of multiheme heme *c*-type cytochromes (Le Gall and Fauque, 1988; Pollock et al., 1991); and periplasmic localization of hydrogenases (Fauque et al., 1989). The pathway of respiratory sulfate reduction is enzymologically distinct from pathways of assimilatory sulfate reduction, which occur in other prokaryotes, algae, and green plants (Peck and Lissolo, 1988).

3.4 Enzymes and Proteins Involved in Respiratory Sulfate Reduction

Respiratory sulfate reduction involves a minimum of four cytoplasmic enzymes (Odom and Peck, 1984): ATP sulfurylase (reaction 3.1); inorganic pyrophosphatase (reaction 3.2); adenylyl sulfate (APS) reductase (reaction 3.3); and one of four bisulfite reductases (reaction 3.4).

$$ATP + SO_4^{2-} \leftrightarrows APS + PP_i \quad \text{ATP sulfurylase} \quad (3.1)$$

$$PP_i + H_2O \rightarrow 2\ P_i \quad \text{inorganic pyrophosphatase} \quad (3.2)$$

$$APS + H^+ + 2e \leftrightarrows HSO_3^- + AMP \quad \text{APS reductase} \quad (3.3)$$

$$HSO_3^- + 6H^+ + 6e \rightarrow HS^- + 3H_2O \quad \text{bisulfite reductase} \quad (3.4)$$

Reduction of sulfate is the only reduction of an inorganic compound in biology that requires ATP, and sulfate reduction has been employed as a measure of the ability of intact cells of sulfate-reducing bacteria to generate ATP by substrate phosphorylation or electron transfer phosphorylation.

Extracts of sulfate-reducing bacteria may contain an enzyme, ADP sulfurylase, which catalyzes a second reaction to convert APS to ADP (reaction 3.5).

$$APS + P_i \rightarrow ADP + SO_4^{2-} \quad \text{ADP sulfurylase} \quad (3.5)$$

This reaction may be important for conservation of a portion of the energy contained in the phosphosulfate bond during oxidation of sulfite to sulfate in phototrophic bacteria and thiobacilli. The reaction catalyzed by ATP sulfurylase is reversible, but the reaction catalyzed by ADP sulfurylase is irreversible and cannot be utilized for the biosynthesis of APS (Peck 1974).

APS reductase APS reductase is a non-heme iron/FAD containing enzyme which catalyzes the reversible reduction of APS to sulfite plus

AMP (Peck and Le Gall, 1982). The reductase also occurs in some phototrophic and sulfur-oxidizing bacteria capable of oxidizing reduced sulfur compounds to sulfate and results in a substrate-level conservation of energy. Neither $NADH_2$ nor $NADPH_2$ directly functions as electron donor for APS reductase, and the in vivo electron donor for this reduction has not been established. Oxidation of $NADH_2$ in crude extracts with APS has been observed, but the electron transfer appears to involve membranes (Kremer and Hansen, 1989) and the specific activity is low. The presence of APS reductase in bacteria obtained from dark, anaerobic environments is believed to be an indicator for the presence of sulfate-reducing bacteria.

Bisulfite reductase Four different molecular species of high-spin bisulfite reductase have been identified, largely on the basis of differences in their optical absorption spectra, and have the following trivial names: desulfoviridin (green); desulforubidin (red); P_{590} (brown); and desulfofucidin (reddish brown) (Fauque et al., 1991). For details of the distribution of these proteins, see Widdel (1988). The high-spin bisulfite reductases are related to assimilatory high-spin sulfite reductase, dissimilatory low-spin sulfite reductases, and nitrite reductases; they all contain siroheme plus non-heme iron, which are probably exchange-coupled through a bridging sulfur atom. For the high-spin assimilatory sulfite reductases (Ostrowski et al., 1989), dissimilatory sulfite reductases (Huang and Barrett, 1991), and nitrite reductase (Peakman et al., 1990), this relationship is supported by sequence data. There is some indirect evidence which suggests that reduction of bisulfite to sulfide is reversible (Schelel and Trüper, 1979); however, this idea requires further documentation.

These bisulfite reductases are distinguished from nitrite and other sulfite reductases by their ability to catalyze the irreversible formation of trithionate and thiosulfate under certain assay conditions, but the role of these reactions in the pathway of respiratory sulfate reduction has not yet been clearly resolved. Neither $NADH_2$ nor $NADPH_2$ directly functions as an electron donor for bisulfite reductase and, as is the case for APS reductase, the in vivo electron donor for the six-electron reduction has not been established. Similarly, oxidation of $NADH_2$ in crude extracts with HSO_3^- has been observed, but the electron transfer appears to involve membranes (Kremer and Hansen, 1989). Extracts of *Dv. vulgaris* contain only traces of ferredoxin, and flavodoxin appears to serve as the electron donor for bisulfite reduction (Irie et al., 1973). Three additional reductases—trithionate reductase, thiosulfate reductase, and a low-spin sulfite reductase (Peck and Le Gall, 1982)—may be involved in the reduction of sulfate to sulfide; however, the roles of these enzymes have not been conclusively documented.

Terminal electron acceptors APS reductase and bisulfite reductase serve as terminal electron acceptor catalysts for respiratory sulfate reduction; as such, they serve cognate biochemical functions with cytochrome oxidase in other respiratory systems. APS reductase and bisulfite reductase are isolated as soluble cytoplasmic proteins, and immunoelectron microscopic studies have, except for one thermophilic species, localized these proteins in the cytoplasm (Kremer et al., 1988b). This location is quite different from that observed with enzymes catalyzing the reduction of other terminal electron acceptors, such as cytochrome oxidase, which is an intrinsic membrane protein.

The cytoplasmic location of the two terminal reductases requires consideration of four unique bioenergetic problems for respiratory sulfate reduction as compared with membrane-bound terminal reductases. First, sulfate must be transported across the cytoplasmic membrane, and this is, in all probability, an energy-requiring reaction involving one or two protons (Cypionka, 1987; Warthmann and Cypionka, 1990). Second, many proteins providing electrons for reduction of sulfate are membrane bound, and electron carriers are required to link these reactions on the membrane with cytoplasmic APS reductase and bisulfite reductase. Third, it is unlikely that cytoplasmic APS reductase and bisulfite reductase directly catalyzed proton or ion translocation across the cell membrane, as they are clearly soluble proteins. Fourth, it is anticipated that sulfide, the reactive and toxic product of sulfate reduction, must be transported out of the cell but may bioenergetically balance sulfate import.

Oxygen can be used as a terminal electron acceptor by some sulfate-reducing bacteria. The physiological significance of this observation is not readily apparent because a cytochrome oxidase is not present. Consequently, some sulfate reducers may be oxygen sensitive or tolerant, and some species are able to reduce oxygen to water (Cypionka et al., 1985; Hardy and Hamilton, 1981; Kobayashi et al., 1982; Fitz and Cypionka, 1989; Postgate, 1984a); however, all lack a classical membrane-bound cytochrome electron-transfer chain terminating in a proton-translocating cytochrome oxidase. Many of the *c*-type cytochromes from these bacteria are autooxidizable and may be responsible for O_2 utilization.

ATP generation, but not growth (Dilling and Cypionka, 1990), has been shown to be coupled to reduction of oxygen, and possible reduction products—water, peroxide, or superoxide—have not been elucidated. Both catalase and superoxide dismutase (Hatchikian et al., 1977) have been found in some sulfate-reducing bacteria. The occurrence of glutathione peroxidase, which has recently been shown to play a significant role in the detoxification of superoxide (Mullenbach et al., 1988), however, has not been reported. Thus, there does not appear to be a generally applicable biochemical or physiological explanation for the strictly anaerobic character of sulfate reducer bacterial growth, but it is clear that these bacteria do not contain a classical electron transfer chain terminating in a cytochrome oxidase.

Cytochrome b Most sulfate-reducing bacteria possess variable amounts of membrane-bound cytochromes *b* that have not been characterized as regards to multiple species and menaquinones with 5 to 7 isoprenoid units (Widdel, 1988). The amount of membrane-bound cytochrome *b* in *Desulfovibrio gigas* is doubled by growth on fumarate over that observed on lactate plus sulfate. Cells of *Dv. vulgaris*, by comparison, do not have fumarate reductase and contain only trace amount of membrane-bound cytochrome *b*, but exhibit usual levels of menaquinone. The presence of cytochrome *b* has been related to fumarate respiration rather than a Q-cycle involved generally in sulfate respiration. There is no direct experimental evidence for the existence of a Q-type cycle or a b-cycle, and information dealing with the bioenergetics and electron transfer components, both types and concentration, does not allow one to confidently infer their functional existence in sulfate-reducing bacteria (Stouthamer, 1988).

c-Type cytochromes These anaerobic bacteria also have a variable assortment of mono- and multi-heme *c*-type cytochromes, many of whose functions still remain obscure (Peck and Le Gall, 1982). These cytochromes may have different cellular operational locations. Tetraheme cytochrome c_3 and the monoheme cytochrome c_{553}, for example, may be operationally localized in the periplasm; hexaheme cytochrome *c* (nitrite reductase, ammonia forming) on the cytoplasmic membrane; and unidentified *c*-type cytochromes of unknown structure and function, in the cytoplasm.

Tetraheme cytochrome c_3 has been shown to function both as a sulfur reductase and as an essential cofactor of [Fe], [NiFe], and [NiFeSe] hydrogenases to reduce low-molecular-mass electron transfer proteins, i.e., ferredoxin, flavodoxin, and rubredoxin (Odom and Peck, 1984). These interactions with tetraheme cytochrome c_3 have been extensively modeled and probable structures formulated (Stewart et al., 1987, 1989). The physiological significance of these models has been in doubt because the tetraheme cytochrome c_3 is periplasmic, whereas ferredoxin, rubredoxin, and flavodoxin are cytoplasmic proteins. In addition, most tetraheme cytochromes c_3 have the catalytic ability to reduce sulfur (S°) to sulfide, i.e., function as a sulfur reductase, and inverted vesicles catalyze anaerobic oxidative phosphorylation coupled to H_2 oxidation and sulfur reduction (Fauque et al., 1980). These observations suggest that the true physiological function of tetraheme cytochrome c_3 may not yet be realized.

The autooxidizable monoheme cytochrome c_{553} [and related monoheme cytochrome c_{553}(550)] exhibit His-Met liganding of iron. The monoheme cytochromes *c* found in sulfate-reducing bacteria are related to the large cytochrome *c* family (Bruschi and Le Gall, 1972) found in most eubacteria and eukaryotes, but the functions of these monoheme *c*-type cytochromes are not established. Transmembrane hexaheme cytochrome *c* biochemically is a nitrite reductase (NH_3-forming) and related to the eubacterial family of respiratory nitrite reductases (NH_3 forming) (Liu and Peck, 1988). It is unrelated to the assimilatory siroheme-non-heme iron-containing nitrite reductases (NH_3 forming) of nutritional importance for many prokaryotes and eukaryotes. Five hemes are low-spin and a single high-spin heme seems to be the site of substrate binding (Costa et al., 1990). In sulfate-reducing bacteria, H_2 oxidation, coupled to reduction of nitrite to NH_3, involves both scalar and vectorial proton generation (Steenkamp and Peck, 1981) and must be regarded as a respiratory process. However, with these bacteria, growth on nitrate is not correlated with the presence of hexaheme nitrite reductase, and the possibility exists that the cytochrome plays some other role in general electron transfer.

Two additional multiple heme-cytochromes c_3 have been isolated from some, but not all, members of the genus *Desulfovibrio*. These include a periplasmic hexadecaheme cytochrome c_3 (65.5 kDa) (from *Dv. vulgaris* Hildenborough) with largely bis-his liganding of the heme iron (Pollock et al., 1991) and a soluble octaheme cytochrome c_3 (25 kDa) from *Dv. desulfuricans* (Norway 4) and *Dv. gigas*. The function and relationships of these two types of multiheme cytochromes are not yet established, and they have thus far been found only in sulfate-reducing bacteria.

In conclusion, sulfate-reducing bacteria contain monoheme and multiheme cytochromes *c*. In the genus *Desulfovibrio*, there is a single monoheme cytochrome *c* but at least three different types of multiple-heme cytochromes c_3 which seem to be unique to sulfate- and possibly sulfur-reducing bacteria (Le Gall and Fauque, 1988).

Hydrogenase Another significant characteristic of sulfate-reducing bacteria belonging to the genus *Desulfovibrio* is the presence of a soluble periplasmic hydrogenase which is postulated to oxidize H_2 for reduction of APS. The hydrogenase may be one of three types: a [12 Fe] hydrogenase; a [NiFe] hydrogenase; or a [NiFeSe] hydrogenase (Fauque et al., 1989).

In bacteria that contain only a single type of hydrogenase, one must suppose that, during growth on H_2 plus sulfate, the single hydrogenase activates hydrogen and provides electrons for each of the major reductases utilized in respiratory sulfate reduction, i.e., APS and bisulfite reductases. Additional hydrogenases, if present, are localized on the cell membrane or in the cytoplasm; *Dv. vulgaris* (for example) contains at least four molecular species of hydrogenase plus nitrogenase (Meyer and Gagnon, 1991).

In bacteria that contain two to four different hydrogenases, evidence has been presented which indicates that hydrogenases may be specifically coupled to APS or bisulfite reductases (Peck and Lissolo, 1988). Thus, in cells of *Dv. vulgaris*, the soluble periplasmic [12 Fe] hydrogenase provides electrons for reduction of APS to sulfite plus AMP, while a membrane-bound [NiFe] hydrogenase, localized on the periplasmic aspect of the cytoplasmic membrane, provides six electrons for reduction of bisulfite to sulfide. The tetraheme cytochrome c_3 (and possibly the octaheme cytochrome c_3) function as a cofactor (Bell et al., 1974) for each of the three soluble periplasmic molecular species of hydrogenases found in these bacteria. Clearly, for different hydrogenases to be required for reduction of APS and bisulfite, pathways of electron transfer through the cell membrane must be different, i.e., only one of the reductive steps can involve the tetraheme cytochrome c_3.

It therefore appears that the soluble periplasmic hydrogenase, regardless of whether it is a [12 Fe], [NiFe], or [NiFeSe] hydrogenase, utilizes the tetraheme cytochrome c_3 as its cofactor and is coupled specifically to APS reductase by a specific transmembrane electron transfer chain and one or more soluble cytoplasmic electron transfer proteins. The membrane-bound periplasmic [NiFe] hydrogenase, which appears to be found in all *Desulfovibrio*, is proposed to be coupled to a specific membrane-bound electron transfer protein, possibly the hexadecaheme cytochrome c_3, which transfers electrons to a soluble electron carrier (probably flavodoxin in *Dv. vulgaris*) for reduction of bisulfite to sulfide. Clearly, details of transmembrane electron transfer are most important for understanding the bioenergetics of respiratory sulfate reduction; however, they have not yet been satisfactorily resolved.

In conclusion, H_2 and hydrogenases seem to have a special role in respiratory sulfate reduction as it occurs in *Desulfovibrio*. Three different hydrogenases that may have unique cellular localizations have been identified and extensively studied. Most species of *Desulfovibrio* possess a soluble periplasmic hydrogenase which may be any one of the three types and requires the tetraheme cytochrome c_3 as a specific cofactor. In *Dv. vulgaris*, the periplasmic [12 Fe] hydrogenase seems to be specific for reduction of APS to sulfite, and it is proposed that this is the general physiological role of soluble periplasmic hydrogenases in other sulfate-reducing bacteria. The bacteria also usually possess a membrane-bound [NiFe] hydrogenase which in *Dv. vulgaris* is localized on the periplasmic surface of the cytoplasmic membrane and produces scalar protons plus electrons to reduce bisulfite to sulfide.

Although these hydrogenases catalyze identical chemical reactions, different cellular locations or electron acceptor specificities would allow the enzymes to feed electrons from H_2 into different electron transfer sequences and thus function as unique dehydrogenases. The ability to couple energy generation to reduction of sulfate to sulfide is not randomly distributed among strictly anaerobic bacteria and, to date, the process has not been reported in any facultative anaerobes. Simple possession of the four enzymes that constitute the pathway of respiratory sulfate reduction (ATP sulfurylase, inorganic pyrophosphatase, APS reductase, and bisulfite reductase) does not confer the ability to use sulfate as a terminal electron acceptor, as evidenced by the presence of the pathway in some sulfur-oxidizing thiobacilli and sulfur-utilizing phototrophic bacteria. On the basis of these two considerations and the absence of mitochondrial type electron transfer sequences, it seems that the ability to employ sulfate as a terminal electron acceptor must involve profound modifications in a bacterium's basic physiology and biochemistry.

3.5 Hydrogen Utilization

Historical development Stephenson and Stickland (1931) first demonstrated anaerobic coupling of H_2 oxidation to reduction of sulfate, plus a number of more reduced sulfur-containing compounds, to sulfide. Their experiments with sulfate-reducing bacteria did not suggest the nature of the respiratory pathway of sulfate reduction or establish whether or not energy conservation was coupled to this oxidation of hydrogen with sulfate. However, they raised questions regarding whether sulfate-reducing bacteria were capable of autotrophic growth on H_2, sulfate, and CO_2.

During the 1950s, considerable research concerning the physiology and biochemistry of sulfate-reducing bacteria had a significant impact on our ideas concerning the bioenergetics of respiratory sulfate reduction. Extracts of *Desulfovibrio* were shown to contain exceptionally high levels of a soluble hydrogenase (Sadana and Jagganathan, 1954) which was not affected by growth conditions. The metabolic role of hydrogenase remained enigmatic, because it could not be associated with any physiological function. In addition, a *c*-type cytochrome generally believed to be associated with aerobic respiration, was isolated from *Dv. vulgaris* (Postgate, 1956; Ishimoto et al., 1954a), and sulfate-reducing bacteria thus became the first nonphotosynthetic, strictly anaerobic bacteria shown to contain a cytochrome.

The outlines of the ATP-dependent pathway for respiratory sulfate reduction were also established, and in 1960 evidence was published (Peck, 1960) that demonstrated that oxidation of H_2 coupled to reduction of sulfate by intact cells of *Dv. vulgaris* involves anaerobic oxidative phosphorylation. The basic observation consisted of the demonstration that H_2 oxidation coupled to ATP-dependent sulfate reduction was inhibited by 2,4-dinitrophenol, a classical inhibitor of oxidative phosphorylation. The conclusion was of general biological importance, because it indicated that oxidative phosphorylation was not restricted to aerobes. Subsequently, anaerobic oxidation of H_2 and concomitant reduction of sulfite in cell-free preparations was shown to support electron transfer-coupled phosphorylation and to require both membrane and soluble components (Peck, 1966). The data indicated that anaerobic oxidative phosphorylation was not significantly different in its fundamental requirements from that reported in aerobic bacteria.

The next significant step in understanding the bioenergetics of H_2 oxidation in sulfate-reducing bacteria was the demonstration that hydrogenase of *Dv. gigas* was largely localized in the bacterial periplasm (Bell et al., 1974). This was the first evidence for the periplasmic localization of a respiratory dehydrogenase in prokaryotes. The observation set the stage to interpret sulfate reducer energy generation in terms of cellular localizations of major respiratory enzymes and the chemiosmotic concept of energy coupling.

Wood (1978) presented the first model that attempted to integrate the then available biochemical and physiological information with chemiosmotic ideas. He took into account the various cellular localizations of enzymes and electron transfer components and postulated the presence of a single membrane-bound electron transfer sequence which was proposed to be responsible for transmembrane electron transfer and proton pumping as occurs in most bacteria. He indicated, but did not discuss, the potential for generation of a proton gradient as a result of periplasmic oxidation of H_2.

At about the same time, it was shown that several strains of *Desulfovibrio* were able to grow with CO_2 and acetate as sole carbon sources, and H_2 plus SO_4^{2-} as their sole energy source (Badziong et al., 1978). From there data, it was inferred that reduction of sulfate to sulfide with 4 H_2 must involve formation of more than the two equivalents of ATP required for activation of sulfate. Evidence gathered from growth yield studies indicated that only H_2 oxidation coupled to reduction of sulfite (or thiosulfate) to sulfide resulted in ATP formation. Thus, it was suggested that H_2 reduction of sulfite to sulfide resulted in generation of three ATPs, while reduction of sulfate to sulfide—in which two ATPs are consumed for sulfate activation—resulted in the net generation of only one ATP (Badziong and Thauer, 1978). To explain their data, one must assume that there must exist some mechanism to avoid scalar production of protons in the periplasm resulting from oxidation of H_2 coupled to reduction of APS.

The bioenergetic importance of scalar proton formation in the periplasm of Gram-negative prokaryotes was first realized in the interpretation of the formate/nitrate redox reaction (reactions 3.6 and 3.7) carried out by *Wolinella succinogenes* (Kröger, 1977).

$$\text{Formate} + \text{fumarate} + H^+ \rightarrow CO_2 + \text{succinate} \quad (3.6)$$

$$\text{Formate} + \text{nitrate} + H^+ \rightarrow CO_2 + \text{nitrite} + H_2O \quad (3.7)$$

Reactions 3.6 and 3.7 consist of two half reactions spatially isolated from each other. In each reaction, formate dehydrogenase is localized on the periplasmic side of the cytoplasmic membrane, while fumarate and nitrate reductases are located as membrane-bound proteins on the cytoplasmic aspect of the membrane. Periplasmic oxidation of formate to CO_2 generates scalar protons and electrons which are transferred across the membrane via a cytochrome *b*/menaquinone electron transfer sequence for fumarate reduction to succinate (or nitrate reduction to nitrite) with consumption of two protons.

Employing the same insights, Badziong and Thauer (1980) proposed a bioenergetic mechanism for sulfate-reducing bacteria involving scalar formation of protons in the periplasm coupled to H_2 oxidation linked to reduction of sulfate and other more reduced compounds of sulfur in the cytoplasm. On the basis of growth yield studies, they proposed that only the six-electron reduction of sulfite to sulfide involved electrogenic or vectorial electron transfer of protons. Transport of SO_4^{2-} into the cell requires at least two protons; however, it is possibly balanced by exit of S^{2-} accompanied by two protons; such a mechanism generates a net production of six protons, which can be utilized for ATP synthesis by ATP synthase (Guarraia and Peck, 1971). However, the utilization of protons for substrate and product transport may be variable and result in variable growth yields.

This concept of vectorial electron transfer was consistent with the cellular localization of enzymes and electron donors and, most interestingly, it suggested that two transmembrane electron transfer sequences were involved in the reduction of sulfate to sulfide: the first coupling oxidation of H_2 to reduction of APS to sulfite; the second coupling oxidation of three hydrogens to reduction of sulfite to sulfide. It is now apparent that reduction of bisulfite with molecular hydrogen is a central focus in the bioenergetics of respiratory sulfate reduction and may have had early evolutionary origins reflecting a biological response to the geological gassings of H_2 and SO_2 in terms of respiratory sulfite reduction.

Enzymology of sulfate respiration The enzymology of respiratory sulfate reduction has been most extensively studied in *Dv. vulgaris*, and results obtained with this organism that have relevance for sulfite reduction will be briefly discussed. A variety of different enzymes and proteins are involved in this process, including various hydrogenases, multiheme cytochromes *c*, and sulfite reductases. These enzymes and proteins have a number of different molecular forms, compositions, and localizations. All of these attributes combine to define and limit the potential role the protein or enzyme can play in the overall process of sulfate respiration.

Three different hydrogenases, termed the [12 Fe], [NiFe], and [NiFeSe] hydrogenases, have been identified and extensively studied (Fauque et al., 1989) by biochemical and molecular biology techniques. The three hydrogenases have been cloned and sequenced, and the [NiFe] and [NiFeSe] hydrogenases belong to the large family of nickel-containing hydrogenases whose structural genes exhibit no homology with structural genes of the [Fe] hydrogenases (Choi, Peck, and Przybyla, unpublished data; Voordouw and Brenner, 1986). Hybridization studies have indicated

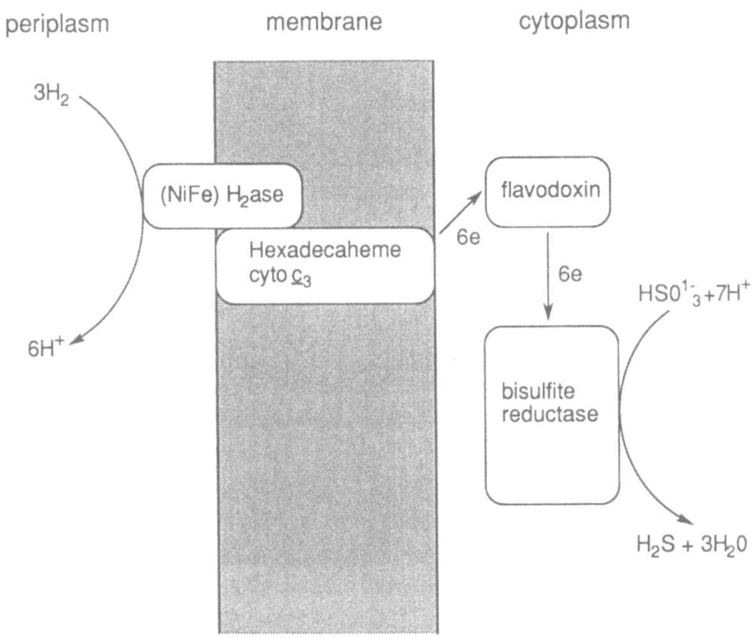

Figure 3.2 Schematic representation of the cellular topology of proteins involved with the H_2/HSO_3^- redox couple.

that only the structural genes for [NiFe] hydrogenase are found in all species of *Desulfovibrio* except one thermophilic species. In *Dv. vulgaris,* the heterodimeric [NiFe] hydrogenase is a membrane-bound protein and is localized on the periplasmic surface of the cytoplasmic membrane. The small subunit possesses a long signal peptide that exhibits homology to the signal peptide of the soluble periplasmic [12 Fe] hydrogenase. Inhibitor studies suggest that this [NiFe] hydrogenase activates H_2 for reduction of bisulfite to sulfide, while the [12 Fe] hydrogenase oxidizes H_2 for reduction of APS to sulfite plus AMP (Peck and Lissolo, 1988).

Inhibition studies with CO (Peck and Lissolo, 1988) on reduction of sulfate with H_2 in *Dv. vulgaris* indicates that the soluble periplasmic [12 Fe] hydrogenase activates H_2 for reduction of APS to sulfite plus AMP, but does not inhibit reduction of sulfite to sulfide. In keeping with the suggestion of Thauer and Badziong (1980), it is proposed in Figures 3.2 and 3.3 that pathways for transmembrane electron transfer to bisulfite reductase and APS reductase are not identical and involve different electron carriers. The [12 Fe] hydrogenase is unique in utilizing tetraheme cytochrome c_3 as its electron acceptor and in its resistance to inactivation by O_2. Although related to the bidirectional [20 Fe] hydrogenase of *C. pasteurianum,* in terms of its nonheme iron active site (H cluster), it contains two ferredoxin

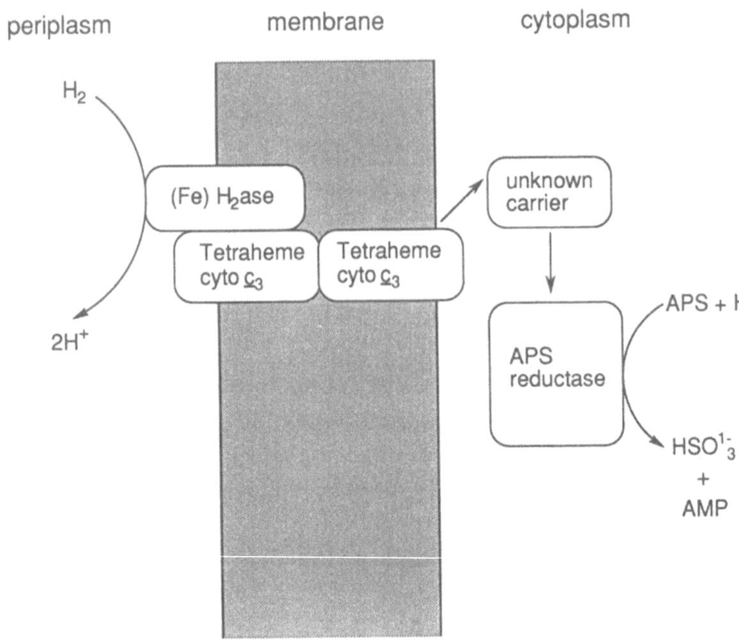

Figure 3.3 Schematic representation of the cellular topology of proteins involved with the H_2/APS redox couple.

type iron sulfur clusters rather than four (Adams, 1990). This conclusion has recently been confirmed by sequencing studies, and it appears that the derived amino acid sequence is more homologous to that of the *hyd* γ gene (Stokkermans et al., 1989) than the [12 Fe] hydrogenase (Meyer and Gagnon, 1991). These observations indicate *Dv. vulgaris* contains at least four different proteins that have hydrogenase activity; however, nothing is yet known about the enzymatic properties of the gene product encoded by *hyd* γ. Hybridization studies of Voordouw et al. (1990a) suggest that the hydrogenase is not widely distributed in the genus *Desulfovibrio*.

It is proposed here that soluble periplasmic hydrogenases regardless of type, i.e., [12 Fe], [NiFe], or [NiFeSe], represent a specific biochemical adaptation for utilization of sulfate as a terminal electron acceptor. Identity of the transmembrane electron carrier coupling the [12 Fe] hydrogenase with APS reduction has not been established, but an attractive candidate is a homodimer of tetraheme cytochrome c_3. It has been proposed that association of two or more molecules of this cytochrome results in the formation of a "electron channel" (Tabushi et al., 1981), which greatly facilitates electron transfer and proton transfer from hydrogenase into liposomes (Tabushi et al., 1984). In addition, the [12 Fe] hydrogenase as well as soluble periplasmic [NiFe] and [NiFeSe] hydrogenases utilize tetraheme cytochrome c_3 specifically as an electron acceptor and, during purification, the cytochrome is tightly associated with each of the hydrogenases. Computer modeling studies suggest that the tetraheme cytochromes c_3 may specifically dock with low-molecular-mass cytoplasmic electron transfer proteins, and a transmembrane localization of tetraheme cytochrome c_3 would resolve this enigma relating to cellular localization. APS reductase (and ATP sulfurylase plus a significant fraction of pyrophosphatase) are soluble cytoplasmic proteins, but identity of the electron transfer protein or proteins (indicated as x) coupling APS reduction with transmembrane electron transfer has not been established.

The [NiFe] hydrogenase co-purifies with an unidentified high-molecular-mass *c*-type cytochrome which has been suggested to be hexadecaheme cytochrome c_3 (about 75 kDa) (Higuchi et al., 1987). *Dv. vulgaris* is not known to contain hexaheme cytochrome c_3. The structural gene (*hmc*) for hexadecaheme cytochrome c_3 has been cloned and sequenced, but the biochemical relationships between this cytochrome and the octaheme cytochrome c_3 have not been satisfactorily resolved. Sequence analysis suggests that the octaheme cytochrome is derived from the hexadecaheme cytochrome c_3 perhaps by proteolytic degradation. Sequence comparisons also indicate that a signal peptide is encoded by the *hmc* gene and are interpreted to confirm the suggested periplasmic localization of the multiheme cytochrome; however, the finding of the hexadecaheme cytochrome c_3 among the periplasmic proteins is not a universal observation (J. LeGall, personal communication), and the presence of a signal peptide is not proof of a periplasmic localization. (See Chapter 5 for further discussion of the molecular biology of these proteins.)

Figure 3.2 indicates the hexadecaheme cytochrome c_3 as a transmembrane protein that interacts with flavodoxin at the cytoplasmic side of the cell membrane. In *Dv. vulgaris* flavodoxin has been shown to be a major low-molecular-weight electron transfer protein (Irie et al., 1973) and to couple H_2 oxidation with bisulfite reduction; however, in *Dv. gigas*, the three iron ferredoxin (ferredoxin II) can also catalyze this reaction (Peck and Le Gall, 1982). The physiological significance of computer modeling studies of the docking of cytoplasmic rubredoxin, ferredoxin, and flavodoxin with the periplasmic tetraheme cytochrome c_3 have been questioned; however, the physiological significance may be in the docking of the three low-molecular-mass electron transfer proteins with the transmembrane hexadecaheme cytochrome c_3 (see Figure 3.2).

Extracts of *Dv. vulgaris* contain two types of sulfite reductases (Lee et al., 1973a): a high-spin bisulfite reductase (desulfoviridin) constitutes about 10% of total soluble protein of *Dv. vulgaris*, and a low-spin sulfite reductase that constitutes less than 1% of the total soluble protein. Because of the extraordinarily high abundance of bisulfite reductase, it is generally considered to be of singular importance in the pathway of respiratory sulfate reduction, and several different types can be identified based on their optical properties, i.e., desulforubidin, P_{582}, and desulfofuscidin (LeGall and Fauque 1988).

Desulfoviridin has an $\alpha_2\beta_2$ structure and a molecular mass of 226 kDa; it contains 0.5 sirohemes coupled to a [4Fe-4S] nonheme iron cluster, 1.5 siroporphyrin, and 3.5 [4Fe-4S] nonheme iron clusters per mole of enzyme. Siroporphyrin, however, does not appear to occur in the other types of bisulfite reductase (Moura et al., 1987b). Under certain conditions of assay with SO_3^{2-} and reduced methyl viologen, desulfoviridin irreversibly produces substantial amounts of thiosulfate and trithionate (Kobayashi et al., 1972). These observations led to the promulgation of the trithionate pathway of respiratory sulfate reduction, which involves trithionate and thiosulfate as intermediates. The majority view, however, favors the direct six-electron reduction of sulfite to sulfide as the actual reaction; however, both pathways may be of physiological significance and may be regulated by different substrates and growth conditions (Peck and Le Gall, 1982). In Figure 3.2, desulfoviridin is shown accepting electrons from the membrane-bound hexadecaheme cytochrome c_3 via flavodoxin, which is probably cycling through the quinone and semiquinone forms of FMN.

The low-spin sulfite reductase consists of a single polypeptide chain (26 kDa) and contains a [4Fe-4S] nonheme iron cluster spin coupled to a siroheme (Moura et al., 1987b). It has been purified from several strictly anaerobic bacteria, and its occurrence does not appear to be related to respiratory metabolism of sulfate (Moura et al., 1986). The amino acid sequence derived from the cloned gene for the *Dv. vulgaris* sulfite reductase (Tan et al.,) exhibits extensive homology with assimilatory sulfite reductase from *Escherichia coli* (Ostrowski et al., 1989), anaerobic sulfite reductase from *Salmonella typhimurium* (Huang and Barrett, 1991), and nitrite reductase from *E. coli* (Peakman et al., 1990).

Energetics of respiration of sulfate and related compounds Growth yields ($Y^{max}_{SO_4^{2-}}$) for *Dv. vulgaris* grown on H_2 plus HSO_3^- are around 35 g/mol (Badziong and Thauer, 1978), which suggests that 3 to 4 moles of ATP are generated per mole of sulfite reduced to sulfide. On the basis of the stoichiometry of 3 H^+/ATP (Maloney, 1983), six scalar protons from the oxidation of H_2 can only support synthesis of 2 ATPs. The formation 12 to 14 H^+ associated with oxidation of H_2 plus sulfite has been reported and indicates that the formation of both scalar and vectorial protons are coupled to H_2 oxidation of sulfite. Because vectorial proton transport involves membrane-associated proteins, this observation suggests that the hexadecaheme cytochrome c_3 may function as a proton pump. Clearly, this is more ATP than indicated by growth yield studies. Such studies, however, may not be an absolute criterion for interpreting ATP yields, and the calculations here may not take into account all of the proton-consuming reactions.

Growth of *Dv. vulgaris* on H_2 plus SO_4^{2-} as sole source of energy in a chemostat generates a growth yield ($Y^{max}_{SO_4^{2-}}$) of 13 g/mol compared with a growth yield (Badziong and Thauer, 1978) on sulfite of 36 g/mol. Clearly, if one assumes that proton import with sulfate is balanced by proton export with sulfide, it does not appear that net proton formation is coupled to oxidation of H_2 with APS. Thauer and Badziong (1980) accommodated this conclusion by proposing that transfer of reducing equivalent involved hydrogen carriers, and thus would not result in net production of vectorial or scalar protons. Figure 3.3 indicates that tetraheme cytochrome c_3 is the electron carrier responsible for transmembrane electron transfer. However, Tabushi et al. (1984), on the basis of studies with artificial systems, suggested that proton influx is accompanied by transfer of electrons resulting in formation of neither scalar nor vectorial protons. This idea has the advantage of providing an essential role for tetraheme cytochrome c_3. This general bioenergetic conclusion is supported by the common observation that with sulfate-reducing bacteria containing periplasmic hydrogenases, proton production associated with the reduction of sulfate with H_2 is low or has not been observed (Kobayashi et al., 1982).

Most sulfate-reducing bacteria can utilize thiosulfate as a terminal electron acceptor, and observed growth yields are similar to those found for sulfite (Badziong and Thauer, 1978). Thiosulfate is believed to be cleaved by reduction to yield sulfite plus sulfide, and proton transport has not generally been observed with oxidation of H_2 in the presence of thiosulfate (Kobayashi et al., 1982). This is generally consistent with the small change in free energy associated with the reaction (reaction 3.8), which is catalyzed by thiosulfate reductase.

$$S_2O_3^{2-} + H_2 \rightarrow HS^- + HSO_3^- \qquad (3.8)$$

Thiosulfate reductase has been identified and purified from extracts of *Dv. gigas* (Hatchikian, 1975) and *Dv. vulgaris* (Haschke and Campbell, 1971; Aketagawa et al., 1985a). It has a molecular mass of around 200 kDa, and redox prosthetic groups appear to be lacking. In extracts of *Dv. gigas*, oxidation of H_2 coupled to reduction of thiosulfate is stimulated by the addition of octaheme cytochrome c_3 but not by addition of tetraheme cytochrome c_3 (Hatchikian et al., 1972). Thiosulfate reductase is usually found in the soluble protein fraction and constitutes less than 1% of soluble protein. When extracts of *Dv. gigas* are prepared below pH 6.0, significant amounts of the reductase are associated with the membrane fraction, suggesting that the enzyme is capable of accepting electrons directly from membrane-bound electron carriers (Peck and Le Gall, 1982). Because of the apparent relation between the octa- and hexadecaheme cytochromes and the role of octaheme cytochrome c_3 in soluble electron transfer to thiosulfate reductase, the transmembrane electron carrier is indicated in Figure 3.4 as the hexadecaheme cytochrome c_3 in an electron transfer pathway similar to that utilized for bisulfite reduction. If this proposed

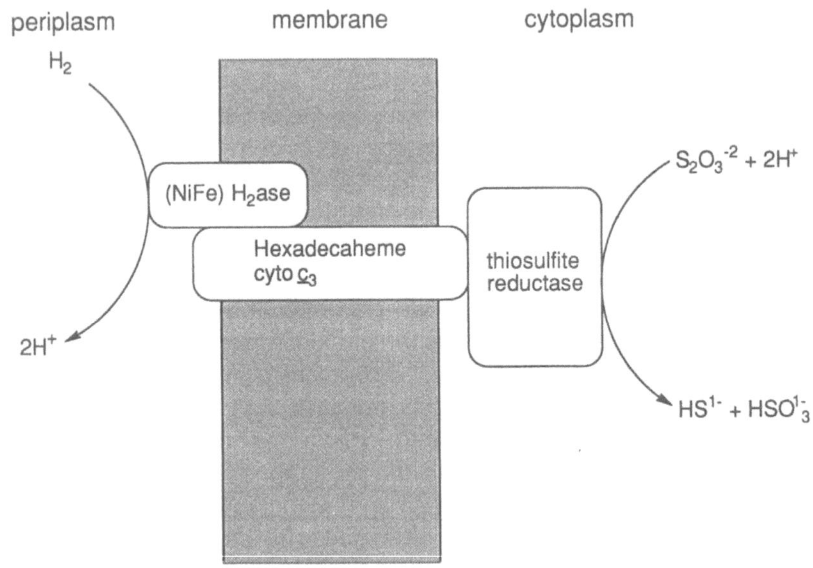

Figure 3.4 Schematic representation of the cellular topology of proteins involved with the $H_2/S_2O_3^{2-}$ redox couple.

relationship is valid, then weak proton translocation should be associated with thiosulfate reduction with H_2; such low level proton translocation associated with this reaction has recently been described (Fitz and Cypionka, 1989).

Many tetraheme cytochromes c_3 are capable of functioning as sulfur reductases, but this hemoprotein from *Dv. vulgaris* is inhibited by H_2S, the product of sulfur reduction (Fauque et al., 1979). The relevance of this important observation regarding tetraheme cytochrome c_3s for sulfur reduction, which occurs in mesophilic and thermophilic Archaebacteria, has not been resolved (Stetter and Gaag, 1983). Inverted membrane vesicles from *Dv. gigas* have been demonstrated to catalyze electron transfer coupled to oxidation of H_2 in the presence of elemental S° (Fauque et al., 1980). The system exhibited P/2e ratios of 0.1, did not require soluble components, and was inhibited by exogenous tetraheme cytochrome c_3, methyl viologen, and pentachlorophenol. The system is difficult to understand in terms of cellular localization of tetraheme cytochrome c_3 in the periplasm. Thus, the sulfur-reducing cytochrome is located inside the membrane vesicles, and the exogenously added S° is located on the outside. Clearly, the system requires further investigation, and it may be that other multiheme cytochromes c_3 are able to carry out reduction of S°.

Energetics of nitrate/nitrite respiration A limited number of sulfate-reducing bacteria are able to utilize nitrate, in addition to sulfate, as

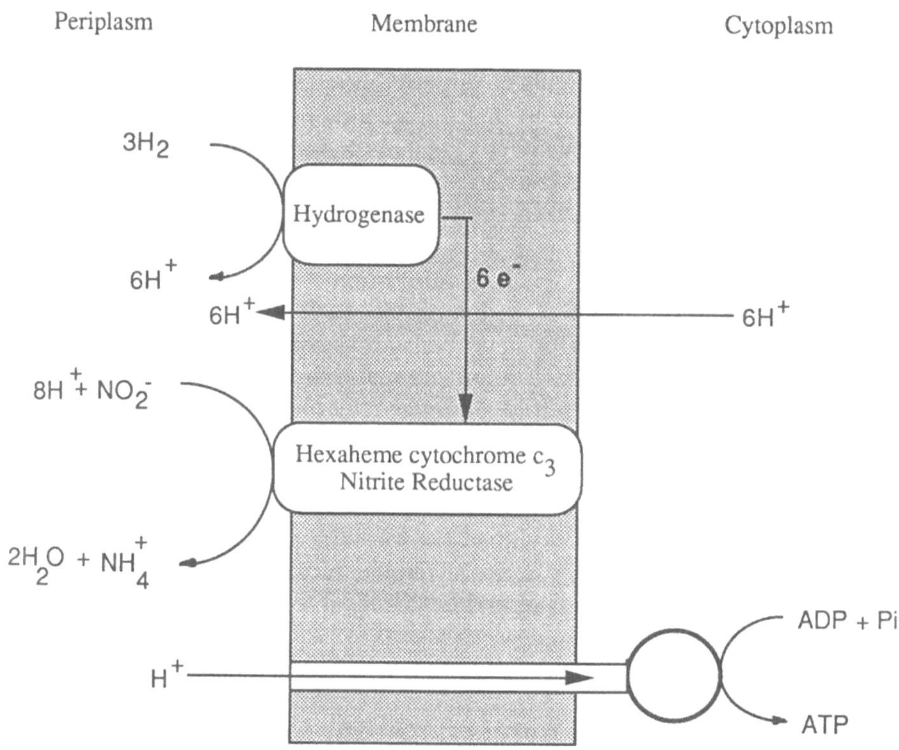

Figure 3.5 Scheme representing scalar and vectorial electron transport coupled to oxidation of hydrogen with nitrite.

their terminal electron acceptor (Keith and Herbert, 1983), and the reaction is coupled to electron transfer phosphorylation (Steenkamp and Peck, 1981). Nitrite reduction is more common in sulfate-reducing bacteria and is not limited to nitrate-reducing species. The reaction is catalyzed by a respiratory nitrite reductase (ammonia-forming) that has a molecular mass of 65 kDa and contains six *c*-type hemes. It is also known as the hexaheme cytochrome c_3 and is widely distributed in strict and facultative aerobes (Liu and Peck, 1988). This nitrite reductase is unrelated to the regulated nitrite reductase (nonheme iron-siroheme containing) found in many plants and bacteria (Vega and Kamin, 1977) where its physiological function is nitrogen assimilation. Oxidation of H_2 with nitrite is coupled to scalar and vectorial proton transport (see Figure 3.5). Nitrite reductase is closely associated with a hydrogenase (Steenkamp and Peck, 1981), and evidence suggests that it is a transmembranous protein. This conclusion is inferred because proton-releasing and nitrite-binding sites are located on the periplasmic aspect of the cytoplasmic membrane, and a benzyl viologen-binding site is on the cytoplasmic side of the membrane.

Hexaheme cytochrome c_3 has not been found in *Dv. vulgaris*; however, cells of *Dv. gigas* catalyze a rapid scalar production of protons as well as ATP (Barton et al., 1983) coupled to the oxidation of H_2 with nitrite. It has not yet been shown that the enzyme responsible for reduction of nitrite in *Dv. gigas* is hexaheme cytochrome c_3. This activity can not be due to bisulfite reductase reduction of nitrite, because it is a soluble, cytoplasmic enzyme and thus could not lead to proton transport coupled to nitrite reduction.

In considering these aspects of nitrite reductase, one can not escape an analogy with the H_2/HSO_3^- redox couple. Both reactions involve scalar and vectorial proton transport, and both apparently utilize unique multiheme *c*-type cytochrome as "proton pumps". Thus, the function of hexaheme cytochrome c_3 as a nitrite reductase may only be fortuitous, and its major function be that of a transmembrane electron carrier. These multiheme cytochromes have been largely relegated to the status of scientific curios, but I propose that these hemeproteins are electron-transferring proton pumps specifically designed for use with periplasmic enzymes of sulfate-reducing bacteria and other anaerobic microorganisms. I anticipate that they will prove to be a central biochemical feature in many types of anaerobic respiration.

3.6 Fermentation of Inorganic Sulfur Compounds

A most exciting contribution to our knowledge of the bioenergetics of sulfate-reducing bacteria has been the discovery (Bak and Cypionka, 1987; Bak and Pfennig, 1987) that certain of these bacteria are able to derive energy and grow by chemolithotrophic fermentations involving disproportionation of the inorganic sulfur compounds, sulfite and thiosulfate, as shown in reactions 3.9 and 3.10.

Sulfite disproportionation: $$4\,HSO_3^- + H^+ \rightarrow 3\,SO_4^{2-} + HS^- \quad (3.9)$$

Thiosulfate disproportionation: $$S_2O_3^{2-} + H_2O \rightarrow SO_4^{2-} + HS^- + H^+ \quad (3.10)$$

Growth yields are low, 2 to 3 gm/mol of substrate, and it is clear that acetate is required as a carbon source for biosynthesis of cell material. The mechanism of energy conservation could involve most of the respiratory sulfate reduction enzymes (Krämer and Cypionka, 1989), and this possible mechanism is summarized below in its simplest forms for sulfite (reactions 3.11 to 3.14) and thiosulfate (reactions 3.15 to 3.18).

Sulfite fermentation

$$3\,HSO_3^- + 3\,AMP \rightarrow 3\,APS + 6e + 6H^+ \quad \text{APS reductase} \quad (3.11)$$

$$3\,APS + 3\,PP_i \rightarrow 3\,SO_4^{2-} + 3\,ATP \quad \text{ATP sulfurylase} \quad (3.12)$$

$$6e + 6\,H^+ + HSO_3^- \rightarrow HS^- + 3H_2O \qquad \text{Bisulfite reductase} \qquad (3.13)$$

$$\text{Sum: } 4\,HSO_3^- + 3\,PP_i + 3\,AMP \rightarrow 3\,SO_4^{2-} + HS^- + 3\,ATP + 3H_2O \qquad (3.14)$$

Thiosulfate Fermentation

$$S_2O_3^{2-} + 2e + 2H^+ \rightarrow HSO_3^- + HS^- \qquad \text{Thiosulfate reductase} \qquad (3.15)$$

$$HSO_3^- + AMP \rightarrow APS + 2e + 2H^+ \qquad \text{APS reductase} \qquad (3.16)$$

$$APS + PP_i \rightarrow SO_4^{2-} + ATP \qquad \text{ATP sulfurylase} \qquad (3.17)$$

$$\text{Sum: } S_2O_3^{2-} + AMP + PP_i \rightarrow SO_4^{2-} + ATP \qquad (3.18)$$

There are two problems with this simple mechanistic view of sulfite disproportionation. The first concerns activities and characteristics of the various enzymes involved in the mechanism outlined above. The second problem involves sources of PP_i necessary for the pathway. First, ATP sulfurylase was not detected, although it is difficult to absolutely prove the absence of any enzymatic activity. Pyrophosphatase activity was decreased, and this observation is consistent with the mechanistic importance of PP_i for the generation of ATP. Hydrogenase activity was increased three to tenfold by growth on sulfite or thiosulfate and suggests the possibility that APS reductase and bisulfite reductase are coupled by means of H_2 and hydrogenases. The high reduction potential of APS reductase ($E'_o = +.060$ mV) compared with hydrogenase ($E'_o = -.414$ mV) is not in favor of such a coupling; however, inhibition of the disproportionation by carboxylcyanide *m*-chlorophenylhydrazone (CCCP) but not *N,N'*-dicyclohexylcarbodiimide (DCCD) might be explained by energy-dependent reverse electron transfer from APS reductase to hydrogenase. If the same electron transfer components are utilized in sulfite disproportionation as respiratory sulfate reduction, tetraheme cytochrome c_3 might be viewed as responsible for this essential energy-dependent electron transfer. Alternatively, electron transfer from APS reductase may involve an entirely different cytoplasmic electron transfer system.

The second problem with the mechanistic view of sulfite disproportionation outlined previously involves the source of inorganic pyrophosphate. Biosynthetic reactions may supply a portion of the required PP_i. It is unlikely, however, that these reactions can meet the needs of the major energy-yielding reaction (reaction 3.12) of sulfate-reducing bacteria when they respire sulfite via disproportionation pathways. Similarly, it is unlikely that reversal of soluble inorganic pyrophosphatase under physiological conditions can generate significant quantities of PP_i (Behreus and DeMeis, 1985). Alternatively, PP_i may be synthesized by a membrane-bound $H^+:PP_i$ synthetase (Maeshima, 1991), and this reaction could account for inhibition of sulfite disproportionation by CCCP.

Clearly, sulfite disproportionation appears to be a variation on the theme of sulfate respiration, but from an evolutionary point of view sulfate respiration should perhaps be viewed as a variation of sulfite disproportionation. If one embraces the view of Skyring and Donnelly (1982) that geological outgassing of SO_2 played an important role in the metabolic evolution of respiratory sulfate reduction, then this geological phenomenon may have led first to respiratory sulfite reduction. This metabolic system, in turn, served as a platform for evolution of APS reductase and perhaps ATP sulfurylase or ADP sulfurylase. Further refinement of sulfite disproportionation may have led to the evolution of microorganisms capable of respiratory sulfate reduction.

3.7 Hydrogen Production by Sulfate-Reducing Bacteria

Sulfate is not an obligatory electron acceptor for growth of sulfate-reducing bacteria on all organic substrates. Postgate (1952) first reported growth of *Dv. vulgaris* on pyruvate in the absence of sulfate with production of acetate, CO_2, and stoichiometric amounts of hydrogen rather that the H_2S produced in the presence of sulfate. These two conversions are compared in reactions 3.19 and 3.20.

$$\text{pyruvate} + P_i \rightarrow \text{acetyl phosphate} + CO_2 + H_2 \quad (3.19)$$

$$\text{pyruvate} + SO_4^{2-} + P_i \rightarrow \text{acetyl phosphate} + S^{2-} + CO_2 \quad (3.20)$$

Energy for growth on pyruvate alone is provided by substrate phosphorylation (reaction 3.19). The possibility exists, however, that cytoplasmic H_2 evolution from pyruvate may function as a proton pump and contribute to overall cellular energy economy. Accurate determination of growth yields on pyruvate has been technically difficult. Thus, experimental evaluation of the role of proton reduction in the bioenergetics of this reaction has not yet been possible. During pyruvate/sulfate respiration (reaction 3.20), little or no hydrogen is usually produced, and growth yield data (Traore et al., 1981) indicate that energy generation is coupled to electron transfer as well as substrate phosphorylation.

During growth of *Desulfovibrio* on lactate plus sulfate, acetate, CO_2, and H_2S are the major products, but variable amounts of H_2, butyrate, and ethanol may be produced depending upon the species under consideration (Hatchikian et al., 1976). Because only two high-energy phosphate groups are produced by substrate phosphorylation during oxidation of lactate, and they are utilized to reduce sulfate (see reactions 3.1 to 3.3), it is necessary to invoke electron transfer-coupled phosphorylation to account for growth during the lactate/sulfate respiration (Peck, 1962). The global involvement of hydrogen and hydrogenases in the metabolism and bioenergetics of sulfate-reducing bacteria (*Desulfovibrio*) is illustrated by numerous studies showing that these bacteria can function as proton-reducing (i.e., H_2-producing) microorganisms in microbial associations.

For example, *Dv. vulgaris* grows well in media containing lactate plus sulfate, forming acetate, CO_2, and H_2S (reactions 3.6 and 3.7), but, in the absence of sulfate, growth on lactate is negligible or absent. This requirement for sulfate can be eliminated by the presence of an H_2-utilizing methanogen, and methane—rather than H_2S—is a major growth product. All of the reducing equivalents generated from oxidation of lactate to acetate and CO_2 appear as methane; thus, H_2 must be regarded as a major product of both lactate and pyruvate oxidation rather than having an incidental or trace involvement in these major reactions (Lupton et al., 1984b).

Many anaerobic bacteria, in addition to sulfate reducers, can completely dispose of electrons originating by oxidation of organic substrates through proton reduction catalyzed by hydrogenase; the efficiency of this process seems to be related to the Gibbs free energy of the H_2-utilizing redox couple rather than the pathways involved in proton reduction (Cord-Ruwisch et al., 1988). These observations project and support a common picture of hydrogenase-containing anaerobic bacteria as enveloped in a "blanket" or "micro-atmosphere" of molecular hydrogen resulting from a steady state between H_2-producing and H_2-utilizing reactions or intraspecies H_2 transfer, i.e., H_2 cycling. Perturbation of this steady-state concentration of hydrogen may occur after exhaustion of an electron acceptor and result in production of an increased amount of H_2 which may be utilized for various reductions by other anaerobic bacteria when present.

In the absence of a *normal* electron acceptor (e.g., SO_4^{2-}), diversion of energy from one bacterium to another by interspecies H_2 transfer to an electron sink is advantageous for the H_2-producing bacterium. Such a process usually allows a net production of ATP by substrate phosphorylation (or possibly H_2 production), but results in loss of ATP production by electron transfer-coupled phosphorylation. The "burst" of H_2 production reported by Tsuji and Yagi (1980) during the early phase of *Dv. vulgaris* growth on lactate plus sulfate may represent simply an "overshoot" in H_2 production prior to establishment of this steady-state concentration of hydrogen. In addition, reduction of intracellularly generated electron acceptors is favored by most organisms.

The specific reactions involved in H_2 evolution have not been studied in great biochemical detail and thus will be discussed only briefly in terms of formate, CO, and pyruvate. Most sulfate-reducing bacteria, which grow in the presence of H_2 and sulfate, also grow in the presence of formate and sulfate (Postgate, 1984a). *Dv. vulgaris* does not grow on formate (except in the presence of sulfate); however, intact cells and extracts catalyze a formate hydrogenlyase reaction in which formate is degraded to CO_2 plus H_2. As shown in reactions 3.21 and 3.22, the reaction seems to require three proteins: formate dehydrogenase, tetraheme cytochrome c_3, and a hydrogenase.

$$HCOOH + 1/2 \text{ cyto } c_{3ox} \leftrightarrows CO_2 + 2H^+ + 1/2 \text{ cyto } c_{3red} \quad (3.21)$$

$$2\,H^+ + 1/2 \text{ cyto } c_{3red} \leftrightarrows H_2 + 1/2 \text{ cyto } c_{3ox} \quad (3.22)$$

Formate dehydrogenase and tetraheme cytochrome c_3 have been shown operationally to be periplasmic proteins (Odom and Peck, 1981b), and, consistent with this cellular localization, the tetraheme cytochrome c_3 gene sequence codes for a positively charged and hydrophobic signal sequence of 21 amino acids (Voordouw and Brenner, 1986). In addition, polyclonal antibodies against purified tetraheme cytochromes c_3 from *Dv. vulgaris* (Miyazaki) inhibited formate hydrogenlyase activity by intact cells and localized the reaction on the cell surface (Tamura et al., 1988). Because cytochrome c_3 is required, a hydrogenase of some sort must certainly be required for the reaction, but the involvement of a specific type of hydrogenase has not been enzymologically demonstrated. A periplasmic [Fe] hydrogenase has been purified from *Dv. vulgaris* (Miyazaki) (Yagi, 1970), and a [NiFe] hydrogenase has been cloned and sequenced (Deckers et al., 1990). Antibodies against the [Fe] hydrogenase do not inhibit the formate hydrogenlyase reaction but do inhibit reduction of sulfite by H_2. Clearly, the enzymology of this situation needs to be resolved. It seems evident that the formate hydrogenlyase system is located in the periplasm and probably constitutes a fortuitous association of enzymes whose reactions are not involved in the bioenergetics of sulfate-reducing bacteria except for providing exogenous H_2.

A second useful and interesting substrate for H_2 evolution is carbon monoxide, which is an intermediate in the Wood autotrophic pathway (Ljungdahl, 1986) of CO_2 fixation and the pathways of methane formation from CO_2 and acetate. In addition, it can also serve as an energy source for anaerobic growth of acetogens and methanogenic bacteria, but these observations can be complicated by the fact that CO is an effective inhibitor of "[Fe]" and nickel-containing hydrogenases (Fauque et al., 1989). The presence of an enzyme capable of oxidizing CO to CO_2 in anaerobic bacteria was first reported by Yagi (1958) studying sulfite metabolism in extracts *Dv. vulgaris*, but most of the enzymological and functional studies have employed other microorganisms.

Generally there are two different types of carbon monoxide dehydrogenases (CODs): (1) nickel and nonheme iron enzymes found in anaerobic and phototrophic bacteria; and (2) a molybdenum and nonheme iron flavoprotein enzyme found in aerobic chemolithotrophic bacteria that utilize CO as a sole energy and carbon source (Ljungdahl, 1986). An example of the former enzyme is found in *Rhodopseudomonas gelatinosa*, which exhibits slow anaerobic growth in the absence of light and in the presence of CO, which is converted to CO_2 plus H_2. The bioenergetic mechanism involved in energy conservation is not immediately evident, but it is suggested that conversion of cytoplasmic protons to H_2 would result in alkalinization of the cytoplasm and, effectively, generation of a membrane potential. This possibility needs to be experimentally verified.

Dv. vulgaris utilizes CO as sole electron donor for growth and sulfate reduction in the presence of acetate plus sulfate and a head-space concentration of CO below 4.0% (Lupton et al., 1984a). Growth is relatively slow, and formation of H_2 from CO was sequentially followed by H_2 utilization to reduce sulfate to sulfide. Growth was apparently due exclusively to oxidation of H_2 coupled to sulfate reduction and not the result of H_2 evolution by the cytoplasmic COD, resulting in alkalization of the cytoplasm, as suggested for *R. gelatinosa*. These growth data were also interpreted to indicate the presence of two types of hydrogenase: a CO-insensitive enzyme coupled to COD and a CO-sensitive enzyme coupled to reduction of sulfate (Lupton et al., 1984a).

Carbon monoxide metabolism can best be understood as involving sequential H_2 cycling. H_2 is evolved in the cytoplasm, as a result of CO oxidation by a nickel-containing hydrogenase, and is subsequently oxidized in the periplasm by the [Fe] hydrogenase, and possibly a nickel-containing hydrogenase, to yield protons. Reducing equivalents generated are used to reduce sulfate. Localization of the CO-metabolizing system in the cytoplasm makes possible the net transfer of scalar protons coupled to reduction of sulfate. It should be noted that substrate phosphorylation does not occur in CO oxidation; consequently, all of the energy required for APS formation, as well as the energy for growth, must be generated by H_2 cycling.

The third H_2-producing activity from sulfate-reducing bacteria is the pyruvate phosphoroclastic system, which reversibly converts pyruvate, in the presence of substrate level concentrations of orthophosphate (P_i), to acetyl phosphate, CO_2, and H_2. The system is cytoplasmic and involves four protein components: a nonheme iron-containing pyruvate:ferredoxin oxidoreductase (PFO) (reaction 3.23), which requires thiamine pyrophosphate (TPP), Coenzyme A, and Mg^{2+} for activity and is oxygen labile; a ferredoxin (or flavodoxin); a hydrogenase capable of oxidizing reduced ferredoxin or flavodoxin (reaction 3.24) with the formation of acetyl phosphate from acetyl CoA by transacetylase (reaction 3.25). Pyruvate:ferredoxin oxidoreductase from *Dv. vulgaris* is stimulated by ATP, but its role in the reaction has not been studied in detail (Yates, 1967).

$$\text{pyruvate} + \text{CoA} + \text{ferredoxin}_{ox} \overset{TPP, Mg^{2+}}{\leftrightarrows} \text{acetyl CoA} + CO_2 + \text{ferredoxin}_{red} \quad (3.23)$$

$$\text{ferredoxin}_{red} + 2\,H^+ \leftrightarrows H_2 + \text{ferredoxin}_{ox} \quad (3.24)$$

$$\text{acetyl CoA} + P_i \leftrightarrows \text{acetyl phosphate} + \text{CoA} \quad (3.25)$$

Most of our information concerning reaction 3.23 has come from studies with the clostridial enzyme and is quite limited (Uyeda and Rabinowitz, 1971). The organization and mechanism of the nonheme iron have not yet been evaluated in detail, and the gene(s) encoding the enzyme system has not been sequenced. Pyruvate:ferredoxin oxidoreductase, in the clostridia, does not have a high level of specificity for its electron acceptors. Many low-molecular-weight compounds, such as FMN, FAD, methyl viologen, benzyl viologen, and tetrazolium derivatives, are reduced but may or may not donate electrons to a specific hydrogenase. Similarly, the oxidoreductase does not exhibit a great deal of specificity towards electron carrier proteins. Rubredoxin is reduced but does not serve as an electron donor for proton reduction. Various ferredoxins and flavodoxin are reduced, and the reduced forms are oxidized by hydrogenase to produce H_2. The specificity of the reaction for ferredoxin, either in terms of oxidoreductase or hydrogenase, is most puzzling. Plant-type ferredoxins (two iron clusters) and bacterial-type ferredoxins (four and eight irons) actively catalyze H_2 formation from pyruvate. Ferredoxin from *Dv. gigas*, which contains a three-iron center, does not catalyze transfer of electrons from PFO to hydrogenase. However, after conversion to a four-iron center protein (Moura et al., 1984), the *Dv. gigas* ferredoxin will catalyze this electron transfer reaction. Tetraheme cytochrome c_3 stimulates the phosphoroclastic reaction of sulfate-reducing bacteria (Akagi, 1967), and the mechanism of this stimulation probably involves a cofactor role of this cytochrome with the sulfate reducer hydrogenases.

Using various known periplasmic and cytoplasmic proteins, a sequence of electron transfer reactions can be constructed as follows:

$$\text{PFO} \rightarrow \begin{matrix}\text{ferredoxin}\\ \text{or}\\ \text{flavodoxin}\end{matrix} \rightarrow \begin{matrix}\text{tetraheme}\\ \text{cytochrome } c_3\end{matrix} \rightarrow \text{hydrogenase}$$

This sequence directs our attention on hydrogenase as a most important enzyme for specificity studies of the phosphoroclastic system. Organisms exhibiting PFO activity may contain one or more of six different nickel-containing and iron-only hydrogenase. One can expect that the presence of iron or nickel at their catalytic sites and differences in organization and numbers of nonheme iron centers will generate interesting displays of electron donor specificities among these enzymes. Indeed, PFOs from different prokaryotes may be expected to contain different combinations of nonheme iron centers, which will be reflected in their electron acceptor specificities. Most acetate-accumulating, sulfate-reducing bacteria (Widdel, 1988) grow well on pyruvate plus sulfate. Many can grow on pyruvate in the absence of sulfate or other anions of sulfur and form H_2 rather than H_2S. In the laboratory, pyruvate is metabolized rapidly and without a lag and is, therefore, the substrate of choice for studying H_2 evolution by sulfate-reducing bacteria.

3.8 Intracellular H_2 Transfer or H_2 Cycling

The availability of an exogenous source of H_2 has played a prime role in the evolution of respiratory sulfate reduction, but sources of H_2 have changed over geological time. Initially, geological outgassing was the sole source of exogenous H_2. Subsequently, however, biological formation of H_2 from phototrophic and other anaerobic bacteria, in terms of interspecies H_2 transfer, assumed increasing importance as exogenous sources of H_2. The ultimate development has been the oxidation of organic substrates in the cytoplasm—or on the cytoplasmic surface of the cell membrane—for evolution of H_2, which is then utilized in the periplasm as an exogenous source of H_2 for respiratory sulfate reduction. This coupling of H_2-evolving and H_2-utilizing reactions has been termed H_2 cycling or intraspecies H_2 transfer.

Close intracellular coupling between H_2-evolving and -utilizing reactions was first proposed to explain the hydrogenase/nitrogenase relationship in nitrogen-fixing bacteria (Dixon, 1972). The basic concept of this H_2 recycling mechanism involves evolution of H_2 by nitrogenase and oxidation of H_2 at the cytoplasmic side of the membrane by a membrane-bound hydrogenase. Reducing equivalents generated are transferred to a proton-translocating electron transfer chain and utilized to reduce O_2 to water. This mechanism was proposed to serve two physiological functions: one was to recover energy lost by generation of H_2; another was to maintain low intracellular partial pressures of H_2, which is an inhibitor of nitrogenase. These ideas, plus those of intraspecies H_2 transfer, strongly support the concept that H_2 plays an essential role in the transfer of energy both between individual microorganisms and among cellular compartments within discrete organisms.

The basic outline of intraspecies H_2 transfer with pyruvate plus SO_4^{2-} in terms of scalar proton transfer is indicated in Figure 3.5 and essentially combines aspects of H_2 evolution and H_2 utilization discussed previously. The notion involves evolution of H_2 from pyruvate by PFO in the cytoplasm utilizing [NiFeSe] hydrogenase localized (Rohde et al., 1989) on the cytoplasmic surface of the membrane or the [20 Fe] hydrogenase. H_2 then diffuses across the cytoplasmic membrane and is oxidized by periplasmic hydrogenases to produce scalar protons. In cells of *Dv. vulgaris*, two separate periplasmic hydrogenases and electron transfer sequences are employed. In one sequence, periplasmic membrane-bound [NiFe] hydrogenase oxidizes H_2, and the reducing equivalents are used to reduce HSO_3^- by bisulfite reductase in the cytoplasm; transmembrane electron transfer to flavodoxin utilizes the hexadecaheme cytochrome c_3. In a second sequence, soluble periplasmic [12 Fe] hydrogenase oxidizes H_2 concomitant with cytoplasmic reduction of APS by APS reductase; it is proposed that the hydrogenase utilizes a dimeric form of tetraheme cytochrome c_3 for transmembrane electron transfer to APS reductase. The scheme is consistent with the specific cellular localizations of electron carriers and enzymes and the presence of a membrane-bound, uncoupler-stimulated ATPase. In particular, it should be noted that the electron transfer pathways from H_2 to bisulfite and H_2 to APS are entirely separate and linked only by molecular hydrogen. The latter is of particular importance for the regulation of energy coupling in sulfate-reducing bacteria.

The first experimental evidence for H_2 cycling was presented by Odom and Peck (1981a), who were studying the ability of spheroplasts of *Dv. vulgaris* to catalyze the lactate/SO_4^{2-} respiration. They observed that washed spheroplast preparations did not catalyze the reaction, but that 70% of the activity could be restored by addition of both purified tetraheme cytochrome c_3 and pure *Dv. gigas* [NiFe] hydrogenase. These unexpected observations were interpreted to indicate that H_2 was an obligatory intermediate in coupling of lactate oxidation to sulfate reduction. Recently, inhibition of [12 Fe] hydrogenase synthesis, employing antisense RNA, has been achieved; inhibition of hydrogenase synthesis is accompanied by inhibition of lactate/sulfate respiration (Berg et al., 1991). These results clearly support an obligatory role for the [12 Fe] hydrogenase in the oxidation of lactate coupled to the reduction of sulfate.

Direct evidence for H_2 cycling has been obtained by comparative studies of pyruvate fermentation and pyruvate/sulfate respiration (Peck et al., 1987). In these studies, production of H_2 into an atmosphere of 20% $^2H_2$ was analyzed under two sets of conditions: one, in the presence of pyruvate alone, which results in net formation of H_2; and two, in the presence of pyruvate plus sulfate, which does not result in net formation of H_2. It was predicted that if H_2 cycling occurred during pyruvate/sulfate respiration, rates of H_2 production from pyruvate would be balanced by H_2 utilization for reduction of sulfate. It should then be possible to "trap" some amount of H_2 in the "pool of $^2H_2$" even if significant isotope effects occurred. When the experiment was performed and the gas phase analyzed for H_2, it was evident that H_2 was produced during the pyruvate/sulfate respiration even in slight excess to the net production of H_2 from pyruvate alone. This result constitutes an unequivocal demonstration of H_2 cycling in *Dv. vulgaris* and strongly suggests that it is of major bioenergetic importance rather than a mechanism for regulating the redox state of electron carriers (Lupton et al., 1984b).

A scheme is presented in Figure 3.6 that incorporates both scalar and vectorial proton production into the hydrogen cycling scheme. Oxidation of H_2 by the [Fe] hydrogenase and transfer of electrons to APS reductase are shown to require protons and not result in net transport of either scalar protons or vectorial protons. Oxidation of H_2 by periplasmic [NiFe] hydrogenase results in transport of six scalar protons; electron transfer by hexadecaheme cytochrome c_3 to bisulfite reductase is proposed to produce six vectorial protons for a total transfer of 12 protons across the membrane. Assuming a stoichiometry of three protons per ATP synthesized by

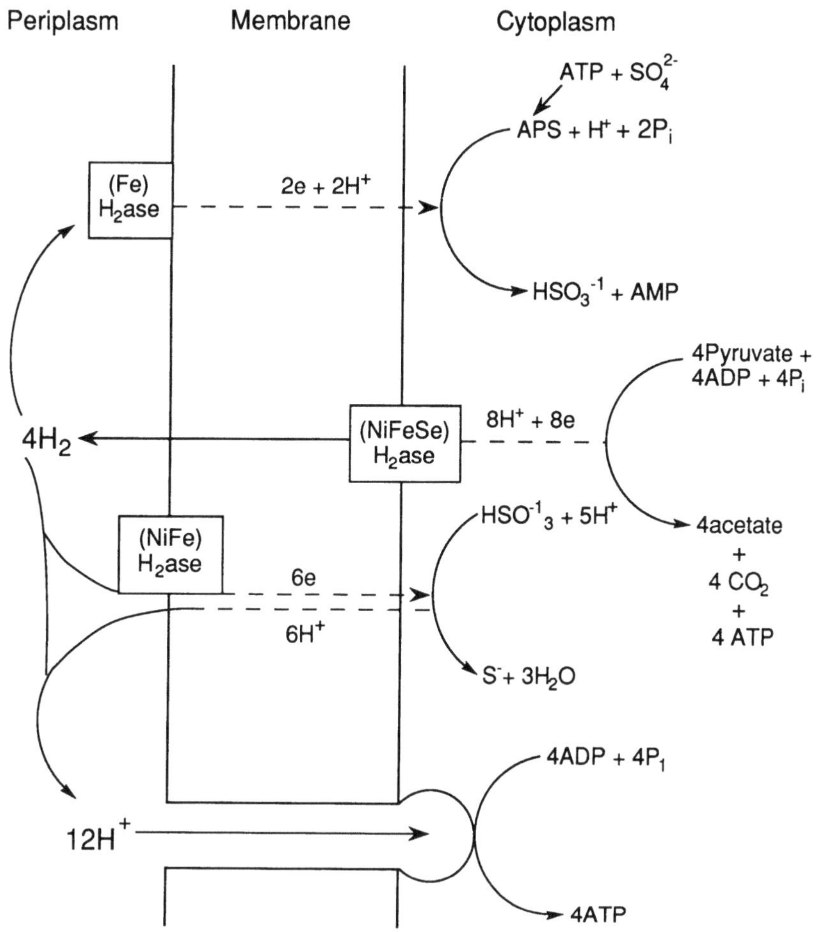

Figure 3.6 Bioenergetics of hydrogen cycling with pyruvate plus sulfate.

ATPase, this scheme suggests the formation of four moles of ATP by reduction of a mole of sulfate; however, two moles of ATP must be utilized for activation of sulfate.

Growth yields are generally consistent with predicted yields of ATP from the proposed scheme for H_2 cycling, but the validity of using growth yields in this manner has been questioned (Tempest, 1978). The term, ($\mathbf{Y}^{\mathrm{max}}_{\mathrm{[substrate]}}$) serves as a measure of the grams of dry cell weight produced by a mole of ATP for a given substrate metabolized by an organism. The term represents a product of all reactions involved in biosynthesis of cell material and thus reflects in a somewhat quantitative fashion the ability of a cell to generate ATP. For example, in pyruvate fermentation by sulfate-reducing bacteria, ($\mathbf{Y}^{\mathrm{max}}_{\mathrm{pyr}}$) is 10 g per mole. This value indicates that one mole of ATP is generated and 10 g of cell materials produced per mole of pyruvate fermented. With H_2, a ($Y^{max}_{SO_4^{2-}}$) of 13 g/mol and a ($Y^{max}_{SO_3^{2-}}$) of 35 g/mol was reported. These values compare with predicted values of 20 g/mol for 4 H_2/SO_4^{2-} and 40 g/mol for 3 H_2/HSO_3^- couples. Observed growth yield for pyruvate/SO_4^{2-} respiration has been determined to be 44 g/mol, which compares with an expected growth yield of 50 from mechanistic considerations (Badziong and Thauer, 1978; Nethe-Jaenchen and Thauer, 1984). It should be noted that the mechanistically predicted formation of ATP is always somewhat greater than that predicted from growth yield. This is probably expected, as we do not as yet know exactly how to correct for the influx of substrates and efflux of products.

The mechanism and bioenergetics of lactate/SO_4^{2-} respiration with regard to H_2 cycling have not yet been satisfactorily resolved. The basic problem concerns thermodynamics in that conversion of lactate to pyruvate plus H_2 is highly endergonic ($\Delta G_o' = +43.2$ KJ/mol) (Pankhania et al., 1988). Nevertheless, studies on intraspecies H_2 transfer indicated that lactate can serve as a source of H_2 as evidenced by the formation of methane (Bryant et al., 1977). This problem was compounded by the observations that H_2 does not inhibit growth on lactate plus SO_4^{2-} and that, similarly, CO, a potent inhibitor of [12Fe] hydrogenase, did not effectively inhibit growth on lactate plus sulfate (Lupton et al., 1984b). The latter difficulty has not been resolved, but it may be, as indicated in Figure 3.6 that H_2 is evolved by a [NiFeSe] hydrogenase which is much less sensitive to inhibition by CO than the [12 Fe] hydrogenase. Pankhania et al. (1988) demonstrated that cell suspensions would quantitatively convert lactate to pyruvate plus H_2 when the H_2 concentration in the gas phase was below 2%. They also observed that protonophores such as CCCP, DCCD, an inhibitor of ATP synthase, and arsenate, which interferes with the metabolism of acetyl phosphate, inhibit H_2 evolution from lactate. It was suggested that this coupling of lactate oxidation with ATP formation might be by substrate activation, reversed electron transfer, or by the active transport of lactate. This phenomenon is reminiscent of the irreversible ATP-dependent evolution of H_2 by nitrogenase, and future progress in understanding this reaction may contribute insights into the evolution of H_2 from lactate.

The mechanism of energy coupling for the lactate/SO_4^{2-} respiration involves the production of 4 ATPs as in the case of the H_2/HSO_3^- redox couple. This value is not supported by growth yield studies, which are lower than predicted by this scheme (($Y^{max}_{SO_4^{2-}}$) = 15.6). The deviation from the expected growth yield may reflect a significant requirement of protons for the import of lactate as well as sulfate or an ATP requirement for the production of H_2. Thus, a possible net requirement of six protons for the import of one sulfate and two lactate would result in a net ATP yield of two, which is more consistent with the observed growth yields. It should be indicated that a stoichiometry of one ATP/H_2 evolved from lactate would consume all the ATP generated by substrate phosphorylation and not allow growth by interspecies H_2 transfer. On the other hand, the lactate/SO_4^{2-} respiration may involve a modified pathway of HSO_3^- reduction such as the "trithionate pathway" or employ the low-spin sulfite reductase with a reduced proton yield but still involving the basic tenets of H_2 cycling. In terms of the trithionate pathway, proton generation might be restricted to only one of the reductive steps, resulting in a predicted value of 1.3 ATPs or two of the reductive steps, giving 2.67 ATPs for the H_2/HSO_3^- redox couple. Similar considerations may apply to the low-spin, assimilatory-type sulfite reductase.

The basic idea of H_2 cycling as a mechanism for the coupling of energy generation to the reduction of sulfate in *Desulfovibrio* offers a framework for understanding many unique aspects of the biochemistry and the physiological and molecular biological observations concerning respiratory sulfate reduction. The concept offers an explanation for the ability of *Desulfovibrio* to act under different conditions as an H_2-evolving or an H_2-utilizing bacterium in interspecies H_2 transfer and is consistent with the bioenergetic importance of H_2 in anaerobic bacteria and ecosystems. The relationship implicit in H_2 cycling should provide insights into the metabolism of other substrates by *Desulfovibrio*, such as ethanol and choline and in the metabolism of other bacteria, for example acetogenesis (Ljungdahl, 1986; Dolfing, 1988) and mixed amino acid fermentations, (the Stickland reaction) (Ljungdahl et al., 1989), and in refining our knowledge of the metabolic and bioenergetic systems of the recently described genera of the sulfate-reducing bacteria (Widdel, 1988). In this regard it should be indicated that not all sulfate-reducing bacteria, such as *Desulfotomaculum*, contain significant amounts of *c*-type cytochromes but are able to couple energy generation to the reduction of sulfate and contain the enzymes of the pathway of respiratory sulfate reduction (Cypionka and Pfennig, 1986).

3.9 Conclusions

Sulfate-reducing bacteria are characterized by the ability to utilize sulfate as their major terminal electron acceptor and to couple the generation of ATP to this reduction. The pathway has been invariably found to involve four soluble cytoplasmic enzymes: ATP sulfurylase, inorganic pyrophosphatase, APS reductase, and a bisulfite reductase. H_2 plays a central role in the bioenergetic mechanisms of energy conservation via the process of H_2 cycling in which H_2 is evolved by a cytoplasmic hydrogenase and utilized for sulfate reduction by periplasmic hydrogenases. The bacteria lack a classical cytochrome electron transfer chain and probably a functional Q cycle. They contain, however, at least two unbranched and independent electron transfer chains that feature multiheme *c*-type cytochromes and are directed individually towards reduction of APS and reduction of bisulfite to sulfide. The two pathways appear to interact and be regulated largely through the agency of molecular H_2. The H_2/HSO_3^- redox couple is solely involved in energy conservation during reduction of sulfate, and both scalar and vectorial protons are produced during oxidation of H_2. In general, mechanistic bioenergetic schemes involving H_2 cycling have been presented that are consistent with the predicted amounts of ATP generated from a consideration of growth yields. H_2 and hydrogenase are of critical importance in understanding these anaerobic bioenergetic systems, and it is enticing to envisage that the two large gene families of hydrogenase reflect the evolution of cytoplasmic H_2-evolving hydrogenases, the [12Fe] hydrogenases, and periplasmic H_2-utilizing hydrogenases, the [NiFe] hydrogenases.

[References, see p. 211]

I acknowledge my continuing collaborations with Dr. Jean Le Gall, Dr. B.H. Huynh, Dr. J.J.G. Moura, Dr. I. Moura, and Dr. D.V. DerVartanian, and I thank each of them for their helpful discussions regarding this chapter. My own research described here has been supported by contracts DEA-5-9-79-ER 10499 from the U.S. Department of Energy, from the National Science Foundation (DMB 9005734), and the National Institutes of Health under a grant (GM 34903).

4

Genetics of the Sulfate-Reducing Bacteria

Judy D. Wall

4.1 Introduction

Genetic studies of sulfate-reducing bacteria have lagged far behind physiological and biochemical investigations. The primary reason for this delay has derived from the strictly anaerobic growth mode of these bacteria and the consequent inability to obtain useful plating efficiencies for quantitation of cell numbers (Postgate et al., 1988). Both the improvement of anaerobic chambers that provide a consistently low O_2 atmosphere and the selection of strains less sensitive to O_2 inhibition have been instrumental in opening sulfate-reducing bacteria to genetic manipulation. Because of the short history of genetics with these bacteria, this review will discuss very basic procedures no longer mentioned in other better analyzed systems.

With the exception of sequence determinations of rRNAs discussed in Chapter 6 of this volume, genetic and molecular biological studies of the sulfate-reducing bacteria have been limited to Group 1 *Desulfovibrio* (Devereux et al., 1989). An important reason for this focus is historical, since members of this genus were the first nonphototrophic anaerobes shown to have cytochromes (Postgate, 1954) and to respire organic substrates (Peck, 1962). In addition, the more rapid growth rate, the extensive metabolic capabilities, the reasonable plating efficiencies, and the resistance or sensitivity to useful antibiotics also have influenced the choice of these strains for genetic development.

Molecular biological tools allow the isolation and manipulation of genes and regulatory elements; however, the exact physiological roles can be learned only from an analysis of these sequences within their original context. Thus, although much has and can be learned from a purely molecular biological approach (see Chapter 5), the full potential of these tools can be realized only when complemented by mutant isolation procedures and genetic exchange techniques.

One report has appeared in which classical enrichments for mutants with non-selectable phenotypes have been demonstrated to be feasible with *Desulfovibrio desulfuricans* ATCC 27774 (Odom and Wall, 1987). This procedure should be equally successful with other strains that are sensitive to penicillin or cycloserine. Recent efforts have concentrated on gene transfer techniques, resulting in reports of transduction (Rapp and Wall, 1987) and conjugation (Powell et al., 1989; van den Berg et al., 1989; Argyle et al., 1992) discussed below. Perhaps the most facile and utilitarian form of gene transfer, transformation, has not been documented for the sulfate-reducing bacteria. However, electroporation has now been successfully applied to one *Desulfovibrio* strain (M. Rousset et al., 1991) and should, in principle, be useful for the introduction of DNA and other molecules into any strain.

4.2 Considerations for Development of a Genetic System

Metabolic properties Energy generation, H_2 metabolism, metal corrosion, and degradation of environmental pollutants top the list of subjects to be explored through genetic tools with sulfate-reducing bacteria. Although these are generic processes, it should be kept in mind that the strain selected must exhibit the metabolic process to be studied and, ideally, have additional growth capabilities allowing that process to be nonessential. The capacity to use alternative substrates for energy generation often allows genetic dissection of the bioenergetic pathway unique to a given substrate. For example, if a bacterium can use either H_2 or organic acids as electron donors, it may be possible to obtain mutants in H_2 utilization among cells growing with organic acids. For this reason, *Dv. desulfuricans* ATCC 27774 was one of the strains selected as a candidate for genetic development. This strain is capable of respiration with nitrate or sulfate as electron acceptor and with either H_2 or organic acids as electron donors and can ferment pyruvate or choline (Peck, 1984). Consequently, it was possible to isolate a mutant of this bacterium that could no longer respire sulfate in the presence of H_2 (Odom and Wall, 1987). Unfortunately *Dv. desulfuricans* ATCC 27774 grew poorly, if at all, on minimal medium. Thus, the facility of obtaining useful nutritional markers is poor. Prototrophic strains capable of rapid growth in defined medium should allow the identification of a wider array of mutations.

Plating efficiencies The strictly anaerobic nature of sulfate-reducing bacterial metabolism has been a problem for genetic manipulation. Anecdotal information suggested that O_2 sensitivity was a major obstacle in obtaining workable efficiencies for plating *Desulfovibrio* as single colonies on the surface of solidified medium (Postgate et al., 1988). However, through the use of the currently available anaerobic chambers and prereduced medium, most steps can now be performed without exposure of cells to significant levels of O_2. Using these procedures, Singleton et al. (1988) calculated efficiencies of plating for *Dv. vulgaris* Hildenborough of between 30% and 80%. Surprisingly, van den Berg et al. (1989) were able to obtain similar efficiencies after plating this same strain aerobically onto prereduced medium that was immediately incubated under anaerobic conditions. Clearly, exposure to O_2 was insufficient to cause detectable damage.

Still, experience with streaking several strains of *Dv. desulfuricans* onto agar surfaces for growth in the anaerobic chamber confirmed that colony formation was more rapid, robust, and more efficient under a layer of prereduced soft agar (Rapp-Giles and Wall, unpublished). Plating efficiencies of *Dv. desulfuricans* ATCC 27774 were between 25% and 50% with either sulfate or nitrate as terminal electron acceptor regardless of the growth stage of the inoculum (Odom and Wall, unpublished). Singleton et al. (1988) showed that plating efficiencies could be improved almost twofold, from about 34% to 56%, by a reduction in calcium concentration. This effect was interpreted to result from a decrease in cellular aggregate formation. In spite of such aggregates, plating efficiencies of 33% are adequate for most genetic manipulations.

Genome structure The DNA base ratios of sulfate reducers, which vary from 34% to 66% GC (Widdel, 1988; Devereux et al., 1989), reflect the heterogeneous nature of bacteria identified by an ability for sulfate respiration. Within the *Desulfovibrio* genus, there are three clusters of DNA composition of approximately 49%, 59%, and 65% GC (Postgate, 1984a); however, there are no obvious correlations between this property and metabolic capacities or phylogenetic relationships of the organisms (Devereux et al., 1990). Sequencing data now accruing are consistent with the reported DNA compositions, although localized compositional aberrations are possible.

The genome sizes of *Dv. vulgaris* Hildenborough and *Dv. gigas* were determined by Postgate and coworkers (1984) from the two-dimensional electrophoresis patterns of DNA fragments derived by sequential digestion with two restriction endoncleases. The sizes of the resulting fragments were estimated and summed to yield the chromosome size. The results showed that the genomes of the two sulfate reducers were substantially smaller than that of *Escherichia coli*; 1.63×10^6 base pairs (bp) for *Dv. gigas* and 1.72×10^6 bp for *Dv. vulgaris* versus 3.95×10^6 bp for *E. coli* (Postgate et al., 1984). In addition, *Dv. gigas* was estimated to contain nine copies of its genome per cell in nitrogen-limited cultures; the copy number increased to 17.2 in a batch culture growing in complete medium. By comparison, *Dv. vulgaris* had only four genome copies per cell versus two for *E. coli*. Obviously, multiple chromosomes could be a major difficulty when attempting to isolate recessive mutations through differential enrichment techniques.

Plasmids During these genome studies, plasmids were first observed in *Desulfovibrio* (Postgate et al., 1984, 1986). Of 16 strains analyzed, 5 had plasmids: *Dv. gigas* had two plasmids of 60 and 105 kbp; *Dv. desulfuricans* Berre sol, one of 135 kbp; and three *Dv. vulgaris* strains—Hildenborough, Wandle, and Brockhurst Hill—each had a plasmid of 195 kbp, and the latter two strains had an additional smaller plasmid of 83 kbp. Although all of the plasmid-containing strains were diazotrophic, a *Klebsiella pneumoniae nifH* gene probe hybridized only to the 195 kbp plasmid of the *Dv. vulgaris* strains. At present, the other plasmids appear to be cryptic. (See Chapter 5 for a discussion of these plasmids and their potential role in nitrogen fixation.)

In their 1988 review Postgate et al. pointed out their "extreme difficulty in extracting manipulable amounts of plasmid DNA" from *Dv. vulgaris* and *Dv. gigas*. Similar difficulties have been encountered upon attempts to recover plasmids introduced by conjugation from *E. coli* (Powell et al., 1989; Rapp and Wall, 1989). Thus, it was pleasantly surprising to observe rather large quantities of a small endogenous plasmid (2.3 kbp) in plasmid minipreparations of *Dv. desulfuricans* strain G200 (Figure 4.1; Wall et al., 1990). This cryptic plasmid, designated pBG1, has been completely sequenced (Rapp-Giles and Wall, in preparation), and some of its structural features are shown in Figure 4.2. It does not appear to function as a replicon in *E. coli*, as no Km^R transformants have been obtained after attempts to introduce a Km^R cassette into pBG1. In addition a recombinant plasmid, pSC1, that has pBG1 cloned into the *EcoRI* site of pTZ18U (United States Biochemical Corporation, Cleveland, Ohio) was unable to be stabilized at high temperature in a temperature-sensitive *polA E. coli* strain. Presently a *mob* site is being added to pSC1 to allow conjugational transfer of pBG1-containing plasmids into *Dv. desulfuricans*. Insight should be gained on the DNA replication machinery of *Dv. desulfuricans* and on the possibility of increasing the efficiency of conjugation into sulfate-reducing bacteria with this small endogenous plasmid.

Antibiotic sensitivities Application of genetic and recombinant DNA techniques requires the availability of a set of selectable markers. Antibiotic resistance (and sensitivity) has played a vital role in genetic manipulation of bacteria, because of the strong selection pressure that can be applied to identify rare events. In general, sulfate-reducing bacteria display resistance to high levels of a number of antibiotics (Postgate, 1984a).

However, strains chosen for genetic analysis to date have some useful sensitivities (Table 4.1). For example, chloramphenicol has proved effective in a selection against *Dv. vulgaris* and for acquisition of a plasmid

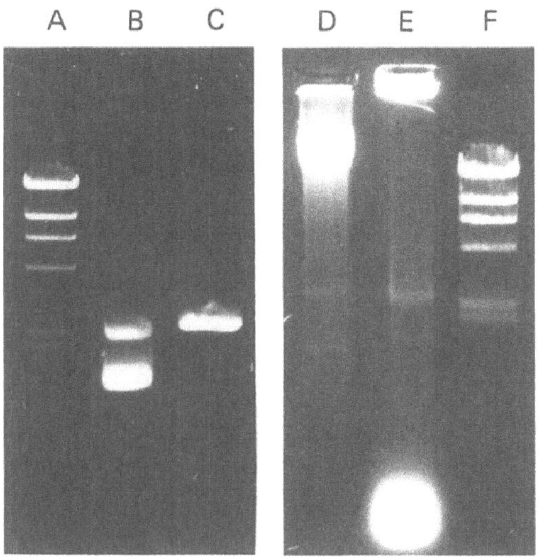

Figure 4.1 Visualization of pBG1 plasmid DNA from *Dv. desulfuricans* strain G200. Ethidium bromide stained agarose gels of: (A) and (F) lambda *Hind*III DNA restriction fragments; (B) uncut CsCl-purified pBG1; (C) *Eco*RI-digested CsCl-purified pBG1; (D) undigested genomic DNA and (E) undigested boiling minipreparation of pBG1 DNA.

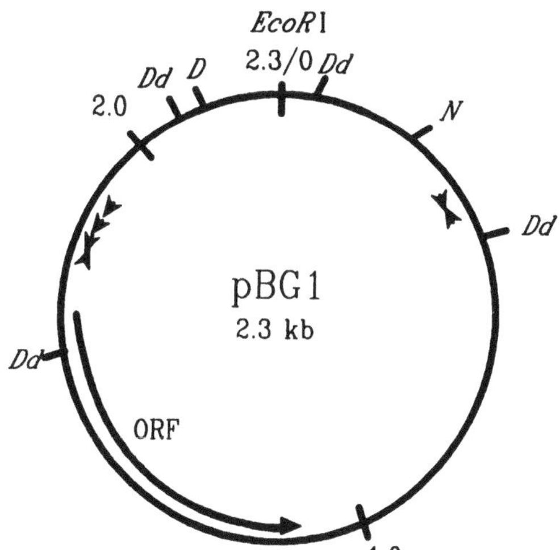

Figure 4.2 Physical map of pBG1. Recognition sites for *Eco*RI, *Dra*I (D), *Dde*I (Dd), and *Nhe*I (N) are shown. A putative open reading frame (ORF) is indicated. A large (19 bp) stem loop structure at 330 bp is indicated by inverted arrows, and a 13-base sequence that is repeated four times is depicted in the region of 1800 bp by arrowheads.

Table 4.1 Antibiotic sensitivity of *Desulfovibrio* strains

Antibiotic[a]	*Dv. vulgaris* Hildenborough[b,c]	*Dv. desulfuricans* ATCC 27774[d]	*Dv. desulfuricans* G100A[d]	*Dv. fructosovorans*[e]
Ap	S(>1)	S(5)	R(10)	R(100)
Sm	R(>100)	R(100)	R(100)	S(100)
Km	R(>100)	S(150)	S(150)	R(50)
Tc	R(12)	S(5)	S(<1)	S(20)
Rf	R(20)	S(50)	R(100)	N.R.
Cm	S(>3)	S(100)	S(50)	S(50)
Sp	N.R.[f]	R(>200)	S(100)	N.R.
Gm	N.R.	S(50)	S(50)	N.R.
Nal	N.R.	S(100)	S(100)	R(200)

[a] Antibiotics were Ap, ampicillin; Sm, streptomycin; Km, kanamycin; Tc, tetracycline; Rf, rifampicin; Cm, chloramphenicol; Sp, spectinomycin; Gm, gentamycin; Nal, nalidixic acid.
[b] Data from van den Berg et al., 1989.
[c] Numbers in parentheses are concentrations of antibiotic in μg per ml at which the cells are either resistant (R) and can grow or sensitive (S) and cannot grow.
[d] Data from Wall and Rapp-Giles, unpublished.
[e] Data from M. Rousset, personal communication.
[f] N.R., not reported.

in a conjugational cross (van den Berg et al., 1989). Classical penicillin enrichments should also be possible with this strain as shown for *Dv. desulfuricans* ATCC 27774 (Odom and Wall, 1987). In contrast, *Dv. desulfuricans* G100A is resistant to ampicillin, cycloserine, and rifampicin (Rapp-Giles and Wall, unpublished), making it necessary to identify nonselectable mutations (such as those resulting in auxotrophy) by screening procedures alone. Powell et al. (1989) took advantage of the sensitivity of *Dv. desulfuricans* strains 8301 and 8312 to streptomycin to select for exconjugants expressing resistance from a plasmid-borne gene. When appropriate antibiotic resistances were limited or unavailable, spontaneously resistant derivatives have been readily isolated from *Dv. desulfuricans* (Rapp and Wall, 1987; Voordouw et al., 1990b).

Restriction systems Among the factors likely to reduce the numbers of transconjugants obtained with recombinant plasmids is the presence of restriction endonucleases, as well as other nonspecific nucleases. *Desulfovibrio* strains are the source of two commercially available restriction endonucleases, *DdeI* and *DdeII*, and the genes for *DdeI* endonuclease and methylase have been sequenced from *Dv. desulfuricans* Norway (Sznyter et al., 1987). Other type II restriction enzymes have not been documented in sulfate-reducing bacteria but are likely to be present. Indirect evidence suggests that nonspecific nucleases may also be present in a number of strains. Chromosomal DNA preparations of *Dv. desulfuricans* often display evidence of significant digestion that may be a factor in the low yields of extractable plasmid DNA. To prepare high-molecular-weight DNA consistently, we have added guanidinium isothiocyanate to the lysis buffer for the immediate inactivation of nucleases (Delgado and Wall, unpublished). This procedure may increase the facility of plasmid isolation.

4.3 *Desulfovibrio* Bacteriophages

Genetic exchange mediated by bacteriophages, transduction, has made enormous contributions to fine structure mapping of mutations in many bacteria. However, few instances of bacteriophages or bacteriophage-like particles have been reported for sulfate-reducing bacteria. As early as 1973, Handley et al. (1973) reported that mitomycin C induced phage-like particles from *Dv. vulgaris* Hildenborough. No plaque-forming ability was observed, and potential gene transfer activity was not explored. Recently, from an enrichment of marine sediment, a lambda-like phage was isolated that could form plaques on *Dv. salexigens* (Kamimura and Araki, 1989). Again, its potential for transduction was not explored. The sensitivity of marine isolates to infection by this phage was proposed as an ecological marker for relatedness.

Only one report has suggested that a bacteriophage-like particle might mediate genetic exchange within a *Desulfovibrio* strain (Rapp and Wall, 1987). A small, apparently defective bacteriophage, Dd1, was found to be released during growth of *Dv. desulfuricans* ATCC 27774. The morphological characteristics of the phage particles resembled those of T7 or T3 coliphages, except that the head was smaller. The head was found to contain double-stranded DNA of a uniform length, ca. 13.5 kbp, that was randomly packaged from the chromosome. The inference that all chromosomal regions have a similar probability of packaging was supported by the occurrence of similar transfer frequencies for each of four different antibiotic markers and by the lack of distinctly reiterated fragments in the restriction endonuclease digestion pattern of the packaged DNA. The DNA in the particles resisted degradation by added nucleases and required divalent cations to be transferred to an appropriate recipient. Routinely, frequencies of 10^{-5} to 10^{-6} were reported in gene transfer experiments. Similar to the *Dv. vulgaris* phage-like particles (Handley et al., 1973), no plaque formation by Dd1 could be detected on any host (Rapp and Wall, 1987). However, mitomycin C treatments did not increase the numbers of the bacteriophage released from *Dv. desulfuricans* as assayed by genetic transfer.

Attempts to demonstrate the ability of this defective bacteriophage to transfer an antibiotic resistance marker from *Dv. desulfuricans* ATCC 27774 to other *Dv. desulfuricans* strains or to other *Desulfovibrio* species were unsuccessful (Rapp and Wall, unpublished). Thus, the practical use of Dd1 is limited to the ATCC strain 27774, although it is a valuable fine-structure mapping tool for that strain. Without an in vitro packaging system, it is not possible to transfer exogenous DNA with this vector. Thus, additional transfer techniques are required for the full analysis of *Desulfovibrio* genetics. The conjugation systems described below fill this need.

4.4 Conjugation of *Desulfovibrio*

Broad host range plasmids are capable of transfer between and stable maintenance in almost all Gram-negative bacterial species. Thus it was reasonable to expect that *Desulfovibrio* species should be no exception once plating conditions and the appropriate selective markers were identified. Almost simultaneously, two reports of successful gene transfer mediated by plasmids belonging to the Q incompatibility group have appeared (Powell et al., 1989; van den Berg et al., 1989) and similar results have been obtained elsewhere (Argyle et al., 1992; Table 4.2). Powell et al. (1989) used two SRB strains as recipients in overnight matings with *E. coli* donors harboring R300B and reported frequencies of 5×10^{-2} and 1 per recipient. Similar frequencies for IncQ plasmid transfer, ca. 1×10^{-2}, were obtained after overnight matings with *Dv. vulgaris* (van den Berg et al.,

Table 4.2 Plasmid transfer by conjugation in *Desulfovibrio*

Recipient	Plasmid stably transferred	Incompatibility group	Antibiotic selection[a]	Reference
Dv. sp.				
NCMIB 8301	R300B	IncQ	Sm_{50}	Powell et al., 1989
Dv. desulfuricans				
NCMIB 8312	R300B	IncQ	Sm_{50}	Powell et al., 1989
Dv. vulgaris				
Hildenborough	pSUP104	IncQ	Cm_{10}	van den Berg et al., 1989
NCMIB 8303	pRK404 Cm	IncP	Cm_5	van den Berg et al., 1989
Dv. desulfuricans				
G200[b]	pKT230[c]	IncQ	Km_{175}	Argyle et al., 1992
	pJRD215[d]	IncQ	Km_{175}	Argyle et al., 1992
	pDSK519[e]	IncQ	Km_{175}	Argyle et al., 1992

[a] Antibiotics used for selection of *Desulfovibrio* transconjugants, with the concentration in $mg \cdot l^{-1}$ as subscript.
[b] A spontaneously nalidixic acid-resistant derivative of wild-type *Dv. desulfuricans* G100A (Weimer et al.,1988).
[c] Bagdasarian et al., 1981.
[d] Davison et al., 1987.
[e] Keen et al., 1988.

1989). Although many rounds of conjugation are possible under such conditions, the frequencies obtained indicated that a very useful level of transfer had occurred. In conjugations of 4 to 6 hours with *Dv. desulfuricans* G200, frequencies as high as 1×10^{-3} were obtained (Argyle et al., 1992).

Transfer of IncP1 plasmids from an *Alcaligenes eutrophus* donor was also reported, but this plasmid was found to be unstable in *Desulfovibrio* sp. 8301 and *Dv. desulfuricans* 8312 (Powell et al., 1989). No instability of IncP1 plasmids was noted with *Dv. vulgaris* Hildenborough (van den Berg et al., 1989). Clear evidence for IncP1 mobilization into *Dv. desulfuricans* G200 was not obtained (Argyle et al., 1992).

The IncQ cloning vector pSUP104 (Priefer et al., 1985) has been used to introduce extra copies of the hydrogenase genes into *Dv. vulgaris* Hildenborough (van den Berg et al., 1989). Although protein levels increased, activity did not increase proportionally, reflecting a complex regulation for this enzyme. When these same genes were transferred into the heterologous background of *Dv. desulfuricans* G200, very low levels but completely processed *Dv. vulgaris* [Fe] hydrogenase was obtained (W.M.A.M. van Dongen, personal communication). In contrast, the introduction of the genes encoding the *Dv. vulgaris* Hildenborough cytochrome c_3 into the G200 host allowed high levels of active cytochrome to be produced that was readily distinguished from the endogenous cytochrome c_3 (Voordouw et al., 1990b). (See Chapter 5 for further discussion.)

Plasmids that are mobilizable but not stable may serve as suicide vectors in marker exchange experiments. van den Berg and colleagues (1989) reported attempts to use pSUP5011, a pMB1 replicon, to construct [Fe] hydrogenase-negative mutants. Although not yet successful, this technique should prove useful once the limitations are identified. One explanation for the inability to obtain the desired mutants was a possibility that the hydrogenase has an essential metabolic function under the growth conditions employed. An alternative approach, also taking advantage of the conjugative capability of the cells, was the reduction of enzyme levels by introducing an IncQ pSUP104 recombinant plasmid that directed synthesis of antisense RNA complementary to hydrogenase mRNA (van den Berg et al., 1991). Antisense RNA produced from the constitutive tetracycline promoter reduced synthesis of the [Fe] hydrogenase two- to threefold and confirmed that this enzyme has a significant role in lactate metabolism in *Dv. vulgaris* Hildenborough.

4.5 Electroporation of *Desulfovibrio*

Electroporation of *Dv. fructosovorans* has been used to create a deletion of the [NiFe] hydrogenase genes by marker exchange (Rousset et al., 1991). The [NiFe] hydrogenase genes were cloned, the coding region replaced by a kanamycin resistance gene from Tn5, and the construct introduced on an IncQ plasmid pGSS33 (Sharpe, 1984). Although the plasmid can be stably maintained, Km^R cells deleted for the hydrogenase genes were obtained from the electroporated cells. The procedure was reported to be rapid and easier than conjugation, and the constructed mutations confirmed that the [NiFe] hydrogenase plays a significant role in the metabolism of *Dv. fructosovorans* (Rousset et al., 1991).

4.6 Summary

In summary, in less than 5 years, the genetics of sulfate-reducing bacteria has moved from considerations of consistent colony formation to sophisticated procedures of gene expression control by antisense RNA. Transduction, conjugation, and electroporation have now been described for *Desulfovibrio* species. Specific problems of high levels of endogenous drug resistance, restriction systems, and unknown nutritional requirements remain for individual strains. However, these limitations should prove to be temporary as additional information is gained and new researchers are attracted to this developing field. Biochemical and physiological questions can now be addressed with all the resources of molecular biology and genetics available.

[References, see p. 211]

I thank M. Rousset and W.M.A.M. van Dongen for communication of results before publication. The patience and technical skill of B.J. Rapp-Giles is gratefully acknowledged. The work was supported by the Basic Energy Research Program of the U.S. Department of Energy through grant DE-FG02-87 ER13713 and by the Missouri Agricultural Experiment Station.

5

Molecular Biology of the Sulfate-Reducing Bacteria

Gerrit Voordouw

5.1 Introduction

The molecular biology of sulfate-reducing bacteria began in 1983 when the gene encoding [Fe] hydrogenase was cloned and sequenced. Since then, a number of genes have been characterized by nucleic acid sequencing. Expression of these genes in functional form was found to be a problem because the host par excellence, *Escherichia coli*, often failed to synthesize functional holo-proteins. However, genetic systems for conjugation and transduction of sulfate-reducing bacteria have now been developed (Chapter 4), and this achievement has largely solved these expression problems.

In this chapter the genes thus far isolated from sulfate-reducing bacteria will be described in detail. The construction of a library of λ-clones for the genome of *Desulfovibrio vulgaris* Hildenborough, which facilitated cloning of genes for flavodoxin and rubredoxin, will also be described. Characteristics of the proteins (e.g., expression as preproteins and consequently likely to reside in the periplasm) deduced from their gene sequence will be discussed, as well as the origin of problems encountered in their functional expression. Finally, without providing a detailed taxonomy (see Chapter 6), attention will be given to the distribution of characterized genes in other *Desulfovibrio* species.

5.2 Cloning, Sequencing, and Expression of *Desulfovibrio* Genes

Introduction and survey Several genes from sulfate-reducing bacteria of the genus *Desulfovibrio* have been cloned in the last 5 years (Table 5.1) and have been studied in detail by nucleic acid sequencing and expression

Table 5.1 Genes characterized for species from the genus *Desulfovibrio*

Gene(s) product	Gene(s) name[1]	References
[Fe]hydrogenase	*hydA,B*	Voordouw et al., 1985 Voordouw and Brenner, 1985 Prickril et al., 1986 Voordouw et al., 1987a Voordouw et al., 1987b van Dongen et al., 1988 Voordouw et al., 1989b
Not identified	*hydC*	Stokkermans et al., 1989 Voordouw et al., 1989b
[NiFe]hydrogenase	*hynA, hynB*	Li et al., 1987 Voordouw et al. 1989a Deckers et al., 1990 Rousset et al., 1990
[NiFeSe]hydrogenase	*hysA, hysB*	Menon et al., 1987 Voordouw et al., 1989a
cytochrome c_3	*cyc*	Voordouw and Brenner, 1986 Voordouw et al., 1987b Pollock et al., 1989 Voordouw et al., 1990
cytochrome c_{553}	*cyf*	van Rooijen et al., 1989
assimilatory sulfite reductase	*asr*	Tan et al., 1991[2]
high molecular weight cytochrome (Hmc)	*hmc*	Pollock et al., 1991
flavodoxin	*fla*	Curley and Voordouw, 1988 Krey et al., 1988 Carr et al., 1990
rubredoxin	*rub*	Voordouw, 1988
rubredoxin oxidoreductase[3]	*rbo*	Brumlik and Voordouw, 1989
desulforedoxin	*dsr*	Brumlik et al., 1990
OMPase	*pyrF*	Li et al., 1986
Nitrogenase, Fe protein	*nifH*	Postgate et al., 1987 Kent et al., 1989
DdeI, restriction endonuclease, methylase	*hsdM,R*	Sznyter et al. 1987
rubrerythrin	*rbr*	Prickril et al., 1991
methyl-accepting chemotaxis protein	*dcrA*	Dolla et al., 1992[2]

[1] Some of the gene names were proposed either by Voordouw and Wall (1992) or Voordouw (1990).
[2] These genes are not further discussed in the text.
[3] The function of this gene product has not been proven.

studies. Many of the cloned genes encode redox proteins. This focus is understandable when we consider the wealth of biochemical data accumulated on redox proteins and redox enzymes of sulfate-reducing bacteria prior to the start of molecular biological studies. Knowledge of the structure or amino acid sequence of proteins such as flavodoxin, rubredoxin, cytochrome c_3, cytochrome c_{553}, and desulforedoxin (Watenpaugh et al., 1972; Herriot et al., 1970; Bruschi, 1976; Adman et al., 1977; Hormel et al., 1986; Sieker et al., 1986; Frey et al., 1987; Ambler, 1968; Trousil and Campbell, 1974; Pierrot et al., 1982; Higuchi et al., 1984; Bruschi and LeGall, 1972; Bruschi et al., 1979) made genes for these proteins obvious targets for cloning.

However, the molecular biology work initially started with the isolation and characterization of genes for [Fe] hydrogenase from *Dv. vulgaris* Hildenborough, for which no structural information was available. The objective of these first studies was to characterize the amino acid sequence of a hydrogen-consuming and -producing enzyme, which had not yet been reported in the literature. The amino acid sequence, deduced from the nucleic acid sequence of the cloned genes (Voordouw et al., 1985; Voordouw and Brenner, 1985) proved to be very interesting, and this success served as a catalyst for further work on related and other genes.

Following the cloning and sequencing of the genes for [Fe] hydrogenase, the gene for its redox-partner cytochrome c_3 was isolated (Voordouw and Brenner, 1986). The nucleic acid sequence clearly indicated the presence of an N-terminal signal sequence for export of this redox-protein to the periplasm. Although this agreed with expectations, it also created some confusion, since a signal sequence had not been identified on the large and small subunits of [Fe] hydrogenase, suggesting the possibility that this enzyme was cytoplasmic (Voordouw and Brenner, 1985). This confusion ended quickly when Prickril et al. (1986), showed that the [Fe] hydrogenase small subunit does contain a complex signal sequence of 34 amino acid residues.

The first evidence that DNA from sulfate-reducing bacteria can be transcribed and translated to a functional protein in *E. coli* was provided when Li et al. (1986) isolated the *pyrF* gene from *Dv. vulgaris* by complementation of an *E. coli* PyrF$^-$ mutant. Further progress on characterization of *Desulfovibrio* hydrogenases was made by Li et al. (1987) and Menon et al. (1987), who cloned and sequenced genes for the [NiFe] hydrogenase from *Dv. gigas* and the [NiFeSe] hydrogenase from *Dv. baculatus*, respectively. The primary structures of the small and large subunits for these hydrogenases were found to be very different from those of [Fe] hydrogenase. Surprisingly, a conserved sequence was found only in the small subunit signal sequences of all of these hydrogenases. This indicates that the export of these redox enzymes to the *Desulfovibrio* periplasm may be accomplished by a special, conserved translocation system that has yet to be characterized.

Genes for enzymes and proteins from *Desulfovibrio*, other than hydrogenase, have been studied by molecular genetic approaches. The gene for the nitrogenase Fe protein was cloned from *Dv. gigas* by exploiting its homology with the *nifH* gene from *Klebsiella pneumoniae* (Postgate et al., 1988). The *rub* and *fla* genes for the well-characterized cytoplasmic redox-carrier proteins, rubredoxin and flavodoxin, were next described (Voordouw, 1988; Curley and Voordouw, 1988; Krey et al., 1988). It was a relief to see that fully functional flavodoxin was formed in *E. coli* transformed with plasmids containing the *fla* gene (Krey et al., 1988; Carr et al., 1990), after attempts to express functional [Fe] hydrogenase and cytochrome c_3 in *E. coli* had failed (Voordouw et al., 1987a; Pollock et al., 1989).

Analysis of the region upstream from the *rub* gene indicated the presence of another reading frame, encoding a possible redox partner of rubredoxin, which was named rubredoxin oxidoreductase (Rbo) (Brumlik and Voordouw, 1989). Despite numerous biochemical and structural studies, the function of rubredoxin in *Desulfovibrio* is presently unknown, and these studies may point the way towards elucidation of this function. *Rbo* has been shown to be present in *Dv. vulgaris* (Brumlik et al., 1990) and has an N-terminus that is homologous to the sequence of desulforedoxin from *Dv. gigas*. The gene for this latter protein (37 amino acids) consists of only 111 nucleotides and is the smallest gene so far isolated from *Desulfovibrio* (Brumlik et al., 1990).

Nucleotide sequence analysis of the cytochrome c_{553} gene indicated the presence of an *N*-terminal signal sequence for export to the periplasm, as was observed for tetraheme cytochrome c_3 (van Rooijen et al., 1989). The observation that the *hmc* gene, encoding the high-molecular-weight hexadecaheme cytochrome *c* (Hmc), also specifies a signal sequence confirms that all known *c*-type cytochromes in the genus *Desulfovibrio* are periplasmic redox proteins. The sequence of Hmc, which covalently binds a total of 16 hemes, suggested the presence of one incomplete and three complete cytochrome c_3 domains; this observation provided an explanation for the confusion in the earlier literature regarding the structure of this cytochrome (Pollock et al., 1991).

We have now entered the "second generation" of *Desulfovibrio* molecular biology. The recent reports of additional hydrogenase gene sequences (Stokkermans et al., 1989; Voordouw et al., 1989b; Deckers et al., 1990; Rousset et al., 1990) now allow us to make sequence comparisons among different genes. Although the number of genes characterized thus far is relatively small, they have had an impact on our understanding of sulfate-reducing bacteria from the genus *Desulfovibrio* at the molecular level. The facts and fiction suggested by these gene sequences are described in this review.

5.3 Hydrogenase Genes

[Fe] Hydrogenase genes Many species of the genus *Desulfovibrio* contain a highly active periplasmic hydrogenase, which was first characterized by Haschke and Campbell (1971). The enzyme, described as a single polypeptide chain with $M_r = 50$ kDa containing three 4Fe-4S clusters (Mayhew and O'Connor, 1982), is easily purified from the *Desulfovibrio* periplasm (van der Westen et al., 1978) in an oxygen-stable form. Antibodies generated against the *Dv. vulgaris* Hildenborough enzyme were used to clone the gene on a 4.7-kb *Eco*RI-*Sal*I fragment of *Dv. vulgaris* DNA, which was entirely sequenced (Voordouw et al., 1985; Voordouw and Brenner, 1985). The nucleic acid sequence indicated that this hydrogenase was a two-subunit enzyme, consisting of a large α-subunit (46 kDa) and a small β-subunit (13.5 kDa). The subunits are encoded by the *hydA* and *hydB* genes and are linked in an operon referred to as *hydA,B*.

Comparison of the nucleotide sequence of the *hydA,B* genes of *Dv. vulgaris* Hildenborough and *Dv. vulgaris* subsp. *oxamicus* Monticello indicated the presence of a conserved sequence, GAC(A/C)G(A/G)TACAAG(C/G)CGG, 80 nucleotides upstream from the translational start of the *hydA* gene (Voordouw et al., 1989b). This sequence could function in transcription initiation, although it does not resemble the *E. coli* consensus promoter sequence (TTGACA and TATAAT for the −35 and −10 elements, respectively). A hairpin loop, which could function as a transcription terminator, is present 200 nucleotides downstream from the translational stop of the *hydB* gene in *Dv. vulgaris* Hildenborough (Voordouw and Brenner, 1985; see also Figure 5.2). These observations suggest that transcription of the *hydA,B* operon could thus result in an mRNA of approximately 2000 nucleotides from which both the α and β subunits can be translated. This transcript size was recently confirmed (van den Berg et al., 1991).

SDS gel electrophoresis of purified hydrogenase confirmed the presence of a small subunit, and the two subunits were shown to be present in a 1:1 molar ratio (Hagen et al., 1986). Since this hydrogenase lacks nickel and two nickel-containing hydrogenases were subsequently discovered in *Desulfovibrio* (Lissolo et al., 1986; Prickril et al., 1987), this enzyme is now referred to as the [Fe] hydrogenase. It is distinct from the [NiFe] and [NiFeSe] hydrogenases, as will be discussed in additional detail below.

Examination of the position of cysteine residues in the α-subunit amino acid sequence, derived from the nucleotide sequence of the *hydA* gene, indicated that two of the three FeS clusters were located at the N-terminus of this subunit (Figure 5.1). The α-subunit sequence contains a total of 18 cysteine residues. Two groups of four at the N-terminus are present in a pattern found also in 8Fe-8S ferredoxins, where these 8 Cys residues are known to coordinate two 4Fe-4S clusters (Adman et al., 1973). These

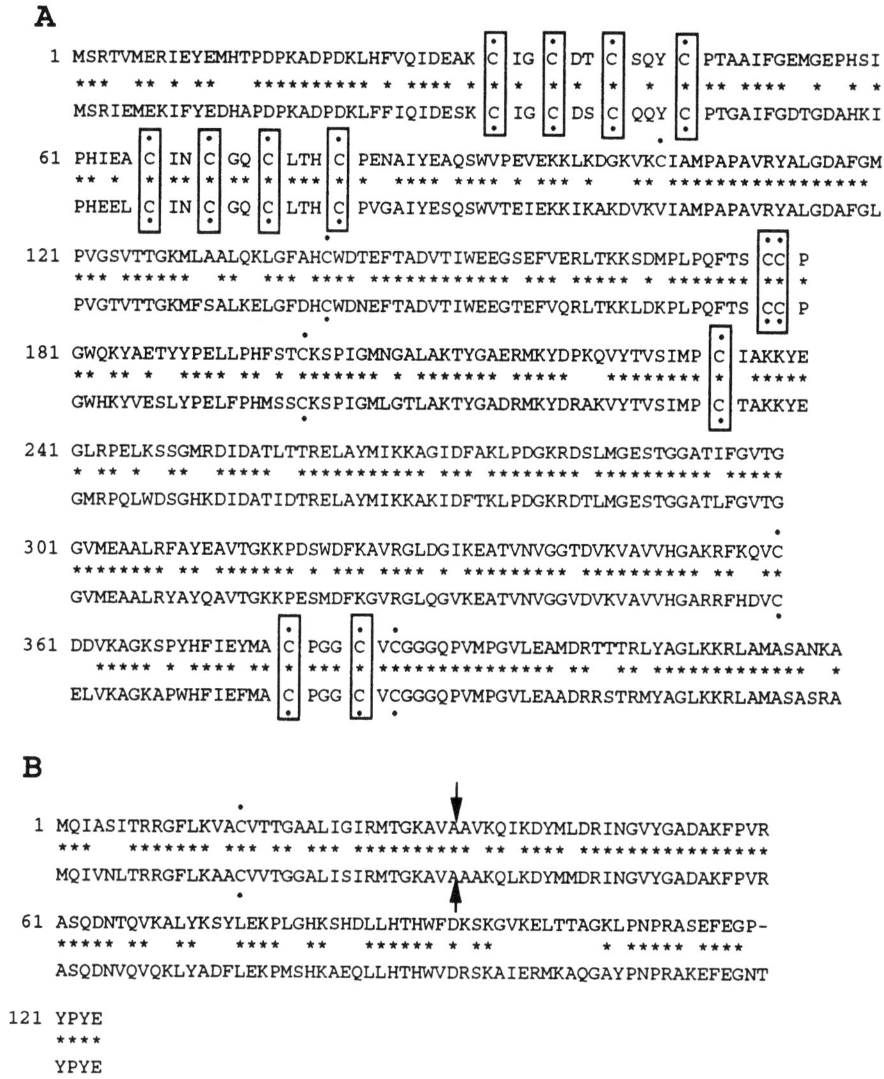

Figure 5.1 Comparison of amino acid sequences of [Fe] hydrogenase subunits. The sequences of the 46 kDa α (A) and the 13.5 kDa β (B) subunits of *Dv. vulgaris* (Hildenborough) (top lines) and *Dv. vulgaris* (Monticello) (bottom lines) are shown. Identical residues are indicated (*). Cysteine residues are indicated (●), while cysteine residues that have been conserved in both [Fe] hydrogenases and in the putative hydC gene product (Stokkermans et al., 1989) are highlighted by boxes. The proposed signal peptide processing site of the small subunits is indicated (↑, ↓).

8Fe-8S ferredoxins function as electron-transport proteins in a variety of bacterial species (Bruschi and Guerlesquin, 1988). The α-subunit of [Fe] hydrogenase thus contains an electron-transferring N-terminal ferredoxin domain (Figure 5.1: residues 1 to 105), which binds two of the three Fe-S clusters. The third cluster is coordinated by some of the nine cysteine residues in the C-terminal domain (residues 106 to 420).

Comparison of the *Dv. vulgaris* Hildenborough [Fe] hydrogenase α-subunit sequence with that of the enzyme from *Dv. vulgaris* subsp. *oxamicus* Monticello (Figure 5.1) indicates that all cysteine residues are conserved with the exception of C-102 at the junction of the two domains, which is replaced by a valine residue in the Monticello sequence. Further definition of essential cysteine residues is possible only by comparing the two α-subunit sequences in Figure 5.1 with that derived of the translated product of the *hydC* gene. This gene, discovered by Stokkermans et al. (1989), is present immediately downstream from the *hydA,B* operon in *Dv. vulgaris* Hildenborough. It encodes a polypeptide of 65.8 kDa, when expressed in *E. coli* minicells, but it is not known whether it is also expressed in *Desulfovibrio*.

The sequence of the *hydC* gene indicates it to be an in-frame fusion of the *hydA* and *hydB* genes. Transcription is in the opposite direction to that of the *hydA,B* operon, as indicated in Figure 5.2. The putative gene product, HydC protein, is also a fusion of α and β polypeptides. Although it is not known whether it serves a function in *Desulfovibrio* (it has been suggested that it could function either as a cytoplasmic hydrogenase or as a helper protein in the assembly of periplasmic hydrogenase), its sequence has been most useful in delineating conserved residues in [Fe] hydrogenase. Recently, Meyer and Gagnon (1991) demonstrated that the sequence of HydC protein is homologous with that of hydrogenase I of *Clostridium pasteurianum*.

Comparison of the Hildenborough α-subunit sequence with that of the putative HydC protein indicates sequence identity of 19% in the N-terminal domain (residues 1 to 105) and a 47% sequence identity in the C-terminal domain (residues 106 to 420). It is significant that five of the nine cysteine residues in the latter domain were conserved (Stokkermans et al., 1989; Voordouw et al., 1989b). These are highlighted in Figure 5.1 and are likely to coordinate to the third FeS cluster, which is thought to represent the H_2 binding site. This cluster is not of the regular 4Fe-4S type and could contain six Fe atoms (Hagen et al., 1986).

A simple structural model can thus be drawn for [Fe] hydrogenase (Figure 5.3). In this scheme, after H_2 reduction, electrons from the H_2 binding cluster, in the C-terminal domain, are transported to the two ferredoxin clusters, in the N-terminal domain, en route to the hydrogenase electron acceptor, which is thought to be the tetraheme cytochrome c_3. Confirmation of this proposed structural model requires a detailed structural analysis of the [Fe] hydrogenase. Although attempts have been made

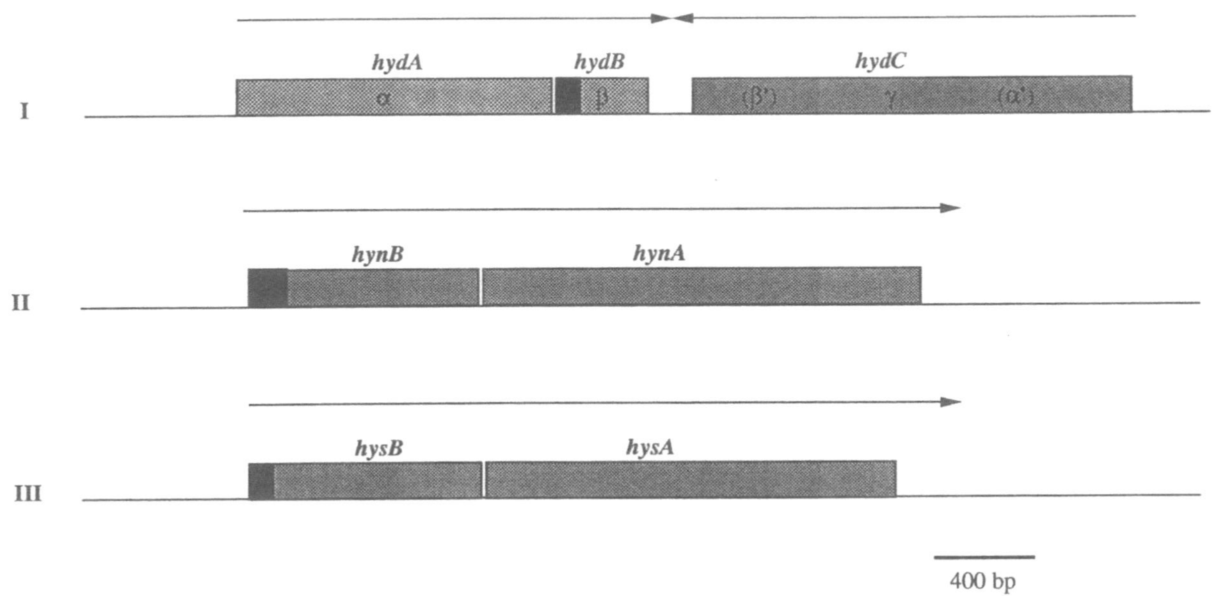

Figure 5.2 Survey of hydrogenase genes that have been cloned and sequenced from *Desulfovibrio*. (I) [Fe] hydrogenase operon from *Dv. vulgaris* Hildenborough. The positions of the *hydA* and *hydB* genes, encoding the large α and small β subunits of the periplasmic [Fe] hydrogenase, are indicated by the shaded boxes. The small subunit signal sequence is represented by the solid black box. The direction of transcription of the *hydA,B* operon is indicated by the arrow. The *hydC* gene is homologous with the *hydA,B* genes as indicated (α', β') and potentially encodes a 66-kDa polypeptide when transcribed as indicated. (II) [NiFe] hydrogenase operon from *Dv. gigas*. The small and large subunits are encoded by the *hynB* and *hynA* genes, respectively. (III) [NiFeSe] hydrogenase operon of *Dv. baculatus*. The small and large subunits are encoded by the *hysB* and *hysA* genes respectively.

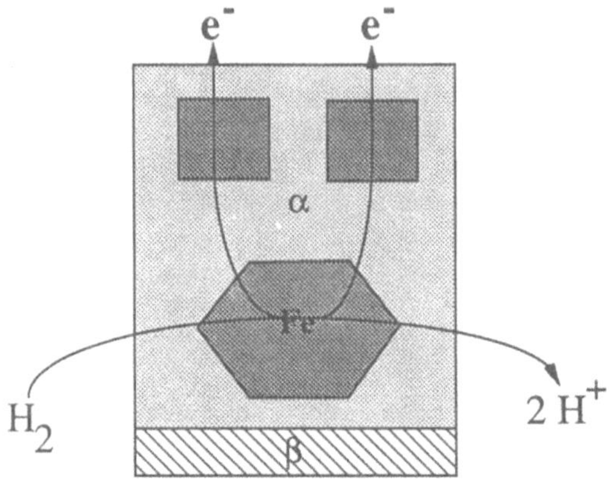

Figure 5.3 Model for [Fe] hydrogenase derived from the sequence data (Figure 5.1), as well as physical and chemical composition data described in the text. The active site cluster (possibly 6Fe-6S, depicted as a hexagon) is coordinated by five conserved cysteine residues of the large α subunit. The two electron-transferring 4Fe-4S clusters (the squares) in the *N*-terminal part of the α subunit are coordinated by two groups of four conserved cysteine residues. This region is homologous to 8Fe-8S ferredoxin. The small β subunit contributes an essential residue other than cysteine (e.g., His) to the active site. For hydrogen consumption electrons flow from the hexagon to the squares, whereas their flow reverses during hydrogen production. The physiological function of this hydrogenase is thought to be hydrogen uptake.

to crystallize the enzyme, no progress has yet been reported towards the elucidation of its structure.

The smaller β-subunit of [Fe] hydrogenase cannot be involved in covalent Fe-S cluster coordination, since the mature protein lacks cysteine residues. As discussed below, the β-subunit plays a role in the translocation of the [Fe] hydrogenase to the periplasm. The initial translation product (Table 5.2: pro-β, 123 residues), was shown to have a 34-amino-acid N-terminal signal sequence. Following cleavage at the signal peptidase cleavage site (Figure 5.1 B), a mature β-subunit of only 89 residues remains in the $\alpha\beta$-complex. The role of this small β polypeptide in [Fe] hydrogenase is not known. It cannot be dissociated with retention of activity and must thus serve some essential function. Comparison of α and β subunit sequences (Figure 5.1 A, B) of the two [Fe] hydrogenases indicates that β is less conserved (71% sequence identity) than α (79% sequence identity).

Studies on the structure-function relationships of the two [Fe] hydrogenase subunits would of course be facilitated by development of a convenient system for expression of the *hydA,B* genes in active form. Since *E. coli* does not express active enzyme, even when grown under

Table 5.2 Properties of three different hydrogenases from *Desulfovibrio*

Hydrogenase	[Fe]	[NiFe]	[NiFeSe]
Organism	*D. vulgaris* Hildenborough	*D. gigas*	*D. baculatus*
Localization	periplasm	periplasm	periplasm or cytoplasm
Nickel (mol/mol)	0	1	1
Selenium (mol/mol)	0	0	1
4Fe-4S cluster	2	2	2
6Fe-6S cluster	1	0	0
3Fe-xS cluster	0	1	0?
M_r (pro-β, β; kDa)	13.6, 9.6	34.0, 28.4	34.2, 30.8
n[1] (pro-β, β)	123, 89	314, 264	315, 283
M_r (α; kDa)	45.8	61.3	56.8
n (α)	420	550	514
Specific activity[2]			
H_2 evolution	4.8	0.44	0.47
H_2 uptake	50.0	1.5	0.12
K_m (H_2), μM	30–300[3]	1[4]	—

[1]Number of amino acid residues per polypeptide chain.
[2]Values are in mmol $min^{-1}mg$ $protein^{-1}$.
[3]Determined for [Fe] hydrogenase from *Clostridium pasteurianum* (Adams and Mortenson, 1984).
[4]Determined for [NiFe] hydrogenase from *Bradyrhizobium japonicum* (Evans et al., 1987).

anaerobic conditions, it cannot be used for this purpose. Nevertheless, it is appropriate to review the results obtained with expression of [Fe] hydrogenase in *E. coli*, since the work illustrates the complexities involved in the expression of a periplasmic redox-enzyme.

The original vector pHV15, containing the *hydA,B* genes of *Dv. vulgaris* Hildenborough on an insert of 4.7 kb, was trimmed to obtain pHV150 with just the two structural genes on a 1.9 kb *Bam*HI insert. Transcription of the *hydA,B* genes from the *lac*-promoter of the pUC-vector led to synthesis of considerable amounts (4% to 6 % of total protein) of both the pro-α and pro-β polypeptides in *E. coli* (Voordouw et al., 1987a). An $\alpha\beta$-complex was purified from *E. coli* cells by the same methods as employed for the isolation of active periplasmic [Fe] hydrogenase from *Desulfovibrio*.

ESR spectroscopy examination of the $\alpha\beta$-complex expressed by *E. coli* indicated that the ferredoxin clusters were present, since addition of dithionite led to nonenzymatic reduction of these clusters (Figure 5.4). However, enzymatic reduction of the ferredoxin clusters by transfer of electrons from hydrogen via the active site could not be observed. In contrast, the enzyme isolated from *Desulfovibrio* is rapidly reduced by

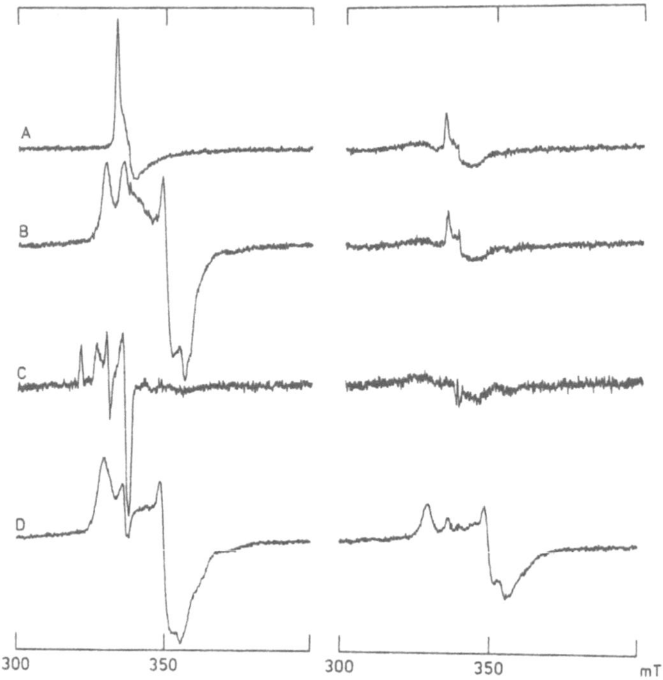

Figure 5.4 Electron flow through [Fe] hydrogenase monitored by ESR spectroscopy. Enzyme isolated from *Dv. vulgaris* Hildenborough (3.55 mg/ml, left-hand traces) is compared with that isolated from *E. coli* (pHV150) (3.05 mg/ml, right-hand traces). (A) Protein as isolated. (B) Spectrum following incubation of protein with H_2. (C) Replacement of H_2 with Ar. (D) Reduction with 50 mM sodium dithionite. Failure of the *E. coli* protein to be reduced with hydrogen points to a defective active site (the hexagon in Figure 5.3). The reduction of this protein with dithionite indicates the presence of the 4Fe-4S clusters. Reprinted with permission from the European Journal of Biochemistry (Voordouw et al.,1987a).

H_2 under anaerobic conditions (Figure 5.4). These results showed that the *E. coli* product had a defective FeS cluster at the active site, which prevented transfer of electrons to the ferredoxin clusters.

Initially, it was thought that the *E. coli* $\alpha\beta$-complex was cytoplasmic (Voordouw et al., 1987a); subsequent work has shown it is periplasmic (van Dongen et al., 1988). However, only a small fraction of the *E. coli* product is present as a soluble, periplasmic $\alpha\beta$-dimer. The majority of the large subunits expressed in *E. coli* are aggregated, while most of the small subunit remains as unprocessed pro-β polypeptide. It thus appears that *E. coli* can assemble only a small fraction of the synthesized [Fe] hydrogenase as a periplasmic, albeit inactive $\alpha\beta$-dimer. In *Desulfovibrio* species, functional expression of periplasmic [Fe] hydrogenase may require action by specific factors (helper proteins) for insertion of Fe-S clusters and translocation of the enzyme. It appears that *E. coli* does not provide a suitable genetic background for functional expression of this protein.

The recent development of genetic conjugation systems for *Dv. desulfuricans* and *Dv. vulgaris* Hildenborough (Powell et al., 1989; van den Berg et al., 1989; Voordouw et al., 1990b) has allowed the reintroduction of cloned *hydA,B* genes into *Dv. vulgaris* Hildenborough. The resulting exconjugant, which had the structural genes for [Fe] hydrogenase on a multicopy plasmid in addition to the copy on the *Dv. vulgaris* chromosome, synthesized both subunits at a 10-fold higher level, but was found to express only 1.5-fold more [Fe] hydrogenase activity. It appears, therefore, that even in the native genetic background it is difficult to overexpress this enzyme, presumably because other factors required for its maturation (see above) become rate limiting. A system for the functional expression of periplasmic [Fe] hydrogenase from the genus *Desulfovibrio* has, unfortunately, not yet been found, and this deficiency hampers further analysis of this interesting enzyme by modern techniques, such as oligonucleotide-directed mutagenesis.

[NiFe] and [NiFeSe] hydrogenase genes Genes for the nickel-containing hydrogenases from *Desulfovibrio* have also been cloned and characterized by nucleic acid sequencing (Menon et al., 1987; Li et al., 1987; Voordouw et al., 1989a; Deckers et al., 1990). As indicated in Table 5.2, these enzymes are very different from [Fe] hydrogenase. The first [NiFe] hydrogenase to be isolated and characterized in detail was the *Dv. gigas* enzyme (Hatchikian et al., 1978). Although the essential presence of nickel in the enzyme was not realized initially, it is now known to contain a single nickel atom and three Fe-S clusters per mole of enzyme (Cammack et al., 1987; Fernandez et al., 1986; Table 5.2). The molecular masses of the small β and large α subunits are much greater than those of the [Fe] hydrogenase; they also have very different properties. The subunits are encoded by an operon in which the gene for the small subunit (*hynB*) precedes that for the large subunit (*hynA*), a gene order different from that in the *hydA,B* operon for [Fe] hydrogenase (Figure 5.2).

The lower activity of the [NiFe] compared with [Fe] hydrogenase (Table 5.2) obscured its presence in *Desulfovibrio* species, such as *Dv. vulgaris* Hildenborough, which contain both of these uptake hydrogenases. Hindsight suggests why [NiFe] hydrogenase was first isolated from *Dv. gigas* and *Dv. vulgaris* Miyazaki F, which lack [Fe] hydrogenase (see below). Absence of a more active enzyme allowed the straightforward assay of the less active [NiFe] hydrogenases in cell extracts from these *Desulfovibrio* species. Lissolo et al. (1986) and Prickril et al. (1987) were the first to show that different classes of hydrogenases can be present in a single *Desulfovibrio* species; for example, *Dv. vulgaris* Hildenborough was shown to contain all three types, i.e., the [Fe], [NiFe], and [NiFeSe] hydrogenases.

Occurrence of these three enzymes in different *Desulfovibrio* species will be discussed in more detail below.

Comparison of nucleotide sequences determined for the *hynB,A* operons from *Dv. gigas* and *Dv. vulgaris* Miyazaki F indicated the presence of two conserved elements at 150 and 180 nucleotides upstream from the translational start of the *hynB* gene. As was found for the *hydA,B* operon, these sites do not resemble an *E. coli* consensus promoter sequence (Deckers et al., 1990). A potential transcription terminator immediately downstream from the *hynA* gene suggests that the *hynB,A* operon is transcribed in *Dv. vulgaris* Miyazaki F as an mRNA of 3000 nucleotides.

This gene organization is quite different from that of *E. coli*, where the structural genes for the small and large subunits of [NiFe] hydrogenase-I are part of a much larger operon, comprising six open reading frames (Menon et al., 1990). The frame immediately following the two hydrogenase structural genes (*hyaC*, in the terminology of Menon et al., 1990) appears to be an integral membrane protein with four hydrophobic membrane-spanning regions. A similar gene is present downstream from the structural genes for [NiFe] hydrogenase from *Bradyrhizobium japonicum* (Sayavedra-Soto et al., 1988; Menon et al., 1990) and *Rhodobacter capsulatus* (Richaud et al., 1990), where this gene has been referred to as *ORFX*. *R. capsulatus* appears to transcribe a 4200 nucleotide mRNA, which includes genes for the membrane-spanning protein and the two structural genes (Richaud et al., 1990). Analysis of the nucleotide sequence of *Dv. vulgaris* Miyazaki DNA indicates the absence of *hyaC/ORFX*-like genes. As noted below, these structural differences may reflect the fact that *Desulfovibrio* hydrogenases react with a periplasmic *c*-type cytochrome electron carrier.

Amino acid sequences of the small and large subunits of [NiFe] hydrogenase from *Dv. gigas* and *Dv. vulgaris* Miyazaki F, deduced from the sequence of their structural genes (Menon et al., 1987; Voordouw et al., 1989a; Deckers et al., 1990), are compared in Figure 5.5. Both subunits are highly homologous; the unprocessed small subunits (314 and 317 residues respectively) share 61%, while the two large subunits (551 and 567 residues respectively), share 68% sequence identity.

The small subunit of [NiFe] hydrogenase is quite different from the much smaller polypeptide found in [Fe] hydrogenase and contains most of the cysteine residues (11 and 12 residues respectively). The large

Figure 5.5 Comparison of the amino acid sequences of the small (A) and large (B) subunits of the *Dv. vulgaris* Miyazaki hydrogenase (top line) with those of the *Dv. gigas* [NiFe] hydrogenase (bottom line). The position of cysteine residues in both subunits is indicated (♠,♥) as well as the position of a conserved element (box) in the small subunit signal sequence. Reprinted with permission from the Journal of General Microbiology (Deckers et al., 1990).

(*a*)

```
                                 ↓
  1 MKISIGLGKEGVEERLAERGVSRRDFLKFCTAIAVTMGMGPAFAPEVARA 50
    ||. || ||:.|||||. ||||||||:|||||:||.|||||||||.|| |
  1 MKCYIGRGKDQVEERLERRGVSRRDFMKFCTAVAVAMGMGPAFAPKVAEA 50
      ↑             ↓  ↓         ↑
 51 LMGPRRPSVVYLHNAECTGCSESVLRAFEPYIDTLILDTLSLDYHETIMA 100
    | :.:||||||||||||||||||:||..:||:|.||||.:|:|||||:||
 51 LTAKKRPSVVYLHNAECTGCSESLLRTVDPYVDELILDVISMDYHETLMA 100
                    ↑  ↑                           ↓
101 AAGDAAEAALEQAVNSPHGFIAVVEGGIPTAANGIYGKVANHTMLDICSR 150
    :||.|.|.||.:|:.:  :|:.|:||||| :..|.:|||:.:.|.|||.
101 GAGHAVEEALHEAIKG..DFVCVIEGGIPMGDGGYWGKVGRRNMYDICAE 148
                 ↓       ↑                         ↑ ↓
151 ILPKAQAVIAYGTCATFGGVQAAKPNPTGAKGVNDALKHLGVKAINIAGC 200
    : |||.||||.|||||:||||||||||||. |||:|| .|||||||||||
149 VAPKAKAVIAIGTCATYGGVQAAKPNPTGTVGVNEALGKLGVKAINIAGC 198
                 ↑                          ↓        ↑
201 PPNPYNLVGTIVYYLKNKAAPELDSLNRPTMFFGQTVHEQCPRLPHFDAG 250
    |||| |:|||:|..| .|: ||||. .||.||||:|||:.||||.||:||
199 PPNPMNFVGTVVHLL.TKGMPELDKQGRPVMFFGETVHDNCPRLKHFEAG 247
                   ↓     ↓        ↓         ↑       ↓
251 EFAPSFESEEARKGWCLYELGCKGPVTMNNCPKIKFNQTNWPVDAGHPCI 300
    |||.||:|.||:||:|||||||||| | |||||  |||.||||:||||||
248 EFATSFGSPEAKKGYCLYELGCKGPDTYNNCPKQLFNQVNWPVQAGHPCI 297
     ↓             ↑     ↑        ↑                 ↑
301 GCSEPDFWDAMTPFYQN 317
    :||||:|||  .||| .
298 ACSEPNFWDLYSPFYSA 314
     ↑
```

(*b*)

```
       ↓
  1 MSGCRAQNAPGGIPVTPKSSYSGPIVVDPVTRIEGHLRIEVEVENGKVKN 50
                     ..  :..|||||:||||||||||||||.||:||
  1 ................MSEMQGNKIVVDPITRIEGHLRIEVEVEGGKIKN 34
                                  ↓  ↓          ↓
 51 AYSSSTLFRGLEIILKGRDPRDAQHFTQRTCGVCTYTHALASTRCVDNAV 100
    |:| ||||||||:||||||||||||||||.||||||.|||||.|.|||.|
 35 AWSMSTLFRGLEMILKGRDPRDAQHFTQRACGVCTYVHALASVRAVDNCV 84
                                  ↑  ↑              ↑
101 GVHIPKNATYIRNLVLGAQYLHDHIVHFYHLHALDFVDVTAALKADPAKA 150
    ||.||.|||.:|||.:||||:|||:||||||||||:|:|..||.||||||
 85 GVKIPENATLMRNLTMGAQYMHDHLVHFYHLHALDWVNVANALNADPAKA 134

151 AKVASSISPRKTTAADLKAVQDKLKTFVETGQLGPFTNAYFLGGHPAYYL 200
    |::|..:||||||...|||||.|:|.:||.|||| ||||||||||||| |
135 ARLANDLSPRKTTTESLKAVQAKVKALVESGQLGIFTNAYFLGGHPAYVL 184
                                                 ↓
201 DPETNLIATAHYLEALRLQVKAARAMAVFGAKNPHTQFTVVGGVTCYDAL 250
    .:|.:||||||||||||:|||||||||:|||||||||||||||.| ||.|
185 PAEVDLIATAHYLEALRVQVKAARAMAIFGAKNPHTQFTVVGGCTNYDSL 234
                                               ↑
251 TPQRIAEFEALWKETKAFVDEVYIPDLLVVAAAYKDWTQYGGTDNFITFG 300
     |:|||||  |:||.:.|:::|||.|||.||: ||:|...| |.||:| |
235 RPERIAEFRKLYKEVREFIEQVYITDLLAVAGFYKNWAGIGKTSNFLTCG 284
                                                    ↑
301 EFPKDEYDLNSRFFKPGVVFKRDFKNIKPFDKMQIEEHVRHSWYEGAEAR 350
    |||.||||||||: ..||:: .|:..:..|:.  |||||:.||||||:|:
285 EFPTDEYDLNSRYTPQGVIWGNDLSKVDDFNPDLIEEHVKYSWYEGADAH 334

351 HPWKGQTQPKYTDLHGDDRYSWMKAPRYMGEPMETGPLAQVLIAYSQGH. 399
    ||:|| |.||:|::||:|||||||||||.||::|.|||| ||:||.. |
335 HPYKGVTKPKWTEFHGEDRYSWMKAPRYKGEAFEVGPLASVLVAYAKKHE 384

400 PKVKAVTDAVLAKLGVGPEALFSTLGRTAARGIETAVIAEYVGVMLQEYK 449
    |.|||| | || .||||||||||||||||||||:. . |: |:| |:...
385 PTVKAV.DLVLKTLGVGPEALFSTLGRTAARGIQCLTAAQEVEVWLDKLE 433
              ↓                       ↑
450 DNIAKGDNVICAPWEMPKQAEGVGFVNAPRGGLSHWIRIEDGKIGNFQLV 499
    .|:  |.: ::..|: |.:.:|||||||||| |||||   :|||:||| |
434 ANVKAGKDDLYTDWQYPTESQGVGFVNAPRGMLSHWIVQRGGKIENFQHV 483
              ↓                                   ↓  ↓
500 VPSTWTLGPRCDKNNVSPVEASLIGTPVADAKRPVEILRTVHSFDPCIAC 549
    |||||.|||||....:|:||..|||||:||:||||||||||||:||||||
484 VPSTWNLGPRCAERKLSAVEQALIGTPIADPKRPVEILRTVHSYDPCIAC 533
              ↑                                   ↑  ↑
550 GVHVIDGHTNEVHKFRIL 567
    ||||||...|:|||||||
534 GVHVIDPESNQVHKFRIL 551
```

subunit has fewer cysteines (9 residues in both sequences). The difference becomes more pronounced when only conserved cysteine residues are considered; 11 cysteines are conserved in the small subunit, whereas only 5 residues are conserved in the large subunit. The large subunit, therefore, can not function as the main site for coordination of FeS clusters, as occurs in [Fe] hydrogenase, and at least two of the three FeS clusters of [NiFe] hydrogenase are bound by the small β subunit. Another difference between the [NiFe] and [Fe] hydrogenases is the absence of a sequence with homology to 8Fe-8S ferredoxin. This difference precludes a straightforward assignment of conserved cysteines in terms of coordination of the active site or the electron-transferring clusters.

Initially, the *Dv. vulgaris* Miyazaki hydrogenase was classified as an [Fe] hydrogenase, because it appeared to lack nickel and was found to have spectroscopic and enzymatic properties (e.g., inhibition of enzyme activity by carbon monoxide) that were incompatible with other [NiFe] hydrogenases (Yagi et al., 1976, 1985). However, the high degree of sequence homology of the *Dv. vulgaris* Miyazaki hydrogenase subunits with those from the [NiFe] hydrogenase of *Dv. gigas* confirmed that the former also belongs to the [NiFe] hydrogenase class.

The nucleotide sequence of the *hynB,A* operon from *Dv. fructosovorans* was also recently determined (Rousset et al., 1990). Comparison of translated amino acid sequences with those of the *Dv. gigas* enzyme demonstrated sequence homologies of 65% and 63% for the pairs of small and large subunits, respectively, and cysteine residues were conserved as described for *Dv. vulgaris* Miyazaki and *Dv. gigas* [NiFe] hydrogenase.

A nickel-containing hydrogenase that also contained one mole of selenium per mole of enzyme was first isolated from *Dv. desulfuricans* strain Norway by Rieder et al. (1984). The enzymatic properties of this [NiFeSe] hydrogenase, and one isolated from *Dv. baculatus* (Texeira et al., 1987), are distinct from those of the *Dv. gigas* [NiFe] hydrogenase. The [NiFeSe] hydrogenase appears more active in H_2 evolution than in H_2 consumption, when compared with either the [NiFe] or [Fe] hydrogenases (Table 5.2). However, the two nickel-containing enzymes are made up of two subunits of similar molecular weight and are encoded by similarly organized operons (Figure 5.2).

The amino acid sequence homology for the small and large subunits, deduced from structural gene sequences, of the *Dv. baculatus* [NiFeSe] hydrogenase and the *Dv. gigas* [NiFe] hydrogenase (Menon et al., 1987; Li et al., 1987; Voordouw et al., 1989a) is much lower than observed for the pairs of [NiFe] hydrogenase sequences. The two pairs of small and large subunit sequences were found to have 38% and 34% identical residues respectively (Figure 5.6), indicating that the [NiFeSe] hydrogenase is a distinctly different enzyme class. The two classes of nickel-containing hydrogenases most likely originated from the evolution of a common

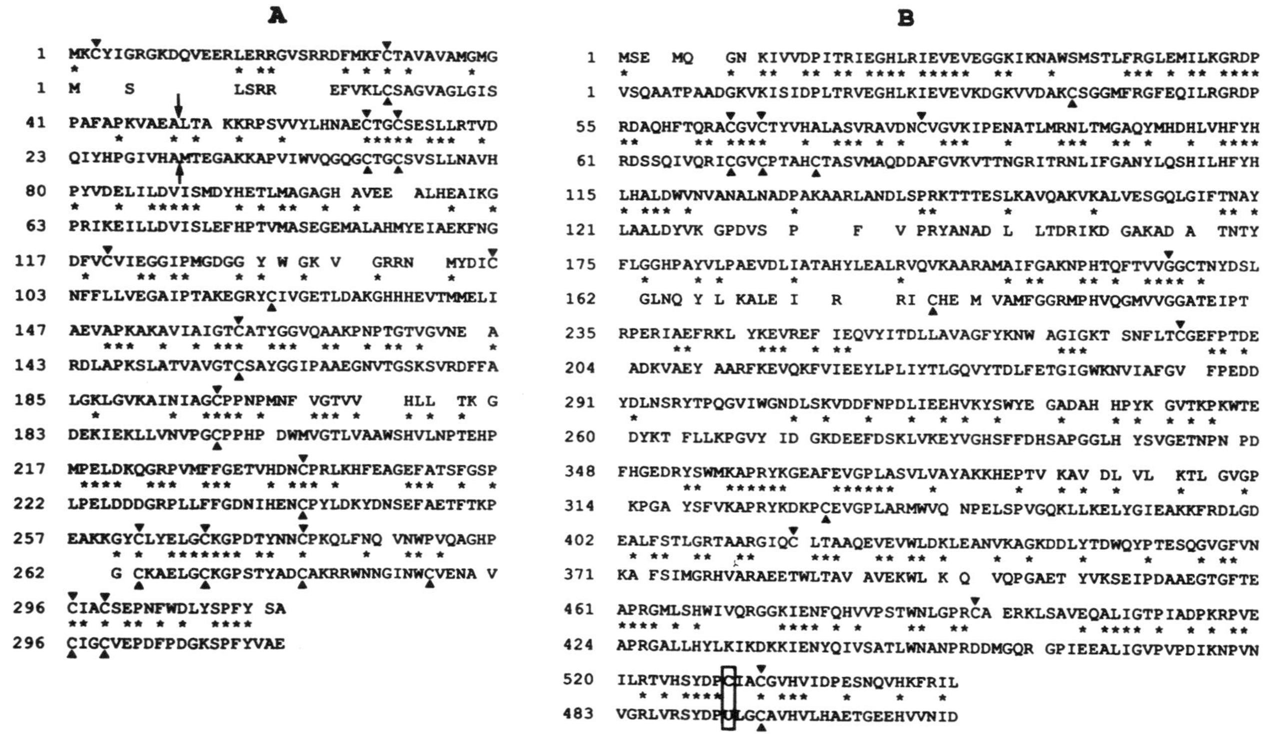

Figure 5.6 Comparison of the amino acid sequences of (A) the small and (B) the large subunits of the *Dv. gigas* [NiFe] hydrogenase (top line) and the *Dv. baculatus* [NiFeSe] hydrogenase (bottom line). The signal peptide processing site is indicated in both small subunits (♠,♥). The location of cysteine residues (▲,▼) and conserved cysteine residues (⧗) is shown in both subunits. The cysteine-selenocysteine (U) homology at the C-terminal end of the large subunit sequence is highlighted by a box. Reprinted with permission from the Journal of Bacteriology (Voordouw et al., 1989b).

ancestral gene, separate from the evolutionary pathway of the [Fe] hydrogenases.

An interesting feature of the *hysB,A* operon nucleotide sequence is the presence of a TGA-codon at the 3′ end of the *hysA* gene (Voordouw et al. , 1989a). This codon has been shown to encode the amino acid selenocysteine in other selenium containing enzymes (Chambers et al., 1986; Zinoni et al., 1986; Sukenaga et al., 1987). Comparison of nucleotide sequences indicates that this codon is matched by a cysteine codon (TGC) in the *hynA* gene for the large subunit of [NiFe] hydrogenase. This comparison illustrates a critical nucleotide in the distinction between [NiFe] and [NiFeSe] hydrogenases. Spectroscopic studies have shown that selenocysteine serves as a ligand to nickel in [NiFeSe] hydrogenase from *Dv. baculatus* (Eidsness et al., 1989; He et al., 1989), and one may expect that the homologous cysteine in [NiFe] hydrogenase serves the same function.

Comparison of amino acid sequences of small and large subunits of a [NiFe] and a [NiFeSe] hydrogenase highlights critical, conserved structural features of a nickel-containing hydrogenase (Figure 5.6) owing to the smaller degree of sequence homology compared to that seen in Figure 5.5. Focusing again on cysteine residues, it appears that ten residues in the small and four residues in the large subunit are conserved. Using the numbering of the *Dv. gigas* sequences, conserved cysteine residues in the small subunit are C-67, C-70, C-162, C-198, C-238, C-263, C-269, C-278, C-296, and C-299, while those in the large subunit are present in two pairs at the N-terminus, C-65 and C-68, and at the C-terminus, C-530 and C-533. The former is selenocysteine in [NiFeSe] hydrogenase of *Dv. baculatus* and serves as a nickel ligand. It thus appears that 13 conserved cysteine residues are available for coordination of nickel and FeS clusters in the nickel-containing hydrogenases, the same number as in [Fe] hydrogenase.

What is the significance of these observations for our understanding of the structure of the three different *Desulfovibrio* hydrogenases? The data in Table 5.2 imply that [NiFe] and [NiFeSe] hydrogenase may coordinate a different number of FeS clusters. Both enzymes are thought to have two 4Fe-4S clusters. The *Dv. gigas* [NiFe] hydrogenase has been reported to coordinate a 3Fe-xS cluster (Kissinger et al., 1989; Hatchikian et al., 1990) that is absent from the [NiFeSe] hydrogenase. Such a major structural difference is incompatible with the considerable sequence homology found for these two hydrogenases. This degree of homology indicates a similar structure for these two enzymes; if three clusters are present in *Dv. gigas* [NiFe] hydrogenase, then three clusters in a conserved three-dimensional arrangement are also expected in the [NiFeSe] hydrogenase. It is peculiar that the presence of the third cluster in the nickel-containing hydrogenases is so controversial. Even for the *Dv. gigas* enzyme it has been stated that its function is unknown, because its reduction potential is too positive (Hatchikian et al., 1990).

Our understanding of the structure of the more complex nickel-

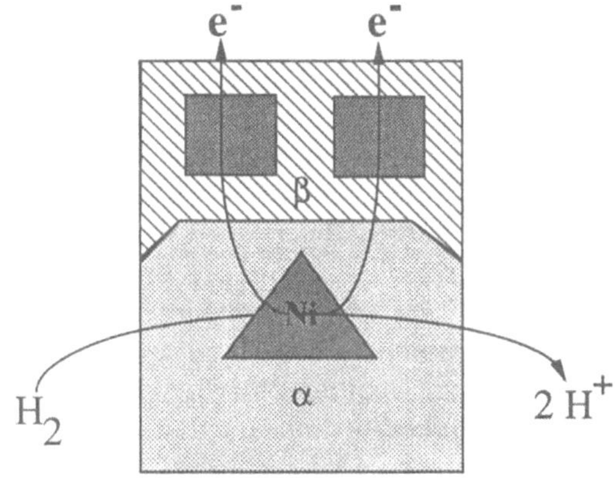

Figure 5.7 Model for [NiFe] and [NiFeSe] hydrogenase derived from the sequence data (Figures 5.5 and 5.6) and drawn in analogy to the model for [Fe] hydrogenase (Figure 5.3). One cluster (possibly 3Fe-xS, depicted as a triangle) is coordinated by conserved cysteine residues from the large α subunit. This cluster is close to the active site nickel. The two electron-transferring 4Fe-4S clusters (the squares) are coordinated by two groups of four conserved cysteine residues in the β subunit. For hydrogen consumption, electrons flow from the nickel-triangle to the squares, whereas their flow reverses during hydrogen production. The physiological function of [NiFe] hydrogenase is thought to be mainly hydrogen uptake, that of [NiFeSe] hydrogenase mainly hydrogen production.

containing hydrogenases is less complete than that of the simpler [Fe] hydrogenase (Figure 5.3), where the third cluster serves as the H_2-binding active site. Although the iron-only and nickel-containing hydrogenase sequences are not homologous, the number of conserved cysteines is similar (13 and 14 residues respectively). The similarity in number of conserved cysteines and number of clusters (even though they may be of a different type, 6Fe-6S vs. 3Fe-xS) suggests the possibility of structural homology between these two different classes of hydrogenase. Figure 5.7 illustrates a possible model for the nickel-containing hydrogenases, derived from that of [Fe] hydrogenase in Figure 5.3 (see also Voordouw, 1990), in which the abnormal Fe-S cluster associated with Ni represents the active site. This site is primarily coordinated by ligands from the large subunit, while the two electron-conducting 4Fe-4S clusters are coordinated by eight conserved small subunit cysteine residues. The merit of the hypothesis that the two structures share common features (Figures 5.3 and 5.7) can be evaluated only by elucidation of the two structures. No progress along that path has been reported beyond the successful crystallization of *Dv. gigas* and *Dv. vulgaris* Miyazaki F [NiFe] hydrogenase (Higuchi et al., 1987b; Niviere et al., 1987). Maroney et al (1991), in EXAFS studies, recently provided evidence for a close spatial relationship between nickel and an FeS cluster.

Export and localization of hydrogenases As indicated in Section 5.1, cloning and sequencing of genes for [Fe] hydrogenase from *Dv. vulgaris* Hildenborough initially raised questions with respect to localization of this enzyme. No signal sequence was found to precede the N-terminus of the large α subunit, determined to be SRTVM etc. (Voordouw et al., 1985; Voordouw and Brenner, 1985). The N-terminal sequence of the small subunit was not directly determined in this work. The fact that the small subunit is synthesized as a pro-β precursor protein was first recognized by Prickril et al. (1986), who demonstrated that the first 34 amino acid residues of pro-β represent a complex signal peptide (Figure 5.8). From the nucleic acid sequences of ten other hydrogenase operons, determined since these first studies, it appears that absence of a signal sequence from the α and presence of a complex signal sequence on the pro-β subunit is a universal feature of these bacterial hydrogenases.

The relevant signal sequences are compared in Figure 5.8. The most interesting observation from these data is that all sequences, irrespective of hydrogenase type and source, possess an RRXFXK consensus box. The strict conservation of this sequence element indicates that the mechanism whereby these enzymes are transported to the bacterial periplasm must have been conserved for these different hydrogenase classes in a variety of bacterial genera (Figure 5.8). A similar observation has never before been made for any other class of N-terminal signal peptides.

The consensus box is separated by 22 residues from the signal peptide cleavage site (A $\downarrow$ X) in all nickel-containing hydrogenases and by 21 residues in [Fe] hydrogenases. The first 13 residues of this intervening region contain no charged amino acids and are largely hydrophobic. The next 8 to 9 residues contain 0 to 2 charged amino acids. Apart from the C-terminal alanine, of the signal peptidase cleavage site, no residue is strictly conserved in the intervening region.

However, if one considers only the 8 [NiFe] hydrogenase signal sequences, extensive sequence homology (7 strictly conserved residues) is evident. It should be pointed out that these [NiFe] hydrogenases share considerable sequence homology in their small and large subunit sequences, and the observation of a conserved small subunit signal sequence within this class is, therefore, less surprising. The consensus box homology between the [Fe] and the nickel-containing hydrogenases is more significant in view of the fact that there is no further sequence homology between these two classes of hydrogenases, as discussed above.

The assembly and export mechanism of both hydrogenase classes is likely to be conserved. Two possibilities for this mechanism have been discussed elsewhere (Voordouw, 1987, 1990). Either the two subunits are exported independently and assembled in the periplasm, or assembly (including insertion of FeS-clusters and Ni, when required) occurs in the cytoplasm. This author favors the latter mechanism, because the complex signal sequence of pro-β can direct export of the $\alpha\beta$-complex via a special,

```
                                                                         cleavage
Sulfate  reducing  bacteria                                                site

                                            ++        +                +   +    +
DvMo  [Fe]                          MQIVNLT RR G F L K AACVVTAAALISIRMTGKAVA AAK
                                            ++        +                +   +    +
DvH   [Fe]                          MQIASIT RR G F L K VACVTTGAALIGIRMTGKAVA AVK
                                            ++ -      +                         -
Db    [NiFeSe]                         MSLS RR E F V K LCSAGVAGLGISQIYHPGIVHA MTE
                +    + +   --+ -++          ++ -      +                    + -
Dg    [NiFe]    MKCYIGRGKNQVEERLERRGVS RR D F M K FCTAVAVAMGMGPAFAPKVAEA LTA
                +    +-   --+  -+           ++ -      +                    - +
DvM   [NiFe]    MKISIGLGKEGVEERLAERGVS RR D F L K FCTAIAVTMGMGPAFAPEVARA LMG
                      +   -++               ++ -      +                    + -
Df    [NiFe]    MNFSVGLGRMNAEKRLVQNGVS RR D F M K FCATVAAAMGMGPAFAPKVAEA LTA

Enterobacteria
                          --        ++      ++        +                    +    -
Ec    [NiFe]          MNNEETFYQAMRRQGVT RR S F L K YCSLAATSLGLGAGMAPKIAWA LEN

Nitrogen  fixing  bacteria
                                   ++       ++        +        +                -
Ac    [NiFe]          ??????????MRRQGIT RR S F L K YCSLT-grpcLGPTFAPQIAHA MET
                           -       ++       ++        +                    +    -
Bj    [NiFe]          MGAATETFYSVIRRQGIT RR S F h K FCSLTATSLGLGPlaAsRIANA LET

Photosynthetic  bacteria
                           - -    -  ++     ++        +   +                +  -  -
Rc    [NiFe]          MMSDIETFYDVMRRQGIT RR S F M K svrspqhvLGLGPSFVPKIGEA MET
                           -     -  ++      ++        +                         -
Rg    [NiFe]              METFYEVMRRQGIS RR S F L K YCSLTATSLGLAPSFVPQIAHA MET


                                            ++        +
                              Consensus box RR   F    K
```

Figure 5.8 Comparison of some hydrogenase small-subunit signal sequences. Sequences were derived from the nucleotide sequences determined for: DvMo, *Dv. vulgaris* subsp. *oxamicus* Monticello (Voordouw et al., 1989b); DvH, *Dv. vulgaris* Hildenborough (Voordouw and Brenner, 1985); Db, *Desulfovibrio baculatus* (Menon et al., 1987); Dg, *Dv. gigas* (Li et al., 1987; Voordouw et al., 1989a); DvM, *Dv. vulgaris* Miyazaki F (Deckers et al., 1990); Df, *Dv. fructosovorans* (Rousset et al., 1990); Ec, *Escherichia coli* (Menon et al., 1990); Ac, *Azotobacter chroococcum* (Ford et al., 1990); Bj, *Bradyrhizobium japonicum* (Sayavedra-Soto et al., 1988); Rc, *Rhodobacter capsulatus* (Leclerc et al., 1988); Rg, *Rhodocyclus gelatinosus* (Uffen et al., 1990). The sequences have been aligned relative to the consensus box (RRXFXK) that is present in every signal sequence. Charged residues D and N (−), and R and K (+) have been indicated. Lower-case residues and question marks indicate deviations from the canonical sequence. These deviations may be caused by nucleotide sequencing errors.

consensus-box-specific export mechanism. Details of this mechanism have yet to be elucidated, and this unusual model, which requires protein export in a folded state, is still far from established.

A further complication of the assembly and export of hydrogenases is that not all of these enzymes are thought to be periplasmic. Localization of the [NiFe] and [NiFeSe] hydrogenase of *Dv. vulgaris* Hildenborough has been studied by immunoelectron microscopy, and the results indicated that both hydrogenases are associated with the cytoplasmic membrane. Data indicate that [NiFe] hydrogenase protrudes into the periplasm and [NiFeSe] hydrogenase into the cytoplasm (Rohde et al., 1990). The latter enzyme thus appears not to be exported. Unfortunately, these results do not allow a satisfactory explanation of the function of the pro-β signal sequence in [NiFeSe] hydrogenases (Figure 5.8). Failure of this hydrogenase to be exported requires cytoplasmic processing of its signal sequence by a protease other than signal peptidase, since the active site of the signal peptidase is in the periplasm.

Localization of hydrogenases and other electron-carriers in *Desulfovibrio* has often been controversial. Biochemical experiments suggest three distinct sites of hydrogenase activity: periplasmic, cytoplasmic, and membrane bound. For instance, whereas the *Dv. gigas* [NiFe] hydrogenase was found in the periplasm, the enzyme from *Dv. vulgaris* Miyazaki F was found to be membrane bound on the periplasmic side and required trypsin treatment for solubilization. The reasons for this different solubilization behavior are not clear.

As is evident from Figure 5.5, both proteins share considerable sequence homology. The large α subunit of *Dv. vulgaris* Miyazaki F has an N-terminal extension of 16 residues. This sequence, however, is not particularly hydrophobic and consequently is unlikely to serve a role in membrane binding. The [NiFe] hydrogenases from genera other than *Desulfovibrio* (Figure 5.8) have, on the contrary, a clear membrane anchor; the small subunit has a C-terminal extension of 40 to 50 residues with a central hydrophobic core of 20 residues. This core helps to anchor these enzymes, which are present on the periplasmic side, to the membrane. As mentioned elsewhere (Voordouw, 1990), this different structure probably reflects the fact that the *Desulfovibrio* hydrogenases interact with a soluble periplasmic electron carrier (cytochrome c_3), whereas hydrogenases from *E. coli*, *B. japonicum*, etc. transfer their electrons to membrane-bound electron carriers. [NiFe] hydrogenases from these organisms are expressed by operons containing additional open reading frames, some of which encode hydrophobic integral membrane proteins (Menon et al., 1990).

Distribution of hydrogenase genes Now that the genes encoding the three main types of hydrogenases have been cloned and sequenced, it is relatively easy to determine their presence in other cultures or newly isolated strains. The conservation of sequences of the *hydA,B*, *hynB,A*, and *hysB,A* operons greatly facilitates this determination. Also, there is no cross-hybridization between hydrogenase gene probes of different classes. Thus, probes for each of these genes can be used to easily assess their presence in an unknown organism by hybridization techniques.

Table 5.3 Distribution of genes for hydrogenase in sulfate reducing baceria

No.	Species	Strain	Source	*hynB,A* [NiFe]	*hysB,A* [NiFeSe]	*hydA,B* [Fe]	*hydC*[1]
1	*Desulfovibrio vulgaris* subsp. *vulgaris*	Hildenborough	NCIMB 8303	+	+	+	+
2	*Dv. vulgaris* subsp. *vulgaris*	Wandle	NCIMB 8305	+	+	+	+
3	*Dv. vulgaris* subsp. *vulgaris*	Brockhurst Hill	NCIMB 8306	+	+	+	+
4	*Dv. vulgaris*	Groningen	NCIMB 11779	+	−	−	−
5	*Dv. vulgaris*	Miyazaki F	T. Yagi	+	+	−	−
6	*Dv. vulgaris* subsp. *oxamicus*	Monticello 2	NCIMB 9442	+	+	+	−
7	*Dv. vulgaris* subsp. *oxamicus*	UofA	D.W.S. Westlake	+	+	−	±
8	*Dv. desulfuricans* subsp. *desulfuricans*	Norway 4	NCIMB 8310	+	+	−	−
9	*Dv. desulfuricans* subsp. *desulfuricans*	Teddington R	NCIMB 8312	+	−	−	−
10	*Dv. desulfuricans* subsp. *desulfuricans*	El Agheila Z	NCIMB 8318	+	−	+	−
11	*Dv. desulfuricans* subsp. *desulfuricans*	Berre Sol	NCIMB 8388	+	+	+	−
12	*Dv. desulfuricans* subsp. *desulfuricans*	Canet 41	NCIMB 8393	+	+	+	−
13	*Dv. desulfuricans* subsp. *desulfuricans*		NCIMB 8307	+	−	+	−
14	*Dv. desulfuricans* subsp. *desulfuricans*	G200	J.D. Wall	+	+	+	−
15	*Dv. gigas*		NCIMB 9332	+	−	−	−
16	*Dv. salexigens*	British Guiana	NCIMB 8403	+	+	−	−
17	*Dv. salexigens*	California	NCIMB 8364	+	+	−	−
18	*Dv. salexigens*		NCIMB 8365	+	+	−	−
19	*Dv. africanus*	Walvis Bay	NCIMB 8397	+	+	+	−
20	*Dv. africanus*	Benghazi	NCIMB 8401	+	+	±	−
21	*Dv. multispirans*		NCIMB 12078	+	−	+	−
22	*Dv. fructosovorans*		DSM 3604	+	−	+	−
23	"*Desulfovibrio*" *thermophilus*[2]		J. Postgate	−	−	−	−
24	*Desulfosarcina variabilis*		DSM 2060	−	−	−	−
25	*Desulfococcus multivorans*		DSM 2059	−	−	−	±
26	*Desulfobulbus propionicus*		DSM 2032	−	−	−	−
27	*Desulfotomaculum ruminis*		ATCC 23193	−	−	±	±
28	*Desulfotomaculum ruminis*		NCIMB 8452	−	−	±	±
29	"*Desulfovibiro*" *baarsii*[2]		DSM 2075	±	−	−	−
30	"*Desulfovibiro*" *sapovorans*[2]		DSM 2055	−	−	−	−

[1]Strong hybridization (+), weak bybridization (±; 100–1000 fold weaker than strong hybridization) or no hybridization (−) is indicated.
[2]Species 23, 29, and 30 have been reclassified (Devereux et al., 1989; Rozanova and Pivovarova, 1988).

Results for 30 different type cultures are summarized in Table 5.3 (Voordouw et al., 1990a). Twenty-two of these cultures belong to the genus *Desulfovibrio*. The data in Table 5.3 indicate that all of these 22 species contain the *hynB,A* operon. The distribution of the *hydA,B* and *hysB,A* operons (for [Fe] and [NiFeSe] hydrogenases) in *Desulfovibrio* is, on the other hand, more limited: chromosomal DNA from 13 and 15 of the 22 *Desulfovibrio* species hybridized with the *hydA,B* and *hysB,A* genes respectively. The *hydC* gene (Figure 5.2), is found only in three related British strains of *Dv. vulgaris* (Hildenborough, Wandle, and Brockhurst Hill). The limited distribution of this gene is a further indication that it presently does not serve an essential function in *Desulfovibrio* physiology. Without considering this rare gene further, it appears that the presence or absence of the structural genes for the three different hydrogenases gives rise to four classes of *Desulfovibrio* species. Class 1, e.g., *Dv. vulgaris* Hildenborough, has genes for all three ([NiFe], [NiFeSe], and [Fe]) hydrogenases. Class 2, e.g., *Dv. fructosovorans*, has genes for two ([NiFe] and [Fe]) hydrogenases, but lacks those for [NiFeSe] hydrogenase. Class 3, e.g., *Dv. desulfuricans* Norway, also has genes for two ([NiFe] and [NiFeSe]) hydrogenases, but lacks those for [Fe] hydrogenase. Finally, class 4, e.g., *Dv. gigas*, has genes for a single [NiFe] hydrogenase, but lacks those for [Fe] and [NiFeSe] hydrogenases.

The occurrence of these four different classes places constraints on the generality of models for hydrogen metabolism in *Desulfovibrio*. For instance, the hydrogen cycling hypothesis (Odom and Peck, 1981a, 1984; see also Chapter 3) postulates cytoplasmic production and periplasmic uptake of H_2. The hypothesis thus requires the presence of two distinct hydrogenases localized in these cellular compartments. Data in Table 5.3 suggest that a universal periplasmic uptake hydrogenase (the [NiFe] enzyme) is indeed present. However, a universal cytoplasmic hydrogenase for H_2 production has not yet been found. Although evidence has been presented for the cytoplasmic localization of [NiFeSe] hydrogenase in *Dv. vulgaris* Hildenborough (Rohde et al., 1990), this enzyme is found in only 15 of 22 type cultures (Table 5.3). Some of the species lacking this enzyme (e.g., *Dv. fructosovorans*; Cord-Ruwisch et al., 1986; Ollivier et al., 1988) have been shown to be H_2 producers. It appears that, unless another as yet undetected cytoplasmic hydrogenase is present in these species, this H_2 is produced by either the periplasmic [NiFe] or the periplasmic [Fe] hydrogenase. On the basis of studies of genetically manipulated strains of *Dv. vulgaris* Hildenborough (growing on lactate-sulfate medium), in which the [Fe] hydrogenase content was reduced by expression of antisense mRNA, van den Berg et al. (1991) recently postulated a role for the latter enzyme in H_2 production.

Despite these comments, it is the author's view that an important physiological function of the latter two hydrogenases, in most *Desulfovibrio* species, is hydrogen uptake. This view raises the question, why two different enzymes are used for this function in the genus *Desulfovibrio*. Although the periplasmic [NiFe] hydrogenase has been found in several bacterial genera (see legend to Figure 5.8), the periplasmic [Fe] hydrogenase appears specific for *Desulfovibrio*. Cytoplasmic [Fe] hydrogenases, which consist of a single polypeptide chain of $M_r = 60$ to 70 kDa binding three FeS clusters, have been isolated from Gram-positive, fermentative bacteria such as *Clostridium pasteurianum* and *Megasphaera elsdenii* (Adams and Mortenson, 1984). This cytoplasmic [Fe] hydrogenase is a low-affinity, high-activity enzyme, while the more complex [NiFe] hydrogenase is a high-affinity, low-activity enzyme. The [NiFe] enzyme may have evolved by selection for high H_2 affinity ($K_m = 1$ μM) at the expense of H_2 turnover.

The presence of this enzyme in all *Desulfovibrio* species allows them to compete for H_2 with each other and with [NiFe] hydrogenase-containing bacteria from other genera (see legend to Figure 5.8) in environments with low average H_2 concentrations (1 μM). The simpler [Fe] hydrogenase has a much higher H_2 turnover (Table 5.2), but also a much lower affinity for H_2 ($K_m = 100$ μM). This enzyme allows *Desulfovibrio* to harvest H_2 at a rapid rate in environments where average H_2 concentrations are high (~100 μM). Species that do not possess this enzyme are not able to take full advantage of H_2-rich environments (but also do not have to invest the energetic cost of synthesizing this enzyme). This view of the function of the two uptake hydrogenases predicts that deletion of either the *hydA,B* or the *hynB,A* operon from strains expressing both operons should lead to viable mutants. Construction of such a mutant has recently been reported by Rousset et al. (1991), who inactivated the *hynB,A* genes by marker exchange mutagenesis. The resulting mutant of *Dv. fructosovorans* expresses only the periplasmic [Fe] hydrogenase. Nevertheless, this strain grows equally well on H_2-sulfate medium as the wild-type strain.

The *hydA,B*, *hynB,A*, and *hysB,A* genes, found throughout the genus *Desulfovibrio*, can be used as gene probes to identify *Desulfovibrio* in environmental samples (Voordouw et al., 1990a). These probes are not suitable, however, to detect sulfate-reducing bacteria from other genera (Table 5.3: 23–30). Some of these organisms have the capacity to use H_2 but have hydrogenase genes that, apparently, are not homologous (or only weakly homologous) with those found in *Desulfovibrio*.

5.4 Genes for *c*-Type Cytochromes

Introduction A variety of *c*-type cytochromes have been purified from sulfate-reducing bacteria. The best known of these is cytochrome c_3 ($M_r = 13$ kDa, 4 hemes). This cytochrome was the first to be described for these bacteria (Postgate, 1954; Ishimoto et al., 1954b) and has, since its early first discovery, been found only in sulfate-reducing bacteria of the genus *Desulfovibrio*. It is considered a diagnostic protein for this genus.

The amino acid sequence of cytochrome c_3 from several *Desulfovibrio* species has been determined, including *Dv. vulgaris* Hildenborough (Trousil and Campbell, 1974), *Dv. vulgaris* Miyazaki (Shinkai et al., 1980), *Dv. gigas* (Ambler et al., 1969), *Dv. salexigens* (Haser et al., 1979), *Dv. desulfuricans* El Agheila Z (Ambler et al., 1971), and *Dv. desulfuricans* Norway (Haser et al., 1979). These studies demonstrated that the amino acid sequence of this cytochrome is not strongly conserved. Sequence homologies with *Dv. vulgaris* Hildenborough cytochrome c_3 (defined as 100%) are: *Dv. vulgaris* Miyazaki (88%), *Dv. gigas* (47%), *Dv. salexigens* (43%), *Dv. desulfuricans* El Agheila Z (40%), and *Dv. desulfuricans* Norway (33%). However, heme binding is conserved in these cytochromes. In each protein, four of *c*-type hemes are covalently bound to cysteine residues in either two or three sequences CXXCH (where X is a variable amino acid) and one or two sequences CXXXXCH. Two histidine residues serve as ligands to the fifth and sixth coordination positions of the iron in all four hemes. Cytochromes c_3 thus have eight cysteine and eight histidine residues for heme binding and coordination (LeGall and Fauque, 1988; Moura et al., 1991).

Despite the low degree of sequence homology, it appears that the three-dimensional structure of this cytochrome has been conserved. The relative positions of the four hemes are almost identical in cytochrome c_3 from *Dv. desulfuricans* Norway (Haser et al., 1979; Pierrot et al., 1982) and cytochrome c_3 from *Dv. vulgaris* Miyazaki (Higuchi et al., 1984). The structure is not symmetric, and the four hemes appear to occupy different positions relative to one another. As determined by a variety of methods, the reduction potential of each of the four hemes is low (−400 to −200 mV).

Odom and Peck (1984) indicate that the physiological role of this cytochrome is most likely to serve as electron carrier for [NiFe] and [Fe] hydrogenase. Its location appears to be largely periplasmic. In vitro stimulation of the phosphoroclastic reaction, catalyzed in the cytoplasm of *Desulfovibrio*, by this cytochrome was considered fortuitous by these authors. As shown below, cloning and sequencing of the cytochrome c_3 gene confirmed its periplasmic location.

A much simpler cytochrome, cytochrome c_{553} ($M_r = 9$ kDa, one heme), has been isolated from *Dv. vulgaris* Hildenborough, *Dv. vulgaris* Miyazaki, and *Dv. desulfuricans* Norway (Odom and Peck, 1984; LeGall and Fauque, 1988; Moura et al., 1991). The single heme has a reduction potential of +10 mV and is coordinated by His and Met ligands like other mono-heme bacterial and mitochondrial cytochromes *c*. The amino acid sequence of the two *Dv. vulgaris* proteins has been reported (Bruschi and Le Gall, 1972; Nakano et al., 1983), and the three-dimensional structure of cytochrome c_{553} from *Dv. vulgaris* Miyazaki has been determined (Nakagawa et al., 1986, 1990). Cytochrome c_{553} is considered to be periplasmic and thought to function as an electron acceptor for oxidation of formate and/or lactate catalyzed by formate and lactate dehydrogenase, respectively (Odom and Peck, 1984; Yagi and Ogata, 1990). This cytochrome is, apparently, not required in all *Desulfovibrio* species for this function, since it is not found in *Dv. gigas* and *Dv. desulfuricans* ATCC 27774 (Odom and Peck, 1984).

Finally, a very interesting but poorly understood and characterized cytochrome of higher molecular weight has been isolated from *Desulfovibrio*. Proteins from this class have been variously referred to as cytochrome cc_3 (Loutfi et al., 1989), octaheme cytochrome c_3 (Liu et al., 1988; Moura et al., 1987a), cytochrome c_3 (M_r 26,000) (Odom and Peck, 1984), and finally, high-molecular-weight cytochrome (Higuchi et al., 1987a; Yagi and Ogata, 1990). Higuchi et al. (1987a) isolated the protein from *Dv. vulgaris* Hildenborough as a 70 kDa protein with an estimated 16 *c*-type hemes. Other cytochromes in this class have lower molecular masses and smaller numbers of hemes per polypeptide. The reduction potentials of this high-molecular-weight class of cytochromes have an average value of −180 mV (Loutfi et al., 1989). Their physiological function is not clear. Odom and Peck (1984) proposed that cytochrome c_3 ($M_r = 26{,}000$) functions as an electron carrier for cytoplasmic hydrogenase. Indeed, in a later review LeGall and Peck (1987) list octaheme cytochrome c_3 from *Dv. baculatus* Norway (*Dv. desulfuricans* Norway) as a cytoplasmic protein.

Cloning, sequencing, and expression of cytochrome c_3 gene The *cyc* gene, which encodes cytochrome c_3, was cloned by Voordouw and Brenner (1986), using two deoxyoligonucleotide probes based on the known amino acid sequence of the protein. Plasmid pCYC3, isolated from a positive recombinant, had the *cyc* gene on a 7.5 kb *Eco*RI-*Hind*III insert of *Dv. vulgaris* DNA. Partial sequencing of this insert indicated that the amino acid sequence of the mature (107 amino acid residues) cytochrome derived from the sequence of the gene is in complete agreement with the amino acid sequence determined for the protein by Ambler (1968) and Trousil and Campbell (1974). The nucleotide sequence also indicated that the cytochrome is synthesized with a preceding typical signal sequence of 22 amino acid residues (Figure 5.9). This sequence has two positively charged residues at the *N*-terminus (RK), is otherwise hydrophobic, and presents an AA signal peptidase cleavage site (Figure 5.9). This sequence strongly resembles those used by *E. coli* for the export of proteins to the periplasm (Benson et al., 1985). The observation of a cytochrome c_3 signal sequence thus confirmed the known periplasmic location of this electron carrier protein.

The nucleotide sequence also indicated the presence of a promoter sequence (TTGACA and TACCAT for the −35 and −10 elements, respec-

	−30 −20 −10 −1	+1
DvH Hmc	MRNGRTLLRWAGVLAATAIIGVGGFWSQGTT	KALP
DvH Cytochrome c_3	MRKLFFCGVLALAVAFALPVVA	APKA
DvH Cytochrome c_{553}	MKRVLLLSSLCAALSFGLAVSGVA	ADGA

Figure 5.9 Comparison of the signal peptide sequences for three periplasmic *c*-type cytochromes from *Dv. vulgaris* Hildenborough. The sequences have been numbered relative to the first residue of the mature protein (+1). Positively charged residues have been underlined. The sequences were derived from nucleotide sequences as described in the text.

tively), resembling the consensus *E. coli* promoter (TTGACA and TATAAT, respectively), 70 nucleotides upstream from the translational start of the *cyc* gene, as well as a potential transcription terminator 50 nucleotides downstream from the translational stop. The nucleotide sequence suggests that the *cyc* gene is transcribed as a monocystronic message of approximately 500 nucleotides. Physical mapping of the *cyc* gene on DNA of λ-clones (see below) did not reveal linkage to any of the hydrogenase genes discussed previously (Voordouw, 1988). These results thus indicate that, although cytochrome c_3 is thought to serve as an electron carrier for one or more of the hydrogenases found in *Desulfovibrio*, its gene is transcribed independently.

The presence of an *E. coli* consensus promoter sequence should facilitate expression of the *cyc* gene in *E. coli*. Expression, as monitored by Western blotting, is indeed observed when *E. coli* TG2 is transformed with plasmid pCYC3. Expression can be improved by limiting the size of the insert from 7.5 kb in pCYC3 to 730 bp in plasmid pJ800, which contains just the *cyc* gene in vector pUC8 (Pollock et al., 1989). Expression appears independent of the orientation of the insert relative to the *lac* promoter of the vector, indicating that transcription is driven by the *cyc* promoter on the insert described above.

E. coli appears to process the cytochrome c_3 signal sequence slowly; both the unprocessed 14 kDa form and the processed 12 kDa form can be distinguished on Western blots. The 12 kDa form was shown to reside in the periplasm, but the 14 kDa form was found to be cytoplasmic or membrane bound (Pollock et al., 1989). One reason for this slow processing could be that *E. coli* appears incapable of covalently inserting the hemes. The periplasmic 12 kDa form is devoid of heme, irrespective of growth conditions chosen for culturing *E. coli* (aerobic, anaerobic, or anaerobic with added nitrate). This could mean that bacterial heme lyase, the enzyme that catalyzes covalent *c*-type heme insertion [e.g., compare with the enzyme isolated from yeast (Dumont et al., 1987)], has a restricted substrate specificity. Alternatively, *E. coli* may lack this catalytic activity entirely, in which case it would not be able to synthesize any *c*-type cytochromes, which have covalently bound hemes.

Recently the 730-bp insert of pJ800 was cloned into broad-host-range vector pJRD215 (Davison et al., 1987), and the resulting plasmid, pJRDC800-1, was transferred to *Dv. desulfuricans* G200 by conjugation with an *E. coli* donor (Voordouw et al., 1990b). The untransformed G200 strain produces cytochrome c_3 with an acidic isoelectric point (pI = 5.8), whereas *Dv. vulgaris* Hildenborough cytochrome c_3 has a basic isoelectric point (pI = 10.5). The recombinant, *Dv. desulfuricans* G200 (pJRDC800-1), produced both the native acidic and the recombinant basic cytochromes c_3. Detailed characterization, by a number of chemical and physical criteria (including NMR spectroscopy) of both proteins demonstrated the recombinant cytochrome c_3 to be indistinguishable from that produced by *Dv. vulgaris* Hildenborough. This result is important for several reasons. First, it opens the route to a more systematic investigation of this cytochrome by site-directed mutagenesis techniques. Second, it also shows that, given the proper "genetic background," the cloned gene is readily expressed in functional form. It is further support for the hypothesis that failure of *E. coli* to synthesize a functional periplasmic cytochrome c_3 must be due to the absence of the proper heme insertion activity. Indeed, synthesis of functional holo-cytochrome c_3 is not restricted to *Desulfovibrio*, but was recently also demonstrated in *Rhodobacter sphaeroides* (Cannack et al., 1991).

With the *Dv. vulgaris* Hildenborough *cyc* gene as a probe in Southern blotting experiments, the presence of the gene could be demonstrated in 12 out of 15 other strains of sulfate-reducing bacteria (Voordouw et al., 1987b). Many of the positive strains, including *Dv. desulfuricans* Norway, exhibited very weak cross-hybridization, in agreement with the low degree of sequence conservation previously discussed. Despite the fact that the presence of this cytochrome is diagnostic for the genus *Desulfovibrio*, the *cyc* gene is not a good probe to identify this genus because of this sequence variability.

Cloning, sequencing, and expression of the cytochrome c_{553} gene
The *cyf* gene, which encodes cytochrome c_{553} from *Dv. vulgaris* Hildenborough, was cloned with deoxyoligonucleotide probes based on the known sequence of the protein (Bruschi and LeGall, 1972). These probes were used to screen a λ-library of *Dv. vulgaris* DNA (discussed below). Two positive clones with identical inserts were isolated, and a 2.2 kb *Pst*I fragment derived from these clones was analyzed by dideoxysequencing (van Rooijen et al., 1989). The amino acid sequence of the mature protein, derived from the sequence of the gene, did not agree with that previously published (Bruschi and LeGall, 1972). It appeared that the order of cyanogen bromide fragments in the latter sequence had been assigned erroneously. This conclusion was confirmed by renewed protein sequencing (van Rooijen et al., 1989). The corrected sequence of the *Dv. vulgaris*

Hildenborough cytochrome is 79 amino acid residues long and is highly homologous to cytochrome c_{553} from *Dv. vulgaris* Miyazaki F, as shown in Figure 5.10.

Like the *cyc* gene discussed above, the sequence of the *cyf* gene indicated the presence of a signal peptide of 24 residues, with typical features for export to the periplasm (Figure 5.9). The *cyf* gene sequence thus confirmed the periplasmic nature of cytochrome c_{553}. An inverted repeat sequence was found 80 nucleotides downstream from the *cyf* gene-coding region. However, a promoter could not be identified on the basis of homology with the *E. coli* consensus sequence, and it is currently not known whether the transcript is mono- or polycystronic.

Cloning, sequencing, and expression of the gene for high-molecular-weight cytochrome As previously discussed, considerable confusion exists in the literature with respect to the molecular nature of a third, higher-molecular-weight *c*-type cytochrome in *Desulfovibrio*. Loutfi et al. (1989) isolated cytochrome cc_3 from *Dv. vulgaris* Hildenborough, for which they reported a partial protein sequence and amino acid composition. The molecular weight of this cytochrome was reported to be 43.3 kDa, and it was thought to be composed of two 20 kDa subunits. SDS-polyacrylamide gel electrophoresis (SDS–PAGE) of purified cytochrome cc_3 indicated the presence of a 70 kDa form (Loutfi et al., 1989). Higuchi et al. (1987a) had previously reported the isolation of a high-molecular-weight cytochrome (Hmc) from *Dv. vulgaris* Hildenborough. The holoform had a molecular mass of 69 kDa by SDS–PAGE and contained 16 *c*-type hemes per polypeptide chain. Recently, Yagi and Ogata (1990) indicated that both cytochrome cc_3 and Hmc have very similar amino acid compositions and thus could be the same protein.

This was proven to be the case by Pollock et al. (1991), who used a deoxyoligonucleotide probe based on the C-terminal sequence of cytochrome cc_3 to clone the cytochrome cc_3 gene. The nucleic acid sequence of this gene indicated that it encodes a polypeptide of 55.7 kDa with 16 sites for covalent binding of *c*-type heme, all of the form CXXCH. These properties are similar to those reported for Hmc by Higuchi et al. (1987a). Thus, cytochrome cc_3, cytochrome c_3 (M_r 26,000), and Hmc are all the same polypeptide, at least for *Dv. vulgaris* Hildenborough. Like the two smaller *c*-type cytochromes discussed above, it appears that Hmc is translated as a pro-protein with an N-terminal signal sequence of 31 residues (Figure 5.9). This observation indicates that this cytochrome is also periplasmic.

The amino acid sequence of mature Hmc is not highly homologous to that of cytochrome c_3. Nevertheless, as proposed by Pollock et al. (1991) and as indicated in Figure 5.11, the structure can be understood in terms of the heme coordination pattern seen in this smaller cytochrome. Beginning at the N-terminus, the elements of this pattern are (i) a sequence

```
                    10         20         30         40         50         60         70
Hildenborough ADGAALYKSC VGCHGADGSK QAMGVGHAVK GQKADELFKK LKGYADGSYG GEKKAVMTNL VKRYSDEEMK AMADYMSKL
              ********** ********* ***     ** ** * ** **  ********* ** ** ***  ** ***** * * *******
Miyazaki      ADGAALYKSC IGCHGADGSK AAMGSAKPVK GQGAEELYKK MKGYADGSYG GERKAMMTNA VKKYSDEELK ALADYMSKL
```

Figure 5.10 Comparison of the amino acid sequences of cytochromes c_{553} from *Dv. vulgaris* Hildenborough, derived from the sequence of the gene (van Rooijen et al., 1989) with that from *Dv. vulgaris* Miyazaki F, determined by direct protein sequencing (Nakano et al., 1983). Sequence identities are indicated with an asterisk.

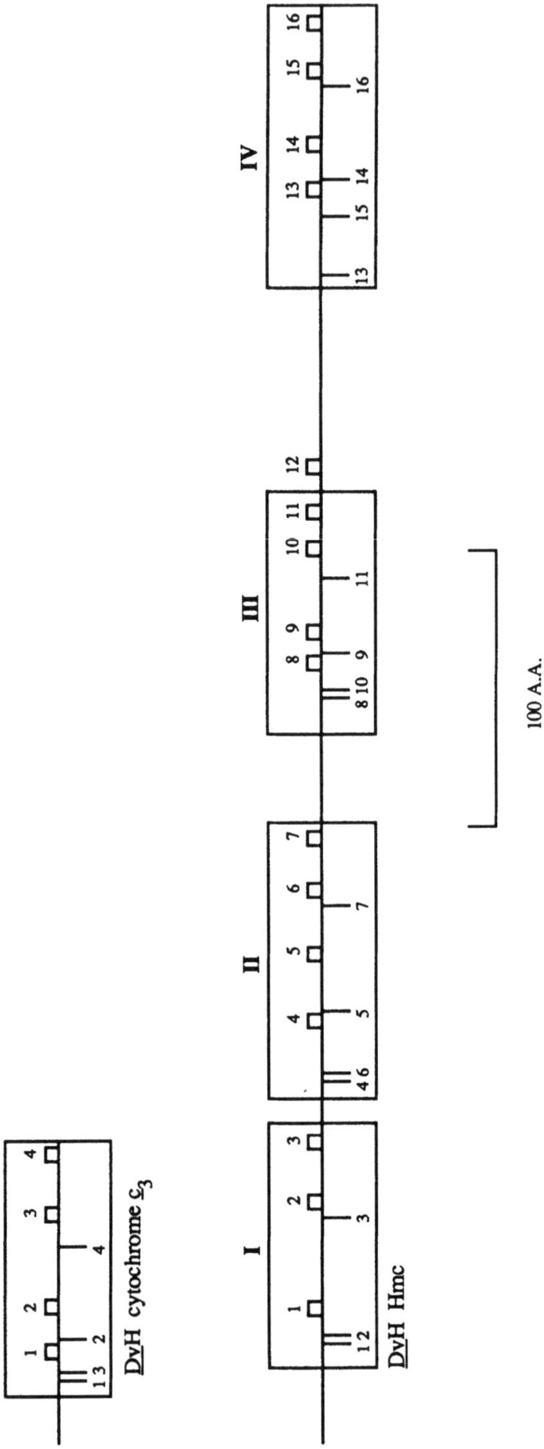

Figure 5.11 Arrangement of heme-binding sites and coordinated histidines in cytochrome c_3 (top) and Hmc cytochrome (bottom). The numbered small boxes are the heme-binding sites CXXCH, except for hemes 2 and 4 of cytochrome c_3, which are covalently bound to CXXXXCH. The downward pointing bars are the sixth coordination position H-residues. The numbers correspond to the heme-binding sites with which these H-residues are associated in cytochrome c_3 and supposedly associated in Hmc. The distribution of heme binding sites and histidine residues in this latter cytochrome suggests the presence of three complete cytochrome c_3 domains (II, III, IV) and one incomplete domain (I), as well as a single heme (heme 12). Heme 12 may not be *bis*-histidinyl coordinated, as described in the text. Reprinted with permission from the Journal of Bacteriology (Pollock et al.,1991).

HXXH, in which the first H coordinates to heme 1 and the second to heme 3; (ii) a CXXCH binding sequence for heme 1; (iii) a single H coordinating heme 2; (iv) a CXXXXCH binding sequence for heme 2; (v) a single H coordinating heme 4; (vi) a CXXCH binding sequence for heme 3; and (vii) a CXXXCH binding sequence for heme 4. All four hemes in cytochrome c_3 have *bis*-histidine coordination. When one recalls that only heme-binding sequences CXXCH are found in Hmc, it appears that this pattern is present three times in the amino acid sequence of Hmc, as shown in Figure 5.11 (domains II, III, and IV). Domain I is an incomplete cytochrome c_3 domain, which lacks elements (iii) and (iv) of the pattern defined above. There is one heme-binding site (heme 12, Figure 5.11), that does not fit into this pattern. Hmc has only 31 histidines in its primary structure. Although spectroscopy (Higuchi et al., 1987a) has shown that most of the hemes must be *bis*-histidine coordinated as in cytochrome c_3, there must thus be at least one heme with histidine-methionine coordination. This is likely to be heme 12, because it does not fit into the cytochrome c_3 pattern.

Thus, Hmc does appear to have a domain structure, and one can easily imagine how proteolysis could lead to apparently smaller cytochromes. As discussed by Yagi and Ogata (1990), proteolysis occurs readily during purification of this cytochrome, especially when the purification is carried out under aerobic conditions. The interpretation of the latter observation is that fully oxidized Hmc assumes a conformation that is more sensitive to proteases.

As in the case of the *cyc* and *cyf* genes, it has not been possible to express the *hmc* gene in functional form in *E. coli*. However, transfer of the gene by conjugation of broad-host-range plasmid pBPHMC-1 from *E. coli* to *Dv. desulfuricans* G200 leads to an exconjugant that overproduces a functional *Dv. vulgaris* Hildenborough Hmc in the periplasm. This abundant source of Hmc will allow more detailed structural and functional studies of this cytochrome in the future, which will hopefully lead to a definition of its function in the periplasmic electron transport chain of *Desulfovibrio*. With Southern blotting it appears that an *hmc* gene, homologous to that of *Dv. vulgaris* Hildenborough, is present in *Dv. vulgaris* Brockhurst Hill, *Dv. vulgaris* Wandle, *Dv. vulgaris* subsp. *oxamicus* Monticello, *Dv. desulfuricans* Berre Sol, *Dv. desulfuricans* Canet 41, *Dv. desulfuricans* Teddington R, *Dv. africanus* Benghazi, and *Dv. africanus* Walvis Bay, while no homologous gene is found in *Dv. gigas*, *Dv. multispirans*, *Dv. salexigens*, *Dv. desulfuricans* El Agheila A, and *Dv. desulfuricans* Norway. If the Hmc sequence has a degree of conservation that is as low as that of cytochrome c_3, then this failure to hybridize with the *Dv. vulgaris* probe does not prove absence of Hmc!

Inspection of the nucleotide sequence upstream from the *hmc* gene suggested the presence of an *E. coli* consensus promoter sequence. However, a transcription terminator was not present downstream from the gene, and hmc may be the first gene in a larger redox protein operon (Pollock, unpublished).

Summary of *c*-type cytochromes In summary of this discussion of the molecular biology of *c*-type cytochromes, gene cloning and sequencing for the three *Desulfovibrio* cytochromes indicate that these genes all encode an N-terminal signal sequence for export of the cytochrome to the periplasm. The *c*-type cytochromes thus participate only in the periplasmic electron transport chain (e.g., electron transport from H_2, via hydrogenase) of *Desulfovibrio*. They do not participate in cytoplasmic electron transport chains [e.g., electron transport to sulfate, via adenosinephosphosulfate (APS) reductase and bisulfite reductase]. In addition, this work has corrected the amino acid sequence for cytochrome c_{553} and elucidated for the first time the sequence of a hexadecaheme cytochrome, Hmc.

5.5 Construction of a Gene Library for *Dv. vulgaris* Hildenborough for Rapid Gene Cloning and Genome Mapping

When multiple gene cloning experiments are planned for a genetically as yet undeveloped organism, it is prudent to construct and maintain a gene library. This has been done for chromosomal DNA from *Dv. vulgaris* Hildenborough in the replacement vector λ-2001 (Karn et al., 1984). For library construction the DNA was partially digested with restriction endonuclease *Sau*3A (↓GATC). Fragments of 15 to 20 kb were isolated by agarose gel electrophoresis and ligated to λ-2001, digested to completion with restriction endonuclease *Bam*HI (G↓GATCC). Following packaging, recombinant phages with a single 15- to 20-kb insert of *Dv. vulgaris* DNA were amplified by lytic growth on *E. coli* Q359, as described elsewhere (Voordouw, 1988). A total of 900 recombinant phages was maintained, initially on plates with *E. coli* Q358 as the lawn, and later as individual phage stocks. DNA was also purified for each of the 900 clones (Voordouw, 1988). Purified DNA from every clone was then denatured by treatment with alkali and spotted on a Hybond-N filter. Following crosslinking of the DNA to the filter by UV-irradiation and neutralization, it was stored at −20°C.

The objective of constructing this gene library is, of course, to accelerate cloning of genes from *Dv. vulgaris* Hildenborough. The filter has been used successfully to isolate the *fla*, *rub*, and *cyf* genes. As an example, λ-clones containing the *rub* gene encoding the electron carrier rubredoxin were identified with a labeled deoxyoligonucleotide probe, as shown in Figure 5.12. Some of these clones had been found to be positive when the filter was hybridized with a *cyc* gene probe. Restriction mapping indicated that the positive λ-clones in Figure 5.12 are overlapping and span

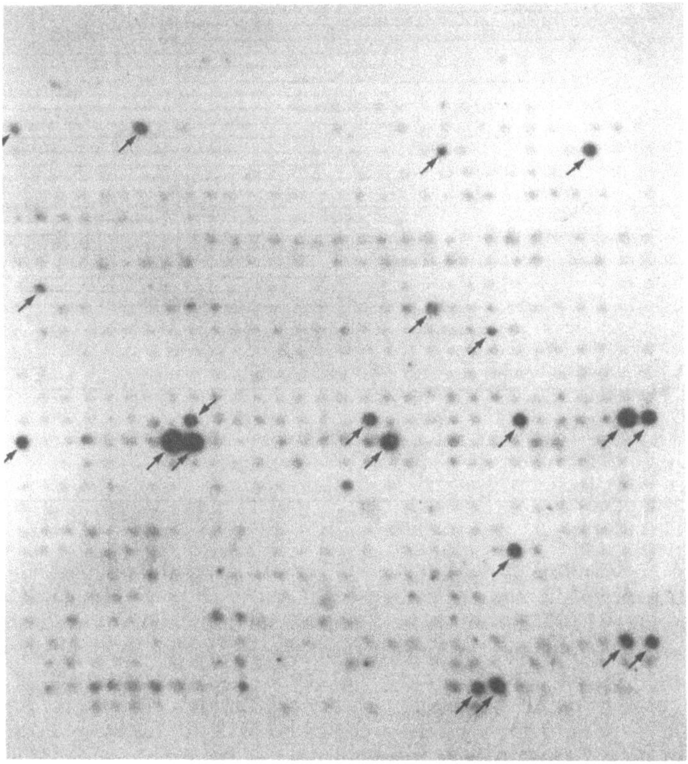

Figure 5.12 Screening of a genomic library, constructed for *Dv. vulgaris* Hildenborough, with a deoxyoligonucleotide probe designed to recognize the *rub* gene, encoding rubredoxin. The arrows point to potential positives. Reprinted with permission from Gene (Voordouw, 1988).

a continuous region of DNA (referred to as a contig) of 35 kb. The *cyc* and *rub* genes are separated by about 17.5 kb in this contig (Voordouw, 1988).

An ideal gene library is completely random, since randomness ensures equal representation of all genome regions. The *Dv. vulgaris* Hildenborough *Sau*3A library is unlikely to approach this ideal, primarily because *Sau*3A sites are not randomly distributed in genomic DNA. This point is confirmed by considering the distribution of λ-clones in the contig covering the *Dv. vulgaris* Hildenborough *hydA,B* operon (Figure 5.13). The region 3′ from this operon is well represented (17 clones), whereas that 5′ from this operon is covered by only a single clone (Figure 5.13: H20). Together these clones span 31 kb of *Dv. vulgaris* DNA.

The size of the *Dv. vulgaris* Hildenborough genome has been estimated to be 1720 kb (Postgate et al., 1984). Accepting this value and

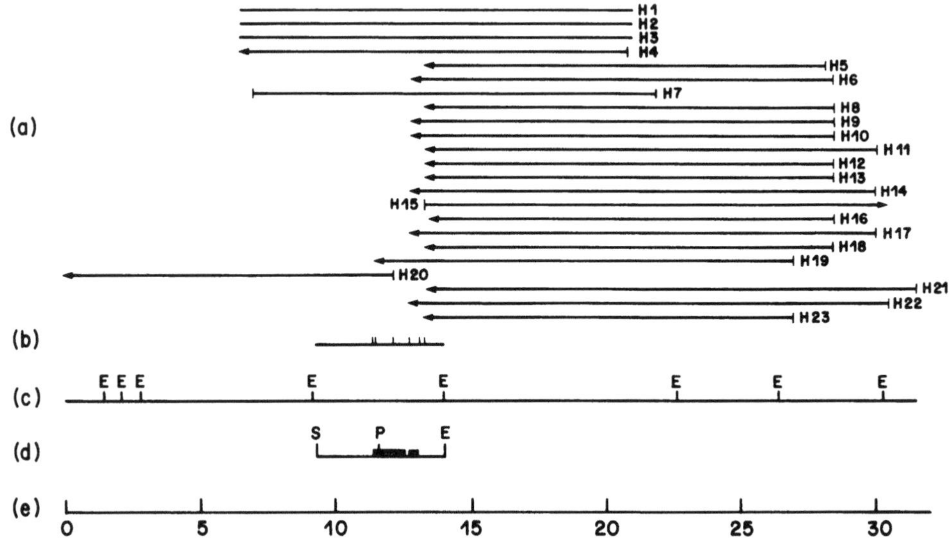

Figure 5.13 Relationship of λ-clones (a) isolated from the *Dv. vulgaris* Hildenborough genomic library as positives with a *hydA,B* probe (b). (a) Position of positive clones H1-H23; the arrows indicate the orientation of the insert in the phage vector λ-2001 and are drawn from the left to the right phage arm. (b) Distribution of *Sau*3A restriction sites in the 4.7 kb *Sal*I-*Eco*RI insert of plasmid pHV15, used to probe the genomic library. (c) Position of *Eco*RI (E) sites in the 31 kb DNA contig defined by the λ-clones H1-H23. (d) Location of the *hydA,B* genes on the 4.7 kb *Sal*I-*Eco*RI fragment used as the probe. (e) Scale in kb. Reprinted with permission from Gene (Voordouw, 1988).

assuming that the 900-λ clones have an average insert size of 17 kb, we can calculate that if the clones were distributed completely randomly, the library would cover 99.9% of the genome (Maniatis et al., 1982). Figure 5.13 refutes this calculation. The conclusion that sections of the genome are absent is also supported by the observation that, although *hydA,B*, *cyc*, *cyf*, *fla*, and *rub* are present (see Table 5.1), *hynB,A*, *hysB,A*, and *hmc* are missing from the library.

5.6 Genes for Cytoplasmic Electron Carriers

Enzymes catalyzing dissimilatory sulfate reduction are present in the cytoplasm of sulfate-reducing bacteria. Cloning and sequencing of their genes has not yet been reported, and this section is, therefore, limited to a description of genes for three well-known (but poorly understood in terms of in vivo function) small electron carrier proteins—rubredoxin, desulforedoxin, and flavodoxin.

The *rbo-rub* operon from *Dv. vulgaris* Hildenborough The gene that encodes rubredoxin in *Dv. vulgaris* Hildenborough was cloned from the λ-library previously discussed. With a deoxyoligonucleotide probe, based on amino acids 1 to 15 of the known rubredoxin sequence (Bruschi, 1976), a large number of positive clones could be identified (Figure 5.12). The sequence of the *rub* gene, derived from these clones, confirmed the previously published amino acid sequence of rubredoxin (with the exception of T-21, for which position a codon for aspartic acid was found). The gene sequence indicated the absence of an *N*-terminal signal sequence, confirming the known cytoplasmic location of rubredoxin (Voordouw, 1988).

Another gene was identified immediately upstream from the *rub* gene (Brumlik and Voordouw, 1989). Nucleic acid sequencing indicated that this gene encoded a 14 kDa protein (Figure 5.14). This was shown to be also an electron carrier protein because its translated N-terminus is homologous with desulforedoxin, an electron carrier isolated from *Dv. gigas* (see below). Direct measurement of the size of *rub* mRNA by Northern blotting and determination of the transcription start site by S1 nuclease mapping indicated that *rub* and its upstream gene form an operon. Since this could mean that the 14 kDa protein exchanges electrons with rubredoxin, it has been tentatively named rubredoxin oxidoreductase (Rbo, product of the *rbo* gene). The transcription start site for the *rbo-rub* operon is approximately at nucleotide 1389. The termination site is at nucleotide 2100 (Figure 5.14).

Since these initial observations the presence of the 14 kDa *rbo* gene product has been confirmed both by immunoblotting studies (Brumlik et al., 1990) and by purification from *Dv. vulgaris* Hildenborough (J. LeGall, personal communication, and Moura et al., 1990) and *Dv. desulfuricans* (Moura et al., 1990). The latter work, especially, has confirmed that Rbo is a redox protein, although its potential interaction with rubredoxin has not yet been characterized. Moura et al. (1990) renamed Rbo as desulfoferrodoxin, largely because the protein does not exchange electrons with pyridine nucleotides. The suggestion by these authors that Brumlik and Voordouw (1989) considered Rbo to be a pyridine-linked oxidoreductase is wrong; the name rubredoxin oxidoreductase was used to indicate a redox protein that exchanged electrons with rubredoxin. Brumlik and Voordouw (1989) did note that an NADH-linked rubredoxin oxidoreductase had been partially purified from *Clostridium acetobutylium*. However, the clostridial enzyme has a flavin prosthetic group and is totally different from the sulfate-reducing bacterial enzyme.

Southern blotting indicated that the *rbo* gene is present in *Dv. vulgaris* species Hildenborough, Wandle, Brockhurst Hill, Groningen, and Miyazaki F, *Dv. vulgaris* subsp. *oxamicus* Monticello 2, *Dv. desulfuricans* Teddington R and Berre Sol, and *Dv. africanus* Benghazi. The Hildenborough *rbo* gene did not hybridize with chromosomal DNA from *Dv. desulfuricans*

```
                                                                                     -35                          -10
GCATACTGTGCCTGAAAAGTGAATCACCCCGGACACATCTTTCCGCTCATGGGCTGGACAAGCCCAGATTAGGATTTACTCGTAAAAAATCACGCCGGTCATTGAACAGACCGCCGACAT
       1310      1320      1330      1340      1350      1360      1370      1380      1390      1400      1410      1420

                                                                     rbs       M  P  N  Q  Y  E  I  Y  K  C  I  H  C  G
GCTGTACACAGGTATAAACAGGTATTTCCGGCGGCCATTCCCGCCTTTCAGACCACCGACAACGAACATGGAGGCCCCATGCCCAACCAGTACGAAATCTACAAATGCATCCACTGTGGC
       1430      1440      1450      1460      1470      1480      1490      1500      1510      1520      1530      1540

N  I  V  E  V  L  H  A  G  G  G  D  L  V  C  C  G  E  P  M  K  L  M  K  E  G  T  S  D  G  A  K  E  K  H  V  P  V  I  E
AACATCGTCGAAGTCCTGCATGCTGGCGGCGGCGACCTCGTGTGCTGCGGCGAACCCATGAAGCTCATGAAGGAAGGCACTTCTGACGGGGCCAAGGAAAAGCACGTGCCGGTCATCGAG
       1550      1560      1570      1580      1590      1600      1610      1620      1630      1640      1650      1660

K  T  A  N  G  Y  K  V  T  V  G  S  V  A  H  P  M  E  E  K  H  W  I  E  W  I  E  L  V  A  D  G  V  S  Y  K  K  F  L  K
AAGACCGCCAACGGCTACAAGGTCACCGTCGGTTCCGTGGCCCACCCCATGGAAGAGAAGCACTGGATCGAATGGATTGAGCTTGTCGCAGACGGTGTGAGCTACAAGAAGTTCCTGAAG
       1670      1680      1690      1700      1710      1720      1730      1740      1750      1760      1770      1780

P  G  D  A  P  E  A  E  F  C  I  K  A  D  K  V  V  A  R  E  Y  C  N  L  H  G  H  W  K  A  E  A  *     rbs           M  K
CCCGGCGATGCGCCCGAAGCCGAGTTCTGCATCAAGGCCGACAAGGTCGTCGCCCGCGAATACTGCAACCTGCACGGCCACTGGAAGGCCGAAGCCTAACCCCGAGGAATCCACGATGAA
       1790      1800      1810      1820      1830      1840      1850      1860      1870      1880 AvaI 1890      1900

 K  Y  V  C  T  V  C  G  Y  E  Y  D  P  A  E  G  D  P  D  N  G  V  K  P  G  T  S  F  D  D  L  P  A  D  W  V  C  P  V  C
AAAGTACGTATGCACCGTCTGCGGTTACGAATACGACCCTGCTGAAGGCGACCCCGACAACGGCGTGAAGCCCGGCACCTCGTTCGACGACCTGCCGGCCGACTGGGTATGCCCCGTGTG
       1910      1920      1930      1940      1950      1960      1970      1980      1990      2000      2010      2020

 G  A  P  K  S  E  F  E  A  A  *  *
CGGCGCCCCCAAGAGCGAATTCGAAGCCGCCTAGTAAGCGCTTTCCATGACGCCGCCGCCGGTTGCCCGTACGGGTTGCCCGCGGCGGCGTTTCTTCGTATAGGGGGGTCATCAGCCCCT
       2030      2040      2050      2060      2070      2080      2090      2100      2110      2120      2130      2140
                 EcoRI
```

Figure 5.14 Nucleotide sequence of the *rbo-rub* operon from *Dv. vulgaris* Hildenborough. The *rbo* gene, encoding a putative rubredoxin oxidoreductase at nucleotides 1499 to 1879, and the *rub* gene, encoding rubredoxin at nucleotides 1896 to 2057, have been translated into protein. Ribosome-binding sites (rbs) are indicated for both genes. The suggested positions for the promoter and transcription terminator are derived from Northern blotting and S1 nuclease mapping experiments. Reprinted with permission from the Journal of Bacteriology (Brumlik et al., 1989).

Norway, El Agheila Z, and Canet 41, *Dv. gigas*, *Dv. salexigens*, and *Dv. africanus* Walvis Bay.

Expression of the *rbo-rub* operon in *E. coli* minicells indicates that both 14- and 6-kDa polypeptides, corresponding to Rbo and rubredoxin respectively, are formed. It is not known whether these polypeptides have the Fe-metal centers (Moura et al., 1990) incorporated.

These results have not led to elucidation of a physiological role for rubredoxin in sulfate-reducing bacteria. Yet, they represent an advance from the situation prior to this work, where almost everything known about rubredoxin was based on in vitro studies [including its three dimensional structure (see references quoted in section 5.2)]; even its potential in vivo redox partner was unknown. Sequencing of the *rbo* gene has provided a new approach to problems of the function of rubredoxin and related redox proteins with poorly understood function. The work demonstrates a way in which molecular genetic studies can help to unravel problems of physiology.

The *dsr* gene from *Dv. gigas* The sequence of the *rbo* gene from *Dv. vulgaris* Hildenborough suggested that *Dv. gigas* desulforedoxin, which is only 36 amino acid residues long, might be synthesized from a larger precursor. There is a homologous region at the *N*-terminus of Rbo that is exactly from residues 1 to 36 (Figure 5.15). Since the *rbo* gene product is isolated from *Dv. vulgaris* as a 14 kDa protein, whereas desulforedoxin isolated from *Dv. gigas* is a 3.9 kDa polypeptide, the question arises whether desulforedoxin is formed by proteolytic cleavage of the *N*-terminus of a larger protein. This question was addressed by cloning the *dsr* gene from *Dv. gigas* and determining its nucleic acid sequence (Brumlik et al., 1990). It appeared that the *dsr* gene has a coding region of only 111 nucleotides, which encodes a polypeptide of 36 amino acid residues, when the initiator methionine is excluded. The amino acid sequence for desulforedoxin, derived from the gene sequence, was in complete agreement with that obtained by protein sequencing (Bruschi et al., 1979).

Southern blotting indicates that the *dsr* gene is found only in *Dv. gigas*, and, as discussed above, this species lacks the *rbo* gene. The *dsr* and *rbo* genes thus do not appear to coexist in a *Desulfovibrio* species at present. An inverted repeat sequence that could serve as a transcription terminator is present downstream from the *dsr* coding region. No promoter has been identified in the upstream region, but also no additional reading frames from which some ideas on function of desulforedoxin in *Dv. gigas* can be formulated have been found.

Desulforedoxin is thus not formed by proteolytic degradation of a larger Rbo-like precursor protein. Instead, it appears likely that the *rbo* gene was formed by fusion of a gene encoding a desulforedoxin domain to a gene for a 10 kDa polypeptide that presently forms the C-terminus of Rbo. As discussed elsewhere (Brumlik and Voordouw, 1989; Brumlik et

```
                     5                        10                       15
pro-asn-gln-tyr-glu-ile-tyr-lys-cys-ile-his-cys-gly-asn-ile-
ala-asn-glu-gly-asp-val-tyr-lys-cys-glu-leu-cys-gly-gln-val-

                    20                        25                       30
val-glu-val-leu-his-ala-gly-gly-gly-asp-leu-val-cys-cys-gly-
val-lys-val-leu-glu-glu-gly-gly-gly-thr-leu-val-cys-cys-gly-

                    35                        40                       45
glu-pro-met-lys-leu-met-lys-glu-gly-thr-ser-asp-gly-ala-lys-
glu-asp-met-val-lys-gln

                    50                        55                       60
glu-lys-his-val-pro-val-ile-glu-lys-thr-ala-asn-gly-tyr-lys-

                    65                        70                       75
val-thr-val-gly-ser-val-ala-his-pro-met-glu-glu-lys-his-trp

                    80                        85                       90
ile-glu-trp-ile-glu-leu-val-ala-asp-gly-val-ser-tyr-lys-lys-

                    95                       100                      105
phe-leu-lys-pro-gly-asp-ala-pro-glu-ala-glu-phe-cys-ile-lys-

                   110                       115                      120
ala-asp-lys-val-val-ala-arg-glu-tyr-cys-asn-leu-his-gly-his-

                   125
trp-lys-ala-glu-ala
```

Figure 5.15 Comparison of the amino acid sequences of Rbo and desulforedoxin from *Dv. vulgaris* Hildenborough and *Dv. gigas* respectively. Identical residues are boxed. Reprinted with permission from the Journal of Bacteriology (Brumlik and Voordouw, 1989).

al., 1990), this finding, as well as the discovery of a ferredoxin domain in [Fe] hydrogenase (see above), indicates that new, larger redox proteins in sulfate-reducing bacteria may have been generated by shuffling and fusion of existing genes for the smaller electron carriers, such as desulforedoxin, rubredoxin, and cytochrome c_3. This mechanism may have led, in addition to Rbo and [Fe] hydrogenase, to rubrerythrin (C-terminal rubredoxin domain; Prickril, 1991) and to Hmc (quadruplication of cytochrome c_3 domain; Figure 5.11). Shuffling and fusion may account for the rich variety of redox proteins found in sulfate-reducing bacteria, and more examples may be found as characterization of these proteins progresses.

The *fla* gene from *Dv. vulgaris* Hildenborough One of the problems in assigning functions to both rubredoxin and desulforedoxin from *Desulfovibrio* is that their reduction potentials are too high (−50 to 0 mV) to function in reduction of sulfate, which requires electrons at potentials of −400 to −200 mV. The FMN prosthetic group of *Dv. vulgaris* flavodoxin has reduction potentials in this range [−450 and −150 mV, (Mayhew and Tollin, 1991)]. Flavodoxin can serve, therefore, as electron donor in sulfate reduction and has been implicated as donor for trithionate reductase in *Dv. vulgaris* Hildenborough (Kim and Akagi, 1985). Flavodoxin is used in lieu of ferredoxin in some organisms and synthesized primarily under conditions of iron deficiency. Laudenbach et al. (1988) have shown that in *Anacystis nidulans* R2, regulation is at the transcriptional level; *fla* mRNA transcription is repressed by the addition of iron. Flavodoxin serves as a specific and sole electron donor to nitrogenase in *Klebsiella pneumoniae*. Deletion of the *nifJ* gene, encoding flavodoxin, leads to loss of nitrogen fixation ability in this organism. Since many *Desulfovibrio* species also have diazotrophic faculties, a similar role for flavodoxin cannot be excluded.

The flavodoxin gene from *Dv. vulgaris* Hildenborough has been cloned and sequenced (Curley and Voordouw, 1988; Krey et al., 1988). Both gene sequences were in good agreement and confirmed the known amino acid sequence of flavodoxin (Dubourdieu and Fox, 1977). The amino acid sequence translated from the nucleotide sequence confirmed the absence of an *N*-terminal signal sequence, in agreement with the cytoplasmic location of flavodoxin. Since flavodoxins are expressed constitutively in *E. coli*, it is not surprising that expression of functional *Dv. vulgaris* flavodoxin is readily achieved in this host. Plasmids, which upon transformation of *E. coli* give a high level of expression of flavodoxin, have been described (Krey et al., 1988; Carr et al., 1990) and gram quantities of fully functional recombinant flavodoxin sufficient for even the most demanding forms of physical study (NMR) can be readily purified. The wild-type protein isolated from *E. coli* has been reported to be indistinguishable from that of *Dv. vulgaris*. Knowledge of the three-dimensional structure of *Dv. vulgaris* flavodoxin (see section 5.2 and the references therein) and the ease with which the cloned gene can be expressed make this protein a prime target for detailed studies on flavoprotein electron transfer reactions by site-directed mutagenesis (Carr et al., 1990).

Potential transcription initiation and termination sequences cannot be readily identified from the *fla* gene sequence. Gene expression in *Dv. vulgaris* may also be iron-regulated (Curley, unpublished), although no elements homologous to the iron-regulated gene from *A. nidulans* (Laudenbach et al., 1988) were found upstream from the *fla* coding region. The *fla* gene is located on the chromosome in *Dv. vulgaris* Hildenborough, whereas the *nifH* gene is present on a large plasmid, which (as discussed below) is easily lost. It thus appears that the *fla* gene in *Dv. vulgaris* is not physically linked to other *nif* genes, as is the case for *Klebsiella*. This fact, together with possible iron regulation of *fla* gene expression, indicates that flavodoxin does not serve as primary electron donor to nitrogenase, but rather, as discussed above, primarily participates in sulfate reduction, e.g., trithionate reduction, under conditions of iron deficiency.

5.7 Nitrogen Fixation by *Desulfovibrio*

Nitrogen fixation by *Desulfovibrio* was first convincingly demonstrated by Riederer-Henderson and Wilson (1970), who refuted an initial report by Postgate (1965) that most strains of *Desulfovibrio* do not fix nitrogen. Since these early studies, nitrogen fixation by this genus has been confirmed both by biochemical (Postgate and Kent, 1984, 1985) and molecular genetic (Postgate et al., 1986) studies. The latter study demonstrated the presence of genes homologous to *nifH* or *nifD* from *K. pneumoniae* in 13 diazotrophic strains of *Desulfovibrio*. Importantly, DNA isolated from three strains incapable of nitrogen fixation appeared not to hybridize with the *Klebsiella nifH* probe. Genes *nifD, K,* and *H* are adjacent in *Klebsiella*. The A3 probe used by Postgate et al. (1986) contained all of the *Klebsiella nifH* gene and part of the *nifD* gene. The *nifH* gene encodes the dimeric Fe-protein, while the *nifD,K* genes specify the tetrameric MoFe-protein of nitrogenase. Cloning and sequencing the hybridizing DNA fragment of *Dv. gigas* (Postgate et al., 1987; Kent et al., 1989) demonstrated that, in this organism, the homologous gene is *nifH*. Hybridization of DNA from probe A3-positive *Desulfovibrio* strains with probe A2, which only contains the *Klebsiella nifD,K* genes, in Southern blotting experiments indicated that these genes must also be closely linked to *nifH* in *Desulfovibrio* species. This conclusion has yet to be demonstrated by nucleic acid sequencing.

The *Dv. gigas nifH* gene was found to be very similar to that of other nitrogen-fixing organisms; in particular, the cysteine residues thought to be involved in binding the [4Fe-4S] cluster of the Fe-protein were conserved (Postgate et al., 1987; Kent et al., 1989). With the *Klebsiella nifH* probe, it was shown that the *Desulfovibrio nif* genes are usually chromosomal, except in some strains (Hildenborough, Wandle, and Brockhurst Hill) of *Dv. vulgaris,* where these genes were found to reside on a large 195-kb plasmid (Postgate et al., 1986). While the plasmids of the latter two strains could be the same, they are different from that of Hildenborough, since the *nifH* gene appears to reside on a differently sized *Eco*RI fragment in the latter strain. It is not known why these sulfate-reducer plasmids that carry *nif* genes are so large. For instance, in *K. pneumoniae,* the 17 genes known to be required for nitrogen fixation (14 in addition to the structural genes *nifH, D,* and *K*) are arranged in a cluster that spans approximately 30 kb.

Unpublished observations from the author's laboratory indicate that the 195-kb plasmid is easily lost from *Dv. vulgaris* Hildenborough. DNA from our laboratory strain failed to hybridize with the *Dv. gigas nifH* probe (kindly provided by Helen Kent, University of Sussex). This Δ*nif* genotype could have arisen by repeated subculturing of our *Dv. vulgaris* Hildenborough strain in Postgate B and C media (Postgate, 1984a), which contain ammonium chloride. To test this hypothesis, the following experiment was conducted. A strain of *Dv. vulgaris* Hildenborough, judged to contain the 195-kb plasmid by hybridization with the *Dv. gigas nifH* probe, was grown in 20 ml Postgate B medium (1 day at 37°C; culture 1). Two ml of culture 1 was then transferred to 20 ml of Postgate B and grown similarly to give culture 2. Following 28 similar transfers, total DNA (chromosomal and plasmid) was extracted and purified from the 29 samples obtained: it was found that *nifH* hybridization was completely lost after 5 transfers, whereas hybridization of the extracted DNA with a chromosomal DNA marker (*hydA,B* which encodes [Fe] hydrogenase) was constant for all samples. The physiological effect of this plasmid loss has not yet been tested, primarily because in our hands it was difficult to grow *Dv. vulgaris* Hildenborough (*nif*) on media lacking ammonium chloride.

These observations have implications for the relationship of flavodoxin and nitrogen fixation in the genus *Desulfovibrio*. Genotypically our laboratory strain of *Dv. vulgaris* Hildenborough is Δ*nif*. The successful cloning of the *fla* gene from this organism indicates that this gene is located on the chromosome and is not closely linked to the structural genes for nitrogenase (*nifH, D,* and *K*), as in *K. pneumoniae.*

5.8 Summary and Future Perspectives

In this chapter, our current state of knowledge of genes from sulfate-reducing bacteria of the genus *Desulfovibrio* has been summarized. The contributions of this work to our understanding of these bacteria has, in the past 5 years, been mainly in the area of consolidation, simplification, and generalization of the extensive biochemical information previously collected for this genus. It is worth while to know that all *c*-type cytochromes in these (and incidentally in all other) bacteria are periplasmic. It is also worth while to know that there are only three different classes of hydrogenases in these bacteria, two of which appear specific for the genus.

These observations set new standards for future biochemical work. The achievements of biochemical work on redox proteins from these bacteria has been extensive. However, the lability of many proteins during isolation often provides the biochemist with a final preparation that may only faintly resemble the macromolecule originally present in the cell. Variability of the final product among different laboratories has led to the perception of a larger variety of redox proteins than actually present. Molecular biology may effectively clear these murky waters. Future proposals for a new cytochrome or hydrogenase in this genus should include the definitive proof of cloning and sequencing of its gene and, with Southern blotting, some documentation of its distribution in the genus.

Future contributions of molecular biology to a better understanding of sulfate-reducing bacteria may be expected. The development of "real" genetics for these bacteria (conjugation, transduction, etc.; see Chapter 4) will facilitate expression and further study of cloned genes. Sequencing of large DNA fragments will allow a level of definition of macromolecular organization in these bacteria that cannot be obtained with biochemical studies alone. In particular, one may hope that the link of the periplasmic to cytoplasmic redox chains through a set of membrane-spanning redox proteins, e.g., as encoded by the *hmc* operon, will be established.

At least two practical applications may emerge from this work. First, one may expect that these results will facilitate development of better gene probes to detect these bacteria in the environment. Second, a clearer understanding of the molecular biology of sulfate-reducing bacteria will contribute to a firmer understanding of their macromolecular constitution and interactions and may lead to a more rational design of agents (biocides) for their selective control.

[References, see p. 211]

I sincerely thank the Alberta Heritage Foundation for Medical Research and the National Science and Engineering Research Council of Canada for their financial support of the research in my laboratory in past years. Michael Brumlik, Mary Carr, Paul Curley, Harm Deckers, Brent Pollock, Ben Prickril, and Gijs van Rooijen contributed hard work and lots of enthusiasm as graduate students or visiting graduate students to the research in my laboratory, while Frankie Wilson and Hansje Voordouw are thanked for providing stability and continuity as laboratory technicians.

6

Phylogeny of Sulfate-Reducing Bacteria and a Perspective for Analyzing Their Natural Communities

Richard Devereux and David A. Stahl

6.1 Introduction

It has been 7 years since publication of the second edition of Postgate's monograph on sulfate-reducing bacteria (Postgate, 1984a). That revision was prompted by the discovery of morphologically and nutritionally varied, sulfate-reducing genera within the bacterial domain (previously named the eubacterial kingdom; Woese et al., 1990; Widdel, 1988). [Note: The taxonomical concept of *domain* is rather new and refers to the highest ranking taxon of a group of organisms.] Until then, sulfate-reducing bacteria had been thought of as a relatively narrow and physiologically specialized collection united by their distinctive production of hydrogen sulfide from reduction of sulfate.

More recently, isolation of extremely thermophilic sulfate reducers of the archaeal domain (previously named the archaebacterial kingdom), such as *Archaeoglobus fulgidus* by Stetter et al. (1987), demonstrated that dissimilatory sulfate reduction is not restricted to the bacterial domain. A recent review by Widdel (1988) summarizes the current appreciation of the diversity among sulfate-reducing bacteria. As is seen throughout this volume, recognition of that diversity has advanced research in the biochemistry, physiology, ecology, and phylogeny of sulfate-reducing bacteria.

One area of fruitful endeavor has been the inference of phylogenetic relationships among the new and classical genera of sulfate-reducing bacteria. This progress has been made possible largely by advances in the ability to rapidly determine complete (or near-complete) nucleotide sequences of small subunit (16S-like) ribosomal RNA (rRNA) molecules (Lane et al., 1985a). Comparison of 16S rRNA sequences has two distinct advantages: first, long tracts of homologous sequence can be rapidly determined; and second, sequence information is easily analyzed. Although 23S rRNA molecules would provide equally valuable comparative data, for historical reasons greater emphasis has been placed on the 16S rRNA molecule; the largest amount of sequence information is now available for this class of molecule. The general use of comparative rRNA sequencing to infer phylogenetic relationships among microorganisms has been well reviewed (Woese, 1987).

Previously, phylogenetic relationships among sulfate-reducing bacteria were inferred by more classical comparative techniques, including cytochrome *c* characterizations, rRNA:DNA homologies, immunological comparisons, and 16S rRNA cataloging. Also, substantial information of biochemistry, physiology, G + C content, pigments, and menaquinone content of many species is available (Pfennig et al., 1981; Peck, 1984; Widdel and Pfennig, 1984; Collins and Widdel, 1986). Comparative sequencing provided a basis for associating known biochemical and physiological properties with natural relationships.

The phylogeny of sulfate-reducing bacteria emerging from 16S rRNA-based studies is generally consistent with the classification based on biochemical and physiological criteria. Inconsistencies between the phylogenetic and earlier classification, especially among species of *Desulfovibrio*, reflected a lack of information upon which to build meaningful distinctions. This will change as the phylogeny, physiology, and ecology of sulfate-reducing bacteria are further delineated.

In this chapter we will summarize recent phylogenetic studies of sulfate-reducing bacteria and the application of 16S rRNA sequence information to environmental studies of these bacteria. A brief overview of the use of 16S rRNA sequences to infer phylogenetic relationships is provided. Where possible, some of the nutritional and biochemical characteristics of sulfate-reducing bacteria have been placed in an evolutionary context. Additional discussions of the physiology and ecology of the sulfate-reducing bacteria are described in separate chapters of this book.

6.2 16S Ribosomal RNA Sequence Comparisons

As will be detailed below, bacteria that share the capacity for dissimilatory sulfate reduction are found within three, possibly four, distantly related phylogenetic groups; the Gram-positive bacteria, the proteobacteria, and within the archaeal domain. Moderately thermophilic sulfate-reducing bacteria, e.g., *Thermodesulfobacterium* (Zeikus et al., 1983), do not appear to be specifically related to Gram-negative mesophilic species within the proteobacteria and likely comprise a fourth, phylogenetically distinct group. The vernacular term for all these organisms, i.e., "sulfate-reducing bacteria", is an appropriate description of common physiology and is useful for determinative purposes. However, the term obscures the evolutionary diversity of the organisms. Although there has never been a formal proposal to place all sulfate reducers within a single taxon, their de facto grouping exemplifies how a descriptive and determinative bacterial classification may overshadow underlying physiological and ecological diversity.

The most desirable classification scheme is one based on evolutionary history (Mayr, 1982), and such a history may be inferred by comparing sequences of homologous biopolymers (Zukerkandl and Pauling, 1965). Homologous biopolymers with a constant rate of random fixed mutations may serve as molecular chronometers to measure evolutionary relationships (Woese, 1987). Since bacteria generally lack an abundance of physiological and morphological characteristics that are phylogenetically interpretable, a molecular approach is generally necessary to infer their phylogeny. Woese pioneered application of 16S rRNA sequence comparisons and with it established the outline of microbial phylogeny (Woese, 1987; Fox et al., 1980). Woese (1987) has recently reviewed the history of microbial phylogenetics.

Why ribosomal RNA? Ribosomal RNA molecules (particularly the 16S rRNA; Figure 6.1) are ideally suited for phylogenetic analyses of bacterial species (Woese, 1987). Ribosomes occur in every cellular organism, and rRNA molecules are functionally and structurally conserved. Sequence conservation of rRNAs was early recognized from studies of heterologous rRNA/DNA hybridization (Pace and Campbell, 1971). This conservation even extends across the highest ranking taxons, the domains Eukarya, Archaea, and Bacteria (Woese et al., 1990), and is particularly striking since these primary evolutionary divergences presumably date to 4 billion years ago (Woese and Olsen, 1986; Woese, 1987). Further, there is no evidence for the transfer of rRNA between species that would obscure their genealogy.

All major classes of rRNA molecules (5S, 16S-like, and 23S-like) have been used as molecular chronometers. The 5S has proven too small for statistically valid estimates of distant relationships. The 16S has been emphasized (relative to the 23S-like rRNA) for historical reasons. The larger 23S molecule was not amenable to sequence determination by older sequencing technologies (below) and is, therefore, something of an untapped resource that will undoubtedly receive greater attention in the coming years.

The first tractable 16S rRNA sequence comparisons, prior to the ad-

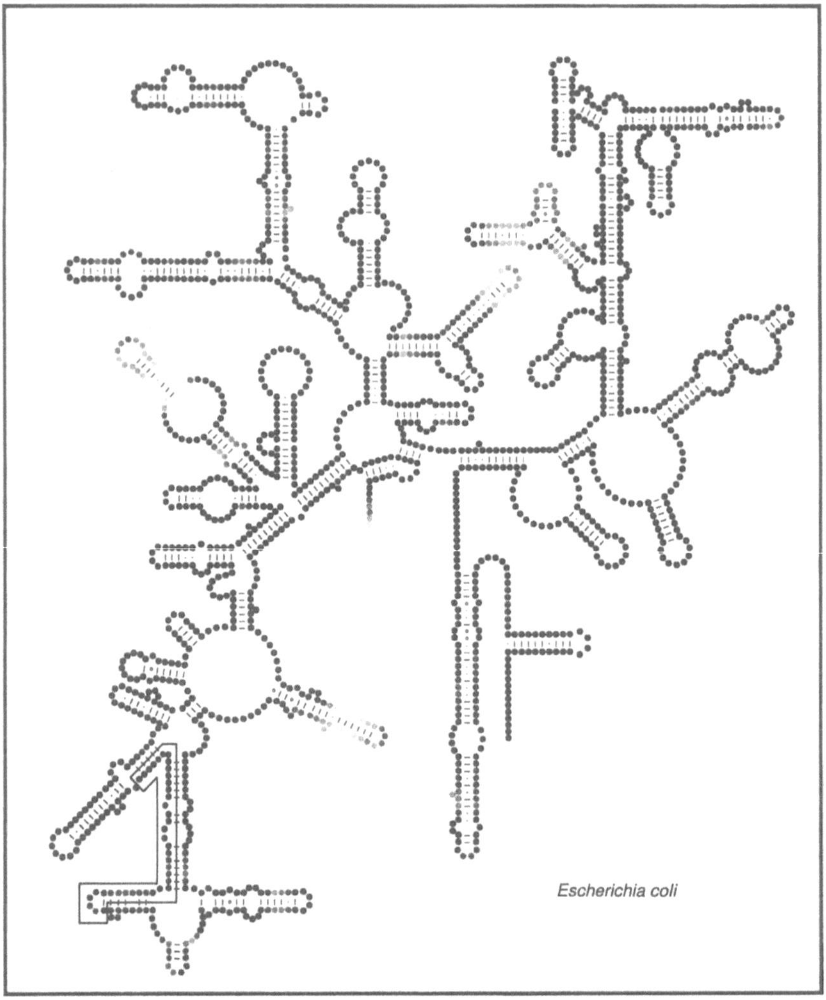

Figure 6.1 Secondary structure of bacterial domain 16S rRNA. Each dot represents one ribonucleotide. The relative sequence conservation for each position is indicated by the degree of shading; black dots correspond to invariant ribonucleotides, lighter dots correspond to variability of ribonucleotides between different bacterial domain sequences. The boxed region (lower left) indicates a target site for an oligonucleotide probe. (Reprinted with permission from Smets et al., 1990. Copyright 1990, American Chemical Society. Figure kindly contributed by R. Guttel.)

vent of recent sequencing technologies, were performed by the procedure of oligonucleotide cataloguing (Fox et al., 1977). Oligonucleotide catalog data provided most of the initial 16S rRNA sequence comparisons that led to the current bacterial phylogenetic scheme, including the discovery of the archaea as a primary evolutionary group distinct from both eukaryotes and "typical" bacteria (Woese and Fox, 1977; Fox et al. 1980; Woese et al. 1985). Initial phylogenetic analysis of sulfate-reducing bacteria was also through comparison of oligonucleotide catalogs (Fowler et al., 1986). The procedure is no longer in general use.

The rRNAs are a significant fraction of cell mass and easily isolated for direct sequence determination. Rapid sequencing is made possible by conserved tracts of nucleotides within the rRNAs which provide sites for sequencing primers (Lane, 1991; Lane et al., 1985a). Cloning of the genes is not necessary, since the rRNA itself serves as template. A primer is annealed directly to the rRNA and, by use of fairly standard nucleic acid sequencing reaction conditions with reverse transcriptase and chain-terminating dideoxynucleotides, sequences of overlapping regions of the rRNA are determined.

Application of the polymerase chain reaction to amplify segments of DNA (Saiki et al., 1988) is beginning to prove useful for rRNA sequence determinations (Medlin et al., 1988; Weisburg et al., 1991; Wisotzkey et al., 1990). Conserved nucleotide sequences near both ends of an rRNA gene are used as priming sites for amplification of the gene. The amplified gene fragment can be cloned into a sequencing vector or sequenced directly. Although considerably more effort may be expended in preparation of the sequencing template, the much better quality of the autoradiogram results in greater overall fidelity.

Phylogenetic trees Phylogenetic trees are graphical representations of the evolutionary relatedness of a group of organisms and can be constructed by comparing the nucleotide sequence of rRNA from different organisms. Base sequences are aligned by using both conserved linear nucleotide sequence tracts and secondary structures as points of reference. Comparison of nucleotide bases at corresponding positions is made for each possible sequence pair. This is directly expressed as sequence similarity or converted to an estimate of evolutionary distance. Sequence similarity indicates the percentage of positions with a common nucleotide base. Evolutionary distance indicates the amount of nucleotide substitutions, or sequence divergence, and includes a correction to account for multiple substitutions at a single position (Hori and Osawa, 1979). Species sharing greater similarity in 16S rRNA sequence are more closely related to each other than they are to species with more divergent (less similar) sequence. The relationships among a set of 16S rRNA sequences may be depicted by using a tree-building or clustering algorithm (Felsenstein, 1984; Olsen, 1985; Woese, 1987).

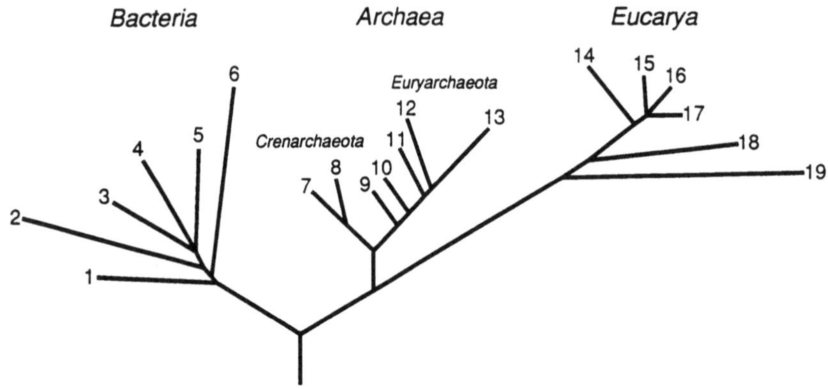

Figure 6.2 Primary evolutionary lineages based on rRNA sequence comparisons illustrating the three domains. Numbers correspond to the groups of organisms as follows (Woese, 1987). **Bacteria**: 1, the Thermotogales; 2, the flavobacteria and relatives; 3, the cyanobacteria; 4, the purple bacteria; 5, the Gram-positive bacteria; and 6, the green nonsulfur bacteria. **Archaea**: kingdom Crenarchaeota: 7, the genus *Pyrodictium*; and 8, the genus *Thermoproteus*; kingdom Euryarchaeota: 9, the Thermococcales; 10, the Methanococcales; 11, the Methanobacteriales; 12, the Methanomicrobiales; and 13, the extreme halophiles. **Eucarya**: 14, the animals; 15, the ciliates; 16, the green plants; 17, the fungi; 18, the flagellates; and 19, the microsporidia. (Figure kindly provided by C.R. Woese. Reprinted with permission of the authors from Woese et al., 1990).

Characterization of small subunit rRNA sequences indicated three primary evolutionary lines that were initially designated the eubacteria, the archaebacteria, and the eukaryotes (Woese and Olsen, 1986). To recognize the primacy of these lines, Woese et al. (1990) proposed that they be ranked above the existing kingdom hierarchy, of traditional taxonomy, as the "domains" Bacteria, Archaea, and Eukarya, respectively. Divergence of the three domains from a common universal ancestor is illustrated in Figure 6.2. Each domain is distinct from the other two; rRNA sequences within a domain are more similar to each other than to sequences from the other domains (Woese et al., 1990). Further, 16S rRNA sequences of each domain exhibit characteristic details in structure while they maintain overall common structure and conserved nucleotide position resemblance among domains. (Woese, 1987).

Ten kingdoms, which were previously ranked as phyla or divisions (Woese et al., 1990), were re-defined within the Bacteria domain (Figure 6.3; Woese, 1987). These groupings initially were based on similarities in oligonucleotide catalogs and unique sequence signatures (Woese et al., 1985); they were later substantiated by comparisons of complete 16S rRNA sequences. As will subsequently be discussed in greater detail, most presently described sulfate-reducing bacteria of the bacterial domain are affili-

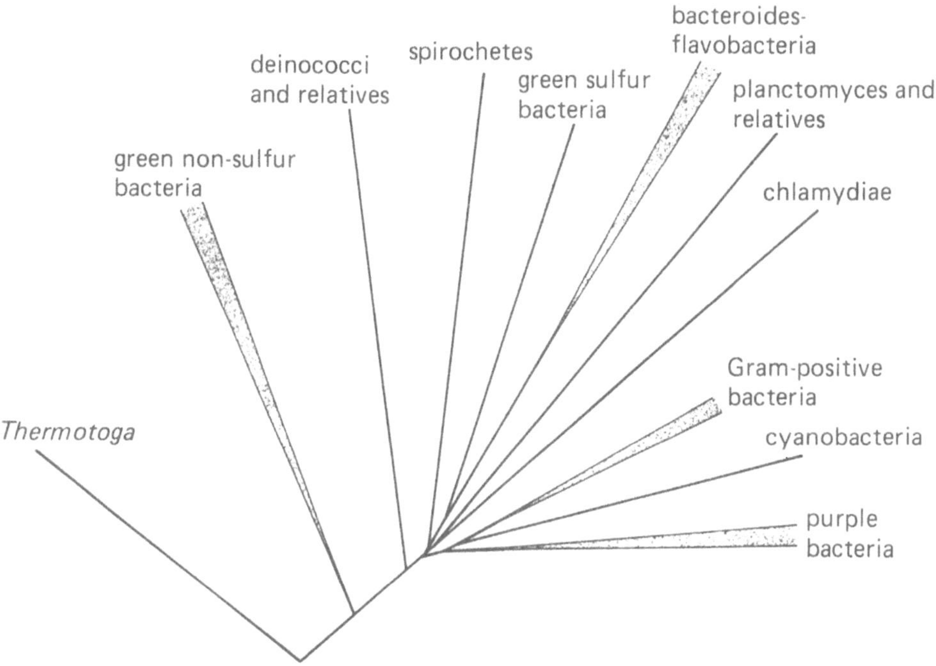

Figure 6.3 Phylogeny of the Bacteria illustrating the kingdoms of the bacterial domain. (From Woese, 1987; reprinted by permission, American Society of Microbiology).

ated with two of these kingdoms, i.e., the Gram-positive bacteria and the proteobacteria—previously designated as the purple bacteria (Stackebrandt et al., 1988; Oyaizu and Woese, 1985; Fowler et al., 1986; Woese, 1987; Devereux et al., 1989).

6.3 Sulfate-Reducing Archaea

Archaeoglobus fulgidus strain VC-16, a hyperthermophilic sulfate-reducing archaea capable of growth between 65°C and 95°C, was isolated from sediments near hydrothermal vents in the vicinities of Vulcano and Stufe di Nerone, Italy (Stetter et al., 1987; Stetter 1988). *Archaeoglobus fulgidus* strain Z, isolated from these same sediments, has also been described (Zellner et al., 1989b), and 16S rRNA cataloging indicated that it was a strain of *A. fulgidus* (Zellner et al., 1989b). An additional species, *A. profundus*, was isolated from the hydrothermal vent systems of Guaymas, Mexico (Burggraf et al., 1990). Subsequently, *Archaeoglobus* was identified as a member of the hyperthermophilic community in volcanic samples taken during the

Macdonald Seamount eruption. These observations indicate that extremely thermophilic sulfate-reducing archaea are common to such ecosystems (Huber et al., 1990). Though not detected in open-ocean seawater during the Macdonald Seamount eruption, hyperthermophilic, sulfate-reducing archaea were obtained by enrichment from the submerged volcanic plume and from a surface slick as far as 1 km from the active center (Huber et al., 1990).

It has been suggested that these archaea may come to inhabit deep oil wells by introduction with sulfate-containing water (Stetter et al., 1987). Although crude oil is not a growth substrate for *A. fulgidus* strain VC-16, the organism is also not inhibited by its presence (Stetter et al., 1987); *A. profundus* exhibited growth on crude oil, perhaps at the expense of trace amounts of acetate (Burggraf et al., 1990). Since *Archaeoglobus* was transported in a surface water volcanic slick during the Macdonald Seamount eruption (Huber et al., 1990), it is a plausible scenario that these thermophilic archaea are introduced into oil formations via injected seawater.

Archaeoglobus fulgidus strain VC-16 uses the same enzymes [adenosine triphosphate sulfurylase, adenylsulfate (APS) reductase, and bisulfite reductase] of the sulfate reduction pathway found in bacterial domain sulfate reducers (Stetter, 1987; Speich and Trüper, 1988). Purified archaeal APS reductase had structural similarities (molecular mass, molar ratios of constituents) with the bacterial reductase (Speich and Trüper, 1988). As pointed out by Speich and Trüper (1988), these observations raise the question of whether sulfate reduction arose just once during early evolution or occurred independently within distantly related phylogenetic groups through lateral gene transfer.

The metabolic capacities of *Archaeoglobus* species have been described and are summarized in Table 6.1 (Stetter et al., 1987; Stetter, 1988; Zellner et al., 1989b; Burggraf et al., 1990). In addition to sulfate, all these archaea utilize sulfite or thiosulfate as terminal electron acceptors. Small amounts of methane are detected during growth. Although sulfur may be reduced, it does not support growth. The *A. fulgidus* strains are capable of both chemolithoautotrophic growth (only with thiosulfate as the electron acceptor) and chemoorganotrophic growth with lactate, pyruvate, or other carbon sources.

Archaeoglobus fulgidus strain VC-16 completely oxidizes lactate to CO_2. The stoichiometry of lactate utilization by *A. fulgidus* strain Z indicates that 65% or more of the acetate generated is further oxidized to CO_2. Little of the carbon from acetate is incorporated into cells; this observation reflects the inability of acetate alone to support growth. In contrast, *A. profundus* is an obligate mixotroph and requires both hydrogen and a carbon source such as acetate, lactate, or pyruvate. No significant amounts of DNA:DNA hybridization occurred between the two species of *Archaeoglobus* (Burggraf et al., 1990).

A unique pattern of cross-reactions between subunits of DNA-

Table 6.1 Substrates supporting growth of *Archaeoglobus* species[a]

Electron donors[b]	Electron acceptors					
	A. fulgidus strain Z			*A. fulgidus strain* VC-16		*A profundus*[c]
	Na_2SO_4	$Na_2S_2O_3$	Na_2SO_3	Na_2SO_4	$Na_2S_2O_3$	Na_2SO_4
H_2/CO_2	−	+	−	−	+	−
Formate	−	+	+	+	+	nr
Acetate	−	−	−	−	−	+
Methanol	−	−	−	+	+	nr
Ethanol	−	−	−	−	+	nr
1-Propanol	−	−	−	−	+	nr
2,3-Butandiol	+	nr	nr	−	+	nr
L(+) or D(−)-Lactate	+	+	−	+	+	+
Pyruvate	+	+	−	+	+	+
Fumarate	−	+	+	−	−	nr

[a] Adapted from Burggraf et al. (1990) and Zellner et al. (1989b).
[b] +, growth; −, no growth; nr, not reported. Other electron donors tested but not supporting growth of *A. fulgidus* strains include propionate, n-butyrate, malate, succinate, and choline.
[c] *A. profundus* additionally requires hydrogen for growth. Thiosulfate and sulfite also serve as electron acceptors.

dependent RNA polymerase from *A. fulgidus* VC-16 and antibodies directed against subunits of a methanogen polymerase suggested that the sulfate reducer represented a new, third branch within the archaea (Stetter et al., 1987). 16S rRNA sequence comparisons were used to investigate the phylogenetic relationship of *A. fulgidus* VC-16 to other archaea (Achenbach-Richter et al., 1987). Initially the VC-16 strain was thought to represent a line intermediate between thermophilic sulfur-reducing archaea (crenarchaeota) and methanogenic and halophilic archaea (euryarchaeota)—the two main branches of the archaeal domain. This phylogenetic placement, thermophilic sulfate-reducing metabolism, and the presence of several methanogenic cofactors (F420 and methanopterin, but absence of others such as CoM and F430) suggested that *Archaeoglobus* could represent a transitional form between the two previously recognized main lineages of the archaea (Yang et al., 1985; Achenbach-Richter et al., 1987; Stetter et al., 1987). However, in describing the substructure of the archaea domain, Woese et al. (1990) have more precisely affiliated *Archaeoglobus* within the euryarchaeota kingdom, comprising the methanogens and their relatives (Figure 6.4).

The isolation and description of these organisms demonstrated that dissimilatory sulfate reduction is not limited to the bacterial domain. Indeed, the occurrence of dissimilatory sulfate reduction in two primary evolutionary lineages suggests that the process is of ancient origins (Stetter et al., 1987).

6.4 Phylogenetic Relationships among Sulfate-Reducing Bacteria

Gram-positive, sulfate-reducing bacteria The genus *Desulfotomaculum* was established for spore-forming species of sulfate-reducing bacteria (Campbell and Postgate, 1965). 16S rRNA oligonucleotide cataloguing of *Desulfotomaculum acetoxidans* and *Dtm. nigrificans* demonstrated the phylogenetic affiliation of this genus with the Gram-positive bacteria (Fowler et al., 1986). The relationship between *Dtm. acetoxidans* and *Dtm. nigrificans* was not very close; their S_{AB} value (which is the measure of similarity between two 16S rRNA oligonucleotide catalogs) is 0.43. This value corresponds to a 16S rRNA sequence similarity (or S) of approximately 86%, based on a correlation developed between the two values (Woese, 1987), and reflects significant physiological differences between the two species. *Dtm. acetoxidans* is mesophilic, has a G + C content of 37 mol%, and completely oxidizes ethanol and acetate. In contrast, *Dtm. nigrificans* is a moderate thermophile with a G + C content of 45 mol% and incompletely oxidizes lactate and ethanol (Postgate, 1984a; Widdel, 1988).

16S rRNA sequence comparisons for two additional species, *Dtm. orientis* and *Dtm. ruminis*, demonstrate affiliation with the Gram-positive

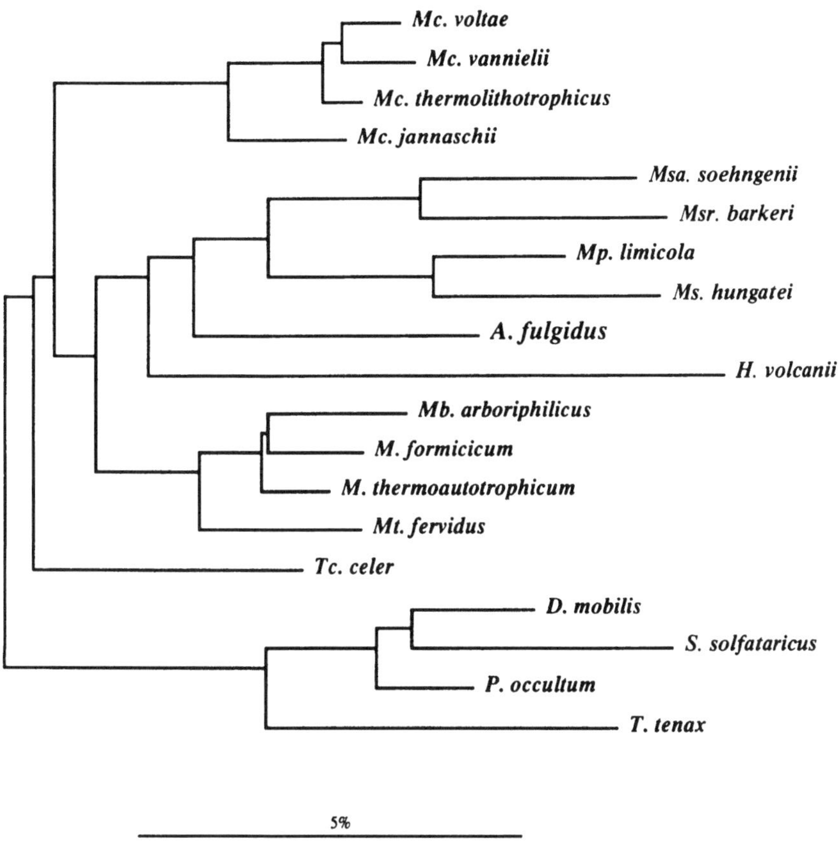

Figure 6.4 Phylogenetic placement of *Archaeoglobus fulgidus* within the archaeal domain. The scale bar represents five estimated changes per 100 16S rRNA sequence positions. (Figure kindly provided by C.R. Woese.)

bacteria and greater diversity within the genus *Desulfotomaculum* (Devereux et al., 1989). The two sequences have only 83% similarity. Indeed, the lineages diverge so deeply within the Gram-positive bacteria (Figure 6.5), it is uncertain whether they are related to each other at the genus taxon. These two species, however, have identical G + C contents and differ only slightly in nutritional capabilities (Postgate, 1984a; Widdel, 1988). Thus, species with similar physiological characteristics may be phylogenetically divergent. These 16S rRNA sequence comparisons suggest that the genus *Desulfotomaculum* is very diverse. This conclusion is in agreement with the range of G + C content for the genus and the nutritional diversity among species (Widdel, 1988).

Gram-negative, mesophilic, sulfate-reducing bacteria Dissimilatory sulfate reduction among Gram-negative species is a phylogenetically res-

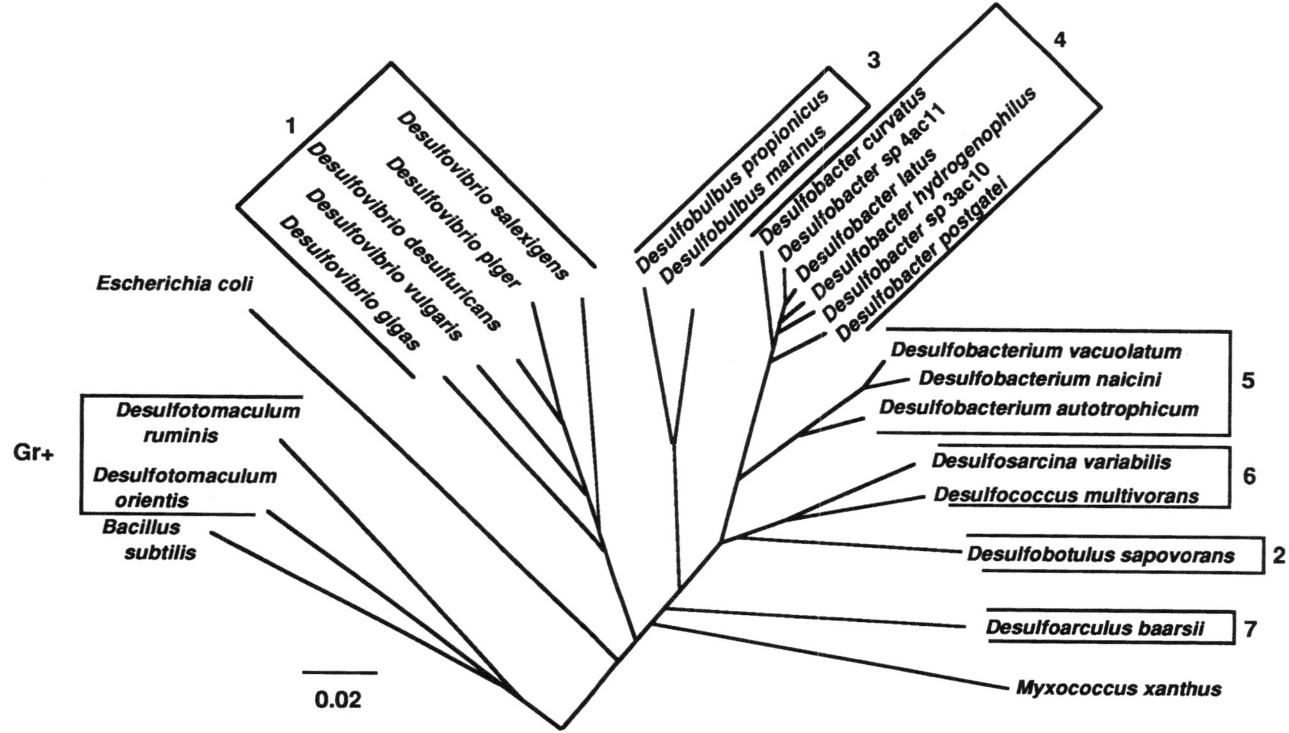

Figure 6.5 Phylogeny of sulfate-reducing bacteria. Numbers correspond to phylogenetic groupings (Devereux et al., 1989); GR+, affiliated with the Gram-positive bacteria. The scale bar is in units of fixed nucleotide substitutions per 16S rRNA nucleotide position.

tricted characteristic in comparison with the ways other groups of bacteria oxidize and reduce inorganic compounds for energy-yielding reactions (e.g., nitrate reduction or sulfur oxidation).

On the basis of 16S rRNA sequence comparisons, Gram-negative mesophilic species of sulfate-reducing bacteria are confined to the delta subdivision (one of the four) of the proteobacteria (Oyaizu and Woese, 1985; Fowler et al., 1986; Woese, 1987; Stackebrandt et al., 1988; Devereux et al., 1989; DeWeerd et al., 1990). The sulfate reducers share their place in the delta subdivision with the sulfur-reducing bacteria, myxobacteria, and bdellovibrios (Hespell et al., 1984; Woese et al., 1985; Fowler et al., 1986; Woese, 1987). Currently, it remains unresolved whether the lines leading to the myxobacteria and bdellovibrios arose from within the group defined by sulfate reducers. However, the myxobacteria and bdellovibrios may represent aerobic adaptations of an ancestral, anaerobic, sulfur-metabolizing phenotype (Woese, 1987)

16S rRNA cataloguing showed that *Desulfovibrio desulfuricans* and *Dv. gigas* are a phylogenetic line distinct from other Gram-negative sulfate reducers (Fowler et al., 1986). The latter group included organisms from the genera *Desulfosarcina, Desulfonema, Desulfobulbus, Desulfococcus,* and *Desulfobacter*. A close relationship (S_{AB} of 0.53; approximately 90% sequence similarity by correlation) was defined only between *Desulfonema limicola* and *Desulfosarcina variabilis*. The study thus indicated considerable phylogenetic breadth among the nonsporing, sulfate-reducing bacteria. Subsequently, comparisons of near-complete 16S rRNAs have defined at least eight phylogenetic groups of Gram-negative, sulfate-reducing bacteria (Devereux et al., 1989; DeWeerd et al., 1990). Evolutionary distance values between eight of the groups are shown in Table 6.2. The values indicate the distinctness of the groups, particularly among *Desulfovibrio* species related to *Dv. desulfuricans*.

The phylogeny of sulfate-reducing bacteria corresponds well with classification based on nutritional and chemical characteristics (Table 6.3), although there were some instances indicating a necessity for taxonomic revision (Devereux et al., 1989, 1990) which has been proposed (Widdel and Bak, 1992). *Dv. sapovorans* and *Dv. baarsii* were found to branch outside the *Desulfovibrionaceae* and have proposed to be renamed as *Desulfobotulus sapovorans* and *Desulfoarculus baarsii*, respectively. *Desulfomonas pigra* was found to be affiliated with the *Desulfovibrionaceae* and was proposed to be renamed *Desulfovibrio piger*. The distribution of menaquinones, in particular, was highly correlated with phylogenetic relationships (Collins and Widdel, 1986). The groups of Gram-negative, mesophilic, sulfate-reducing bacteria are described below.

***Desulfovibrio* species** By far the largest collection of Gram-negative, sulfate-reducing bacteria described are those species of the genus *Desulfovibrio*. The genus was initially established for sulfate-

Table 6.2 16S rRNA sequence similarities between representatives of the sulfate-reducing bacteria phylogenetic groups[a]

Organism	1.	2.	3.	4.	5.	6.	7.	8.
1. *Desulfovibrio desulfuricans*	—							
2. *Desulfobotulus sapovorans*	0.820	—						
3. *Desulfobulbus propionicus*	0.821	0.832	—					
4. *Desulfobacter postgatei*	0.823	0.847	0.841	—				
5. *Desulfobacterium autotrophicum*	0.824	0.875	0.840	0.888	—			
6. *Desulfococcus multivorans*	0.822	0.882	0.832	0.853	0.868	—		
7. *Desulfoarculus baarsii*	0.827	0.849	0.841	0.818	0.829	0.867	—	
8. *Desulfomonile tiedjei*	0.805	0.835	0.846	0.819	0.832	0.860	0.873	—

[a]Sequence comparisons made over 1100 unambiguous nucleotide positions.

Table 6.3 Selected properties of sulfate-reducing bacteria including some strains without species designations[a]

								Electron Donors[i]									
Species[b]	16S rRNA[c] group	Form	Mol % G + C	Desulfoviridin[e]	Motility[f]	Major menaquinone[g]	Oxidation[h]	Hydrogen	Formate	Lactate	Ethanol	Acetate	Fatty acids (carbon atoms)	Fumarate	Malate	Benzoate
Desulfovibrio																
desulfuricans	1	vibrio	59	+	+	MK-6	i	+	+	+	+	–	–	+	+	–
vulgaris	1	vibrio	65	+	+	MK-6	i	+	+	+	(+)	–	–	+	+	–
gigas	1	large vibrio	65	+	+	MK-6	i	+	+	+	(+)	–	–	–	–	–
africanus		vibrio	65	+	+	MK-6 (H_2)	i	+	+	+	+	–	–	–	+	–
salexigens	1	vibrio	49	+	+	MK-6 (H_2)	i	+	+	+	+	–	–	–	+	–
sulfodismutans		vibrio	64	+	+	nr	i	(+)	–	+	+	–	–	–	–	–
carbinolicus		rod	65	+	–	nr	i	+	+	+	+	–	–	+	+	nr
piger	1	rod	66	+	–	MK-6	i	+	–	+	+	–	–	–	nr	nr
Desulfomicrobium																
baculatus		short rod	57	–	+	nr	nr	(+)	+	+	–	–	–	–	+	nr
apsheronum		rod	52	–	+	nr	nr	+	+	+	–	–	–	+	+	nr
Desulfobotulus																
sapovorans	2	vibrio	53	–	+	MK-7	i	–	–	+	–	–	4–16	–	–	–
Desulfobulbus																
propionicus	3	oval or onion	60	–	+/–	MK-5 (H_2)	i	+	–	+	+	–	3	–	–	nr
marinus	3	ellipsoid	nr[d]	–	+	MK-5 (H_2)	i	+	+	+	+	–	3	nr	nr	nr
elongatus		rod	59	–	+/–	MK-5 (H_2)	i	+	–	+	+	–	3	–	nr	nr

Table 6.3 Continued

Species[b]	16S rRNA[c] group	Form	Mol % G + C	Desulfoviridine[e]	Motility[f]	Major menaquinone[g]	Oxidation[h]	Electron Donors[i]: Hydrogen	Formate	Lactate	Ethanol	Acetate	Fatty acids (carbon atoms)	Fumarate	Malate	Benzoate
Desulfobacter																
postgatei	4	oval rod	46	–	+/–	MK-7	c	–	–	–	–	+	–	–	–	–
hydrogenophilus	4	rod	45	–	–	MK-7	c	+	–	–	(+)	+	–	–	–	–
latus	4	large oval rod	44	–	–	nr	c	–	–	–	–	+	–	–	–	–
curvatus	4	vibrio	46	–	+	MK-7	c	+	–	–	+	+	–	–	–	–
sp. 3ac10	4	rod	nr	–	+/–	MK-7	c	–	–	(+)	+	+	–	nr	nr	nr
sp. 4ac11	4	rod	nr	–	+/–	MK-7	c	–	–	(+)	–	+	–	nr	nr	nr
Desulfococcus																
multivorans	6	sphere	57	+	+	MK-7	c	–	+	+	+	(+)	3–16	–	–	–
Desulfosarcina																
variabilis	6	oval rod, packages	51	–	+/–	MK-7	c	+	+	+	+	(+)	3–14	+	–	+
Desulfobacterium																
autotrophicum	5	oval rod	48	–	+	MK-7	c	+	+	+	+	(+)	(3)–16	+	+	–
vacuolatum	5	oval rod/sphere	45	–	–	MK-7 (H_2)	c	+	+	+	(+)	(+)	(3)–16	+	+	–
phenolicum		oval rod	41	–	+	MK-7 (H_2)	c	–	(+)	–	(+)	(+)	(4)	(+)	(+)	+
indolicum		oval rod	47	–	+	MK-7 (H_2)	c	–	(+)	–	(+)	(+)	(3)	(+)	(+)	–
catecholicum		lemon shape	52	–	–	nr	c	(+)	(+)	(+)	(+)	(+)	(3–20)	(+)	(+)	+
niacini (formerly Desulfcoccus)	5	irregular sphere	46	–	–	MK-7	c	+	+	–	+	(+)	(3)–16	+	+	–

Desulfonema																
limicola		filament	35	+	g	MK-7	c	+	+	+	−	(+)	3–14	+	−	−
magnum		filament	42	−	g	MK-9	c	−	+	−	−	(+)	3–10	+	(+)	+
Desulfoarculus																
baarsii	7	vibrio	66	−	+	MK-7 (H_2)	c	−	+	−	−	+	(3)–18	−	−	−
Desulfotomaculum																
nigrificans		rod	49	−	+	MK-7	i	+	+	+	+	−	−	−	−	−
orientis	Gr+	slightly curved rod	45	−	+	MK-7	i	+	+	+	+	−	−	−	−	−
ruminis	Gr+	rod	49	−	+	MK-7	i	+	+	+	+	−	−	−	−	−
antarticum		rod	nr	−	+	nr	i	nr	−	+	nr	−	−	nr	nr	nr
acetoxidans		slightly curved rod	38	−	+	MK-7	c	−	−	−	+	+	4–5	−	−	−
guttoideum		rod	52	−	+	nr	i	+	−	+	−	−	nr	nr	−	nr
sapomandens		rod	48	−	+	nr	c	nr	+	+	+	(+)	4–18	(+)	(+)	+
kuznetsovii		rod	49	−	+	nr	c	+	+	+	+	+	3–18	+	+	−
Thermodesulfobacterium																
commune		rod	34	−	−	MK-7	i	+	nr	+	−	−	−	nr	−	nr
mobilis		rod	38	−	+	MK-7	i	+	+	+	−	−	−	nr	nr	nr

[a] For references see text and Widdel (1988).
[b] Rearrangements in the classification as inferred from phylogenetic studies are included in this table. Because "*Desulfovibrio*" *sapovorans* and "*Desulfovibrio*" *baarsii* represent separate lineages, they were removed from the genus *Desulfovibrio*. "*Desulfococcus*" *niacini* is affiliated with the genus *Desulfobacterium*.
[c] 1–7, defined in Devereux et al. (1988) and Figure 6.5.
[d] nr, not reported.
[e] +, present; −, absent.
[f] +, motile; −, nonmotile; g, gliding motility.
[g] After Collins et al. (1986); (H_2) indicate a terminal saturation in the isoprenoid side chain.
[h] c, complete; i, incomplete.
[i] +, utilized; (+), poorly or slowly utilized; −, not utilized; nr, not reported.

reducing bacteria that do not produce spores, that contain cytochrome c_3 and desulfoviridin, and that primarily utilize lactate or hydrogen as electron donors (Postgate, 1984a; Postgate and Campbell, 1966; Widdel, 1988). Most of the species contain a menaquinone with six isoprene units as the major menaquinone (Collins and Widdel, 1986). Phylogenetic diversity of the genus was suggested by a number of observations: (1) rRNA:DNA hybridizations showed *Dv. desulfuricans* rRNA to be only 57% related to *Dv. vulgaris* rRNA (Pace and Campbell, 1971); (2) the G + C content of the genus ranges from 49 to 66 mol% (Widdel, 1988); (3) surface antigens of the same species demonstrated a high degree of specificity (Abdollahi and Nedwell, 1980; Singleton et al., 1985); and (4) the cytochromes c_3 of the species were divergent, as evidenced from immunological and sequence comparisons (Singleton et al., 1984; LeGall and Fauque, 1988).

Species that are physiologically similar to the type strain of *Dv. desulfuricans* (i.e., those that incompletely oxidize lactate to acetate and may utilize hydrogen as electron donor) were found to comprise a coherent but phylogenetically diverse group, distinct from the other groups of sulfate-reducing bacteria (Devereux et al., 1989, 1990). A new family, *Desulfovibrionaceae*, was proposed to describe the phylogenetic depth and cohesiveness of the group (Devereux et al., 1990). In addition to similar physiology, the family is united by a characteristic sequence signature of conserved 16S rRNA nucleotides that distinguish the *Desulfovibrionaceae* from other lines of sulfate-reducing bacteria.

Within the family are at least five "*Desulfovibrio*" lines with 16S rRNA distance measurements that are as deep as those of other recognized genera of sulfate-reducing bacteria. The recognized lines of genus depth within the proposed family are represented by (1) *Dv. salexigens* and *Dv. desulfuricans* strain El Agheila Z; (2) *Dv. africanus*; (3) *Dv. desulfuricans* ATCC 27774, *Dv. piger* and *Dv. vulgaris* Hildenborough; (4) *Dv. gigas*; and (5) *Desulfomicrobium baculatus* and *Dv. desulfuricans* strain Norway 4 (Figure 6.6).

Two species, "*Desulfovibrio*" *baarsii* and "*Desulfovibrio*" *sapovorans*, are not affiliated with the *Desulfovibrionaceae* (Devereux et al., 1989, 1990; Widdel and Bak, 1992). Rather, they both represent distinct lines that originate within the remaining genera of sulfate-reducing bacteria. This conclusion agrees with metabolic and biochemical features of the two species (Widdel, 1988). Neither organism contains desulfoviridin, and both utilize fatty acids. *Dv. baarsii* is capable of growth on acetate (but not lactate), and *Dv. sapovorans* is an incomplete oxidizer of lactate. These species are considered for reclassification as separate genera, as mentioned above.

Comparison of partial 16S rRNA and 23S rRNA sequences showed *Dv. desulfuricans* ATCC 27774 and *Dv. "multispirans"* to be very closely related to the type strain of *Dv. desulfuricans* Essex 6, whereas *Dv. desulfuricans* ATCC 7757 was more closely related to *Dv. vulgaris* Hildenborough

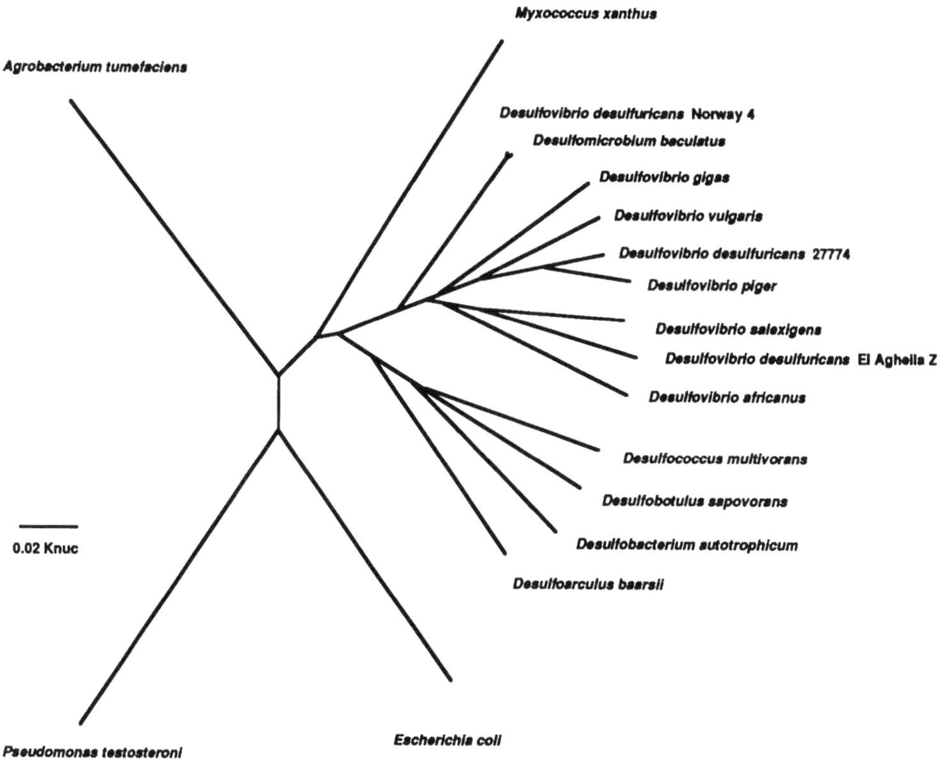

Figure 6.6 Phylogeny of *Desulfovibrio* species. The scale bar is in units of nucleotide substitutions per 16S rRNA sequence position (from Devereux et al., 1990; reprinted by permission, American Society of Microbiology).

(Devereux et al., 1990). The strains most closely related to *Dv. desulfuricans* Essex 6 exhibited the ability to utilize nitrate as an alternative electron acceptor during growth on lactate, which may be useful to distinguish this lineage from others within the family. The relatively closely related *Dv. piger* (95% 16S rRNA sequence similarity) does not reduce nitrate (Moore et al., 1976).

Desulfomicrobium baculatus, which was previously classified as *Desulfovibrio* (Rozanova et al., 1988), and *Dv. desulfuricans* Norway 4 shared near-complete sequence identity of 16S rRNAs (Devereux et al., 1990). They are distinct among the *Desulfovibrionaceae* because of rod-shaped morphology, use of desulforubidin instead of desulfoviridin as bisulfite reductase (Lee et al., 1973b), and in the amino acid sequence divergence of their cytochromes c_3 (Moura et al., 1988).

In other studies, Brune et al. (1983) isolated a *Desulfovibrio* (strain F-1) that contains both cytochrome c_3 and desulfoviridin and is capable of growth on furfural and sulfate. They also showed that *Dv. gigas* possesses the capability for growth on furfural. These observations indicated that *Desulfovibrio* species are perhaps more nutritionally versatile than previously recognized. The strain F-1 was later placed between *Dv. gigas* and *Dv. vulgaris* and renamed as a new species, *Dv. furfuralis*, based on 16S rRNA oligonucleotide cataloguing (Folkerts et al., 1989).

Additional 16S rRNA sequencing has demonstrated that *Dv. desulfuricans* subsp. *aesturaii* is unrelated to the type species and may represent an additional lineage within the *Desulfovibrionaceae* (Devereux and Stahl, unpublished). The characteristics of many organisms presently recognized as *Desulfovibrio* species (Widdel and Pfennig, 1984) may be sufficient to require their reclassification as new genera within the family *Desulfovibrionaceae*. However, since the *Desulfovibrionaceae* lineages are currently defined primarily by rRNA sequence divergence, it is not presently desirable to reclassify them as new genera (Wayne et al., 1987). This problem is a challenge that will benefit from future investigations of the physiology and ecology of the *Desulfovibrionaceae*.

Remaining genera Because *Desulfovibrio* species comprise a monophylogenetic group within the delta subdivision of the purple bacteria, the remaining genera of sulfate-reducing bacteria—*Desulfobacterium, Desulfobulbus, Desulfococcus, Desulfobacter*, and *Desulfosarcina*—which have been analyzed by 16S rRNA sequence comparisons, may be representative of another such assemblage. That is, their phylogenetic diversity is equivalent to that of the *Desulfovibrionaceae*. Also, there is a small but distinctive 16S rRNA sequence signature to unite them and distinguish them from the *Desulfovibrionaceae* (Devereux et al., 1990). Thus, the designation *Desulfobacteriaceae* has been proposed (Widdel and Bak, 1992). Nonetheless, further 16S rRNA sequence comparisons should be made between additional sulfate-reducing bacteria (particularly with additional fatty acid-utilizing species and the gliding *Desulfonema* species), myxobacteria, bdellovibrios, and sulfur-reducing bacteria to establish the cohesiveness of the sulfate-reducing genera comprising this assemblage.

The oval- to lemon-shaped *Desulfobulbus* species incompletely oxidize lactate to acetate and may be enriched for on their characteristic growth substrate, propionate (Widdel and Pfennig, 1982; Widdel and Bak, 1992). The genus is a phylogenetically distinct lineage within the sulfate-reducing bacteria. *Dbu. propionicus* and *Dbu. marinus* (formerly sp. 3pr10) have 16S rRNA sequences that are 92% similar and share no more than 86% sequence similarity with those of other genera (Devereux et al., 1989). *Desulfobulbus* species are phenotypically distinguishable from other genera of sulfate reducers by their nutrition, incomplete oxidation of lactate, growth with propionate (but not higher carbon-chain fatty acids), and in their possession of a dehydrogenated menaquinone with five isoprene units as the major menaquinone (Collins and Widdel, 1986; Widdel and Pfennig, 1982; Widdel, 1988). The branching point for the *Desulfobulbus* line, upon reexamination of the 16S rRNA sequence data, was found to differ from the one depicted earlier (Devereux et al., 1989) and has been corrected for the tree shown in Figure 6.5.

Desulfobacter species are characterized by their ability to grow well on acetate but lack, or have a limited ability to utilize other substrates (Widdel and Pfennig, 1981a; Widdel, 1987). They oxidize acetyl coenzyme A via the citric acid cycle (Gebhardt et al., 1983; Schauder et al., 1987) in contrast to other completely oxidizing species of sulfate-reducing bacteria thus far investigated, which use the carbon monoxide dehydrogenase pathway (see Chapter 2 for details) (Schauder et al., 1986). The species have similar G + C content (44 to 46 mol%) (Widdel and Pfennig, 1981a; Widdel, 1987), which reflects their close phylogenetic relationship (95% 16S rRNA sequence similarity; Devereux et al., 1989).

Desulfobacter species are related to the *Desulfobacterium* species by as much as 90% similarity in 16S rRNA sequence, and the two groups radiate from a common line (Devereux et al., 1989). The range of G + C content for *Desulfobacterium* species is 41 to 52 mol% (Widdel, 1988) and encompasses that of *Desulfobacter* species. *Desulfobacterium* species are, however, nutritionally much more versatile and utilize formate, fumarate, malate, butyrate, higher carbon-chain fatty acids; some species even utilize indole, benzoate, and phenol (Bak and Widdel, 1986a,b; Brysch et al., 1987; Imhoff-Stuckle and Pfennig, 1983; Widdel, 1988). The 16S rRNAs of *Dbt. vacuolatum*, *Dbt. autotrophicum*, and *Dbt.* (formerly *Desulfococcus*) *niacini* share at least 95% sequence similarity, indicating close phylogenetic relationships among these species (Devereux et al., 1989).

Desulfococcus multivorans and *Ds. variabilis* comprise an additional phylogenetic group and share 92% similarity in 16S rRNA sequence. "*Desulfovibrio*" *sapovorans* appears to be related to this group, sharing 90% 16S rRNA sequence similarity with *Dc. multivorans* (Devereux et al., 1989).

DeWeerd et al. (1990) compared the 16S rRNA sequence of *Desulfomonile tiedjei* with those from other sulfate-reducing bacteria and the sulfur-reducing bacterium *Desulfuromonas acetoxidans*. *Dmo. tiedjei* represents a new genus of unusual sulfate-reducing bacteria that perform reductive dehalogenation of halobenzoates and have a unique morphological structure resembling a collar from which the otherwise rod-shaped cell divides (Shelton and Tiedje, 1984; DeWeerd et al., 1986, 1990; Mohn et al., 1990). Although *Dmo. tiedjei* contains both cytochrome c_3 and desulfoviridin, it does not group with the *Desulfovibrionaceae* (DeWeerd et al., 1990), on the basis of 16S rRNA sequence analysis.

6.5 Application of 16S rRNA Sequences to Determinative and Ecological Studies

As noted in Chapters 7 and 8, enumeration and identification of sulfate-reducing bacteria are of both ecological and economic concern. However, culture enrichment techniques may underestimate population densities of sulfate reducers by a thousandfold (Jørgensen, 1978; Gibson et al., 1987). Isolation of pure cultures of sulfate-reducing bacteria is complicated by requirements for anaerobic techniques and further constrained by the long generation times of many species. Identification of isolates, particularly *Desulfovibrio*-like species, currently relies on a limited number of biochemical characters, which is the root of existing imprecise classification.

Ribosomal RNA, particularly 16S rRNA, sequence data now provide a means to circumvent many of these difficulties by providing unambiguous characters (nucleotide sequences) of identity. Within the context of comparative 16S rRNA sequencing, techniques have been developed (discussed below) that use rRNA sequence information for studies in both determinative and environmental microbiology.

Characterization of microbial populations by rRNA sequence comparisons Pace and colleagues provided the outlines for application of 16S rRNA sequence comparisons to characterize natural microbial populations (Olsen et al., 1986; Pace et al., 1986; Stahl, 1986). The technique is based on the direct extraction of rRNAs, or their genes, from environmental biomass without culturing individual population members. Indeed, there is a great likelihood that the rRNAs, or their genes, so obtained represent species that have never been brought into culture. As such, the methodology is well suited to analyze microbial populations whose members, such as sulfate-reducing bacteria, are difficult to isolate or grow in the laboratory.

The smaller 5S rRNAs have been used to characterize relatively simple microbial communities. These molecules may be separated by high-resolution gel electrophoresis and sequenced directly. 16S rRNA genes from complex microbial systems may be identified in recombinant DNA libraries and sequenced. Recombinant libraries are derived from total DNA (isolated from available biomass) or following selective amplification of 16S rRNA sequences by the polymerase chain reaction (PCR). Weller and Ward (1989) developed an alternative procedure in which the cDNA of 16S rRNA, produced by reverse transcription, may be cloned. The rRNA sequences from the microbial population are compared with sequences of well-characterized strains in culture collections. These data provide a phylogenetic assessment of the microbial population structure and, by inference, some indication of physiological types that may be present (Stahl, 1986).

Determinative analysis of 5S rRNA sequences from unculturable bacteria have been used to infer phylogenetic affiliations of endosymbionts associated with invertebrates near hydrothermal vents (Stahl et al., 1984), microbial community members of a hot spring (Stahl et al., 1985) and a copper leaching site (Lane et al., 1985b), and an Antarctic endolithic *Vibrio* species (Colwell et al., 1989).

Giovannoni and colleagues (1990a) analyzed sequences of 16S rRNA genes derived from bacterioplankton, and Ward and colleagues (1990) characterized 16S rRNA sequences from a hot spring microbial mat. Both studies revealed new phylogenetic lines in the respective populations and indicated an abundance of previously unrecognized bacterial diversity. In fact, none of the cloned sequences matched the putative community members previously obtained in pure culture.

The collection of 16S rRNA sequences now represents most of the described genera of sulfate-reducing bacteria (Devereux et al., 1989, 1990; DeWeerd et al., 1990). These data now serve to facilitate comparisons with sequences derived from environmental samples. Since the physiology of sulfate-reducing bacteria corresponds with their phylogenetic groupings (Table 6.3), such sequence comparisons would be expected to give a good indication of the physiological types of sulfate-reducing bacteria present in a natural population.

The utility of using 16S rRNA sequences to characterize sulfate-reducing bacteria within a complex microbial community has been demonstrated (Amann et al., 1992). A diverse microbial community from an inoculum of groundwater was established within a fixed-bed bioreactor operating under anaerobic, sulfidogenic conditions. Nucleic acids were extracted from the community, and 16S rRNA gene sequences were obtained with the polymerase chain reaction (PCR). Primers for PCR were designed to selectively amplify 16S rRNA genes of delta-subdivision eubacteria. Since the primers hybridize to conserved nucleotide tracts, PCR should result in amplification of 16S rRNA genes from a phylogenetically diverse group of delta subdivision (including sulfate-reducing bacteria) organisms. Amplified gene products were cloned, sequenced, and compared with the 16S rRNA sequences from sulfate-reducers and other well-characterized bacteria.

Data in Figure 6.7 show the phylogenetic placement of two bioreactor species inferred by sequence comparison of their cloned 16S rRNA genes. One organism, designated as Bioreactor 2, was most closely related to *Dv. vulgaris* and another, designated as Bioreactor 1, was most closely related to the sulfur-reducing bacterium *Desulfuromonas acetoxidans*. An additional clone (not shown in the figure) was most closely related to a spirochaete. The spirochaete gene was amplified because the primers used in the PCR were targeted to highly conserved 16S rRNA nucleotide tracts and are not completely exclusive. As further indication of the diversity in natural microbial populations, no sequence from the community exactly matched any that had been previously determined. Nonetheless, the phylogenetic affiliations allow reasonable inferences concerning the physiology of species present in the bioreactor. For example, the species related to *Dv. vulgaris* most likely utilizes sulfate as a terminal electron acceptor for oxidation of simple organic compounds (such as lactate, formate, and ethanol) or hydrogen. In contrast, the species related to *Dmn. acetoxidans* may utilize sulfur as a terminal electron acceptor for oxidation of acetate. These in-

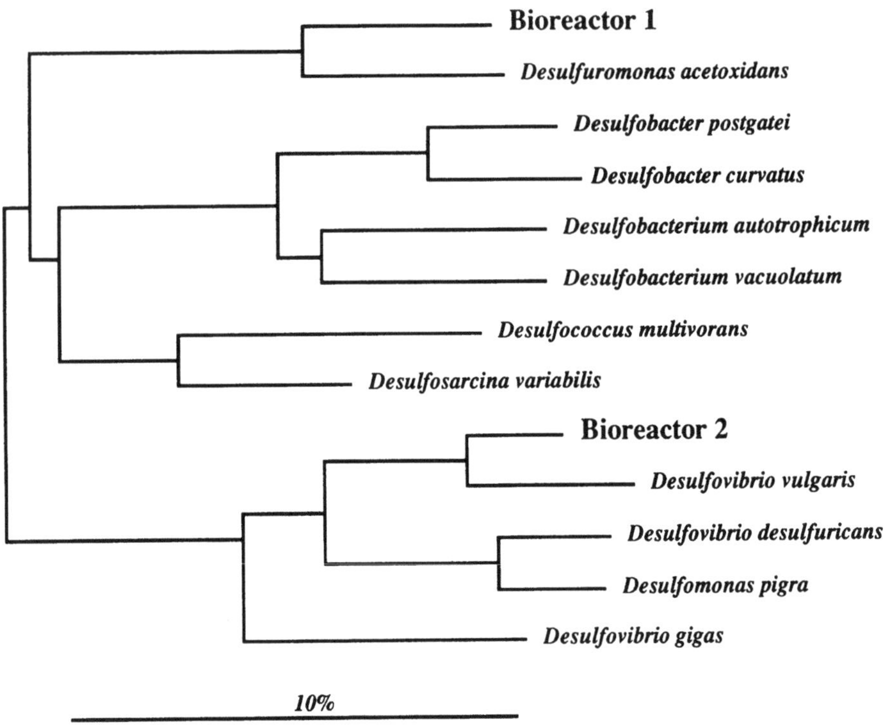

Figure 6.7 Phylogenetic analysis of 16S rRNA genes obtained from the microbial community of a sulfidogenic bioreactor. Bioreactor 2 was found to be most closely related to *Desulfovibrio vulgaris,* and Bioreactor 1 to *Desulfuromonas acetoxidans*. The scale bar represents 10 estimated changes per 100 16S rRNA sequence positions.

ferences also suggest enrichment schemes by which the bioreactor species may be isolated and further studied in the laboratory.

rRNA-targeted hybridization probes It is generally possible to define tracts of nucleotides within 16S or 23S rRNAs that are unique to a phylogenetically coherent microbial group. Synthetic DNA oligonucleotides complementary to these tracts can then be used as hybridization probes to detect and quantify rRNA of a particular phylogenetic group within a natural microbial community (Stahl et al., 1988). Hybridization probes may be synthesized that are specific for closely related assemblages of organisms, such as a species, or that circumscribe larger phylogenetic groups such as genera, families, or kingdoms (Giovannoni et al., 1988; Stahl et al., 1988; Betzl et al., 1990; Hahn et al., 1990; Tsien et al., 1990). Alternatively, longer probes of similar specificities may be obtained from cloned rRNA gene fragments or by isotopically labeling reverse transcripts of rRNA templates (Festl et al., 1986; Regensburger et al., 1988; Giovannoni et al., 1990b). Since actively growing cells may contain 10^4 ribosomes

(Maaloe and Kjellgard, 1966), rRNAs offer a naturally amplified target for hybridization. This amplification provides for sensitive detection (by hybridization to isolated nucleic acids) as well as a means to directly observe single cells in environmental samples (below). The design and application of determinative nucleic acid probes have been reviewed by Stahl and Amann (1991).

Oligonucleotide probes for specific rRNAs are typically about 20 nucleotides long. A satisfactory probe must be about nine nucleotides long and should contain four G-C base pairs for duplex stability (Szostak et al., 1979). A probe's specificity for a defined target molecule (or gene) increases with nucleotide length. Further, a single mismatch between oligonucleotide probe and nontarget sequences may result in a significant decrease in duplex stability relative to target sequence with no mismatches (Wallace et al., 1979). The probe will, therefore, dissociate much more rapidly from the nontarget sequence than the target sequence during the washes following hybridization. Two such mispairs between target and nontarget sequences further decrease duplex stability and provide even greater discrimination. For example, a 12-nucleotide-long probe with one mismatch is sufficient to correctly bind to a fragment within a restriction digest of phage lambda DNA, and a probe of 15 nucleotides can be used to detect a single, unique gene in a yeast genomic library (Szostak et al., 1979). These factors permit oligonucleotide hybridizations that detect a single base-pair difference within the nucleotide sequence complexity of the human genome (Conner et al., 1983; Miyada and Wallace, 1987). Finally, use of the rRNA nucleic acid fraction for hybridizations will greatly reduce sequence complexity of the nucleic acid pool. Thus, under carefully controlled hybridization and wash conditions, oligonucleotide probes offer a high degree of specificity.

On the basis of our phylogenetic studies, we constructed a set of six oligonucleotide probes, which were complementary to conserved tracts of 16S rRNA from phylogenetically defined groups of sulfate-reducing bacteria, to use in determinative and environmental studies (Devereux and Stahl, unpublished). Four probes are directed to specific genera and identify: (1) *Desulfobacterium* spp., (2) *Desulfobacter* spp., (3) *Desulfobulbus* spp., or (4) *Desulfovibrio* spp. The remaining two probes encompass more diverse assemblages. One probe is specific for the phylogenetic lineage composed of *Dc. multivorans*, *Ds. variabilis*, and *"Desulfovibrio" sapovorans*. The remaining probe encompasses this lineage and also those comprising *Desulfobacter* spp. and *Desulfobacterium* spp. Sequences of the oligonucleotide probes are given in Table 6.4. Probe specificity was demonstrated by hybridizations against membrane-bound nucleic acids from target and nontarget species with probes labeled at their 5′ ends with ^{32}P. As is shown in Figure 6.8, the probes combine excellent specificity with a strong hybridization signal.

These probes have great potential for applications to environmental studies of natural microbial communities of sulfate-reducing bacteria. One

Table 6.4 Sequences of sulfate-reducing bacteria-specific rRNA probes

Species	probe sequence	16S rRNA taget site[a]
Desulfovibrio spp.	TACGGATTTCACTCCT	687–702
Desulfobulbus spp.	GAATTCCACTTTCCCCTCTG	660–679
Desulfobacterium spp.	TGCGCGGACTCATCTTCAAA	221–240
Desulfobacter spp.	CAGGCTTGAAGGCAGATT	129–146
Desulfococcus multivorans		
Desulfosarcina variabilis		
Desulfobotulus supovorans	ACCTAGTGATCAACGTTT	814–831
Desulfobacter spp.		
Desulfobacterium spp.		
Desulfococcus multivorans		
Desulfosarcina variabilis	CAACGTTTACTGCGTGGA	804–821
Desulfobotulus sapovorans		

[a]Nucleotide positions by *Escherichia coli* numbering.

experimental approach is through hybridizations to membrane-bound nucleic acids (extracted from environmental samples), as was done in studies of ruminal microbial populations (Stahl et al., 1988). Quantitative and unbiased recovery of rRNA is required for detailed community characterizations of sulfate-reducing bacteria in aquatic sediments. Several procedures to extract rRNA from environmental samples have been described which are amenable to sediment studies (Stahl et al., 1988; Weller and Ward, 1989; Hahn et al., 1990). Hybridizations of the *Desulfovibrio*-specific probe to rRNA extracted from sediment samples is shown in Figure 6.9. Such hybridizations, however, do not give estimates of bacterial densities (e.g., cells/ml or cells/g sediment), since the rRNA content of bacterial cells in the environment is unknown. Rather, the results are expressed as the proportional representation (relative abundance) of a specific target rRNA in the total rRNA of the microbial community (Giovannoni et al., 1990a; Stahl et al., 1988). Nonetheless, relative abundance of specific rRNA may provide a better indication of potential metabolic activity (in contrast to cell numbers) of a target species, since the rRNA content of a cell is proportional to growth rate (Bremer and Dennis, 1987; Maaloe, 1979).

Use of fluorescent DNA probes for single-cell identification Thus far, this discussion of using rRNA for determinative and environmental studies has been limited to radioactive oligonucleotide probes for quantification of specific rRNAs present in environmental samples. It should not go unnoticed that this general experimental format is also amenable to nonisotopic detection systems. There are, however, many alternatives to radioisotopic detection, and this is an area of rapid development. Thus, the reader is referred to specific articles and reviews for more detailed discussion of these techniques (Jablonski et al., 1986; Kumar et

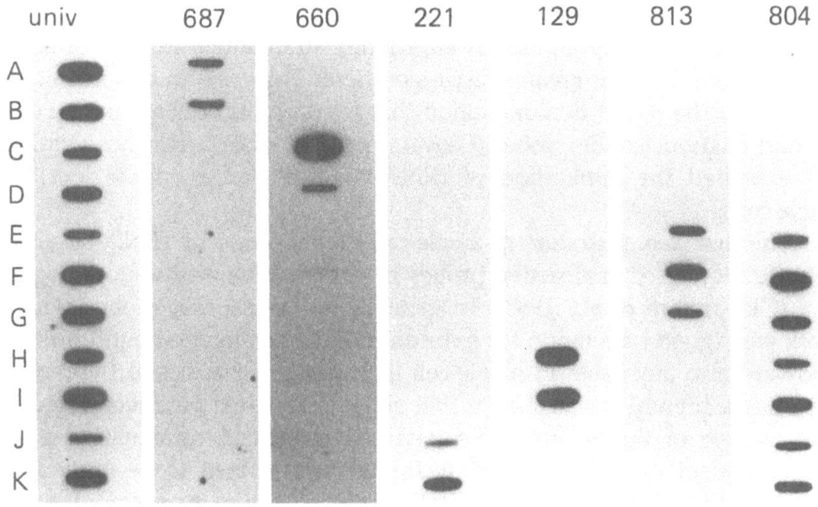

Figure 6.8 Autoradiograms of ^{32}P-labeled oligonucleotide probes hybridized to nucleic acids of sulfate-reducing bacteria. Nucleic acids (by row) were from the corresponding sulfate-reducing bacteria: A, *Desulfomonas pigra*; B, *Desulfovibrio salexigens*; C, *Desulfobulbus propionicus*; D, *Desulfobulbus* sp. 3pr10; E, *Desulfococcus multivorans*; F, *Desulfosarcina variabilis*; G, *Desulfovibrio sapovorans*; H, *Desulfobacterium vacuolatum*; I, *Desulfobacterium niacini*; J, *Desulfobacter latus*; K, *Desulfobacter* sp. 4ac11. Hybridization probes used for strips of nucleic acids by column: univ., universal probe for virtually every 16S rRNA; other probes used are designated by 16S rRNA target site, and their sequences and specificities are given in Table 6.4.

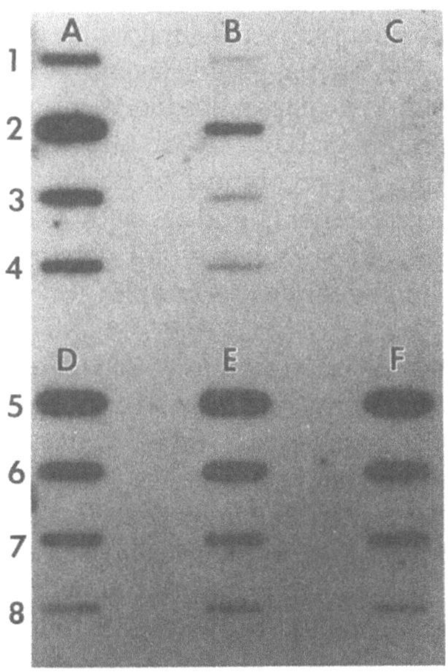

Figure 6.9 Autoradiogram showing results from hybridization of the *Desulfovibrionaceae*-specific rRNA probe to nucleic acids purified from sediment that was spiked with cells of *Desulfovibrio salexigens*. Nucleic acids in rows 1 to 4 are from unamended sediment (row 1) and sediment amended with 10^8, 10^7, or 10^6 *Dv. salexigens* cells/gram sediment (rows 2 to 4). Amounts of nucleic acids applied are 800 ng (column A), 80 ng (column B), and 8 ng (column C). Columns D to F respectively contain nucleic acids from *Dv. salexigens*, *Desulfovibrio vulgaris*, and *Desulfomonas pigra* applied in amounts of 10, 5, 2.5, and 1.25 ng (rows 5 to 8, respectively).

al., 1988; Matthews and Kricka, 1988; DeLong et al., 1989; Stahl and Amann 1991). One application is sufficiently straightforward and of such general utility to merit greater discussion here. The value of the technique arises from the recent demonstration that fixed whole cells are permeable to short oligonucleotide probes (Giovannoni et al., 1988). This observation has extended the application of rRNA-targeted probes to the level of single cells.

The first demonstration of single-cell identification by rRNA-targeted hybridization used radioactive probes in combination with autoradiography (Giovannoni et al., 1988). In general, we found that probes which work well (good sensitivity) for hybridization to membrane-bound nucleic acids are also amenable to single-cell identification via in situ hybridization. Subsequently it was shown that single cells could be directly visualized by use of fluorescent-dye-conjugated probes in combination with epi-fluorescent microscopy. This technique was shown to be applicable for direct identification of single cells both in culture or environmental samples (DeLong et al., 1989; Amann et al., 1990a; Tsien et al., 1990; Devereux, unpublished). More recently, fluorescent probes were used in combination with flow cytometry (Amann et al., 1990b). Identification of selected sulfate-reducing bacteria by probe-conferred fluorescence is shown in Figure 6.10.

Work now in progress suggests that the technique should be amenable to visual inspection of specific populations (or assemblages of populations) in sediment samples and surface-associated consortia (Amann, unpublished; Devereux, unpublished). Thus, the approach potentially has broad application.

Much of our development of whole-cell hybridization and detection protocols focused on sulfate-reducing bacteria. Given the often limited physiological criteria available to unambiguously resolve many of the bacterial species (e.g., *Desulfovibrio* spp.), this new approach should be of interest to investigators working either with the comparative biochemistry of these organisms or studying their environmental distribution and activity. As mentioned previously, growth rate is proportional to ribosomal content and to the amount of fluorescent probe that hybridizes with a cell (DeLong et al., 1989). Thus, the technique offers an avenue for determining the "metabolic status" of single cells in environmental samples.

Fluorescent probe length and mismatch discrimination Whole-cell hybridization specificity appears comparable to hybridization with nucleic acids immobilized on nylon membranes (Amann et al., 1990a). Thus, as discussed earlier, at least two mismatches are generally required for good discrimination between target and nontarget sequences. Similarly, the length of the fluorescent DNA probes ranges from about 16 to 25 nucleotides. This average length offers excellent specificity at wash temperatures compatible with whole-cell studies. The inclusion of formamide in the whole-cell hybri-

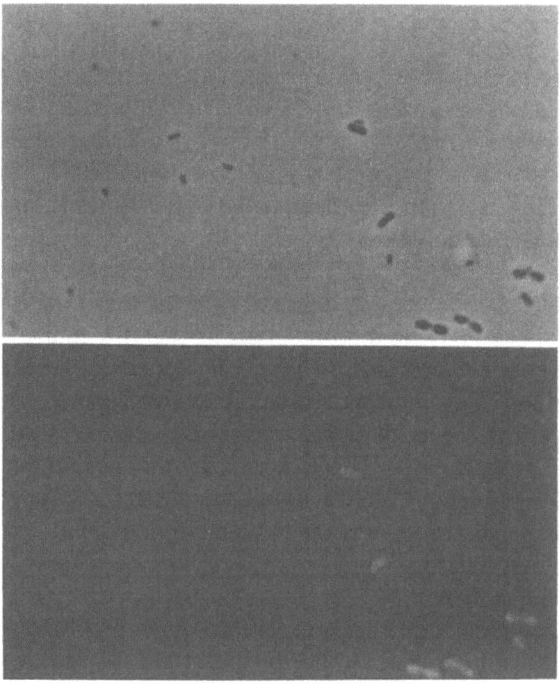

Figure 6.10 Identification of selected sulfate-reducing bacteria by probe-conferred fluorescence. (Top) Phase contrast light micrograph of *Desulfobulbus* cells (ovals) and *Desulfobacter* cells (rods) hybridized with a rhodamine-labeled oligonucleotide probe specific for *Desulfobacter* spp. 16S rRNA. (Bottom) The same field observed by fluorescence microscopy, demonstrating specific identification of the *Desulfobacter* cells.

dization buffer has been examined for several fluorescent probes and observed in some cases to increase sensitivity (Stahl, unpublished).

6.6 Summary

The phylogenetic diversity of dissimilatory sulfate-reducing bacteria far exceeds what was envisaged a decade ago. Indeed, this diversity is so great that it is comparable to the primary evolutionary lineages of all living organisms. An analogy with the evolutionary distance separating the archaeal and bacterial domains is appropriate to understand this diversity. The archaea are no more related to typical bacteria than are the eukarya (Woese and Fox, 1977; Fox et al., 1980; Woese et al., 1990). By comparison, organisms of the archaeal domain that perform dissimilatory sulfate reduction (e.g. *Archaeoglobus*) are no more related to other sulfate-reducing bacteria (e.g., *Desulfovibrio* and *Desulfotomaculum*) than either is related to eukaryotes.

Considerable diversity also exists among many typical sulfate-reducing bacteria. Spore-forming *Desulfotomaculum* species are separated from Gram-negative mesophilic species at a level used to distinguish between bacterial phyla (or kingdoms; Woese et al., 1990). Among Gram-negative species, isolation of nutritionally varied types leads to realization of considerable physiological diversity. In support of these biochemical observations, currently at least eight distinct phylogenetic lineages of Gram-negative species can be defined. One of these lineages comprises *Desulfovibrio* species having the phylogenetic breadth equivalent to a bacterial family. The *Desulfovibrionaceae* presently consist of recognized *Desulfovibrio* species that may represent as many as five separate genera. The remaining genera of Gram-negative mesophilic species of sulfate-reducing bacteria are equally diverse. This is not to say that the notion of specialization by sulfate-reducing bacteria should be altogether discarded. Indeed, the trait is of such comparatively limited phylogenetic occurrence as to suggest that dissimilatory sulfate reduction can occur only within certain highly adapted groups.

Ribosomal RNA sequence comparisons have provided a phylogenetic framework for comparative studies in the molecular biology, physiology, and biochemistry of sulfate-reducing bacteria. The sequences also have ecological applications. Environmental studies of sulfate-reducing bacteria must begin with an understanding of their natural diversity. These organisms occur within complex microbial communities, such as those of anaerobic marine sediments, from which their enumeration, isolation, and identification has been difficult. The 16S rRNA sequence collection offers a data set against which additional sequences, obtained either from newly isolated species or through genes cloned from environmental samples, can be compared. Oligonucleotide probes provide a means to rapidly identify members of specific groups of sulfate-reducing bacteria, in either pure cultures or environmental samples, and to quantify the contribution of their rRNA to the total rRNA of a microbial community. The technique may also lend itself to development of selective enrichments to bring previously unobtainable strains into culture.

[References, see p. 211]

We thank C.R. Woese for providing figures of the universal rRNA tree and the *Archaeoglobus* tree; R.I. Amann for whole cell hybridizations with fluorescent-labeled oligonucleotide probes; and V. Caseo for preparing the tables. This work was supported by grants EPA-CR-812496 from the United States Environmental Protection Agency and N00014-88-K-0093 from the Office of Naval Research to D.A.S. Mention of commercial products or company names does not imply endorsement by the U.S. Environmental Protection Agency.

7

Ecological Actions of Sulfate-Reducing Bacteria

David W. Smith

7.1 Introduction

Few groups of bacteria have undergone such enormous and rapid changes in the way microbiologists perceive them as have sulfate-reducing bacteria in the last 20 years. Consider, for example, the following excerpt from the basic description of this group and its properties from Stanier et al. (1970), the leading general microbiology text of the time:

> The sulfate-reducing bacteria constitute a distinctive physiological group. They are dependent on the reduction of sulfate as a mode of anaerobic energy-yielding metabolism. . . . The range of utilizable organic substrates is narrow, being confined for most strains to a few organic acids.

As noted in Chapter 1, there were only two recognized genera at that time, the Gram-negative *Desulfovibrio* and the Gram-positive, spore-forming *Desulfotomaculum* (Postgate and Campbell, 1966).

Postgate (1979), in the first edition of the defining treatise on sulfate-reducing bacteria, did not substantially alter this picture of bacteria which are limited in nutritional options and simple in characterization. However, by the time of Postgate's second edition 5 years later (Postgate, 1984a), almost all of the comfortingly simple notions I had learned in graduate school in the early 1970s had disappeared. The bacteria we call sulfate reducers are now assigned to more than a dozen different genera, some of which are known to use sulfur compounds other than sulfate as their terminal electron acceptors and some of which are even known to oxidize sulfur compounds. The range of demonstrated electron donors has exploded, owing primarily to the efforts of Pfennig and Widdel (see Widdel, 1988 for summary). The evolutionary relations of these several genera are very interesting and have been discussed by Devereux in Chapter 6. The present chapter will not concentrate on taxonomic questions of individual sulfate-reducing bacterial species. What is of greater importance here is an examination of how our increased physiological understanding of these bacteria has affected the ecological interpretations we make of their activities in nature.

Concomitant with these large changes in the understanding of taxonomy and physiology has been a proliferation of ecological studies of sulfate-reducing bacteria (Widdel, 1988). Two related motives may be discerned for this ecological interest. One motive involves microbial mineralization processes; the second is concerned with the pollution role of sulfate-reducing bacteria.

First, increased attention to these organisms has come about as part of a general increase of interest in the microbiology of anaerobic environments, especially the overall process termed mineralization, in which organic material is decomposed to simpler, inorganic molecules. Second, these activities have frequently been studied as key, often limiting, steps in the cycles of molecules of carbon, nitrogen, sulfur, and phosphorus needed to sustain biological productivity. However, the specific rates and controlling factors of these conversions in the absence of O_2 have historically received much less attention (Howarth and Teal, 1980).

A traditional view of microbial mineralization activity in anaerobic environments is summarized by Capone and Kiene (1988). Here we see that respiration of organic compounds in ultimately anaerobic environments proceeds through consumption of ever less preferable electron acceptors, from O_2 to NO_3 to SO_4^{2-} to CO_2 with the formation of H_2O, N_2, H_2S, and CH_4, respectively. Fermentation reactions occur in different organisms independently of respiratory activities (see Chapter 1 and later in this chapter for comments about the changing definitions of fermentation). In some cases fermenting organisms, primarily bacteria, are essential intermediaries in mineralization owing to the formation of products that serve as electron donors for the various respirers. Acetate, for example, is seen increasingly as a key organic intermediate of this type, while H_2 gas produced fermentatively is of major energetic consequence in anaerobic environments.

The standard model systems for this sequence are the sediments of lakes or shallow nearshore marine areas or intertidal marsh areas, which are very common along the Atlantic coast of North America. It is usually the case that oxidation and mineralization do not go to completion in these anaerobic sediments. There are four general reasons for this incompleteness: (1) there is often a limited amount of electron acceptor; (2) there is an accumulation of products that require molecular O_2 for degradation, owing to enzymatic and thermodynamic limitations (Gottschalk, 1986); (3) some products, such as humic acids, aromatic acids, tannins, and lignin derivatives, are toxic and inhibit further metabolism (Doetsch and Cook, 1973); and (4) physicochemical conditions, primarily pH and temperature, limit rates of degradation. The inevitable conclusion of these restrictions is accumulation of partially oxidized organic material, generally leading to peat formation in the relatively short term (a period of decades in marshes) and coal or oil in the very long term (hundreds of thousands of years in ancient lake basins). The photograph of Tollund Man (Doetsch and Cook, 1973) certainly makes the point that even 2000 years in an anoxic bog may not result in much damage to complex organic molecules such as those which comprise a human body.

The most important implication of this general overview of anaerobic environments for the present discussion is that in the short term of an annual cycle, waterlogged sediments are extremely active in oxidizing organic matter they receive. What emerges is a view of dynamic, complicated interactions among bacterial communities carrying out sulfate reduction, methane production, and various fermentations. The bulk of this chapter will deal with sulfate reduction activity (SRA), but it must be emphasized at the outset that no single microbial activity can be studied meaningfully in isolation. Therefore, methane production activity (MPA) and fermentation will be indispensable aspects of the presentation.

The second motive for the current interest in sulfate-reducing bacteria is that they are significant as nuisances in anthropogenic environments. Postgate (1984a) briefly describes several examples of sulfate reducer-mediated pollution: corrosion of pipes, contamination of oil drilling facilities, contamination of petroleum storage facilities, blackening of paper pulp during processing, blackening of paint containing metal pigments (see also Hamilton, 1985; Ford and Mitchell, 1990a). It is clear that the negative effects of sulfate-reducing bacteria in these environments do not result from new activities, but rather from enhancement of certain reactions that are normal for the group. The emphasis of such studies has been very pragmatic and largely directed to control or elimination of sulfate reducers from the afflicted area. It is probably more appropriate to refer to their study as environmental microbiology, in which management is an important concern, rather than microbial ecology, in which mechanisms and relations are most important. As a result there is limited application of the results of these studies to investigations that pursue more conventional ecological questions in natural environments; therefore, the reader is referred to Chapter 8 for discussion of the pollution aspects of sulfate-reducing bacterial activity.

7.2 Experimental Approaches to Microbial Ecology

Before details of SRA and sulfate-reducing bacteria in situ are addressed, it will be instructive to visit briefly a long-standing dichotomy in fundamental approaches to microbial ecology. To be specific, the question is whether one should measure the number of cells of a specific type (the biomass approach) or whether one should instead measure some aspect of the specific activity of interest (the functional approach). To be sure, the division is not so clearly drawn in actual studies, and many investigators will carry out some combination of the two. Nonetheless it is usually true that a given report will be primarily of one type or the other. The question has many ramifications, some going back to the pioneers of microbiology in the 19th century. For example, Bull and Slater (1982) give a comprehensive and highly readable account of the history of using pure cultures in microbiological research. On the one hand, it is not possible to ascribe specific effects to specific bacteria unless pure cultures are used. On the other hand, pure cultures are by definition unnatural populations, and it is quite reasonable to suspect that a cell's metabolism may well be different depending on what neighbors it has at any given time.

The ecological emphasis of the present chapter forces attention on this central dichotomy. Sulfate-reducing bacteria have been defined largely on the basis of their physiology as elucidated in laboratory culture. As a result there have been many investigations in which these bacteria have been enumerated from natural samples by use of specific selective media. Attempts have been made to extrapolate from such population counts to presumed environmental impact, but in general these conversions are fraught with difficulty and should be accompanied by substantial disclaimers (Wright and Coffin, 1984). The basic problem is that there are many gradations of activity so that the enumerated cells might have been in any condition from dormancy to vigorous activity prior to sampling (Paul and Voroney, 1984; Nannipieri, 1984).

Many procedures have been devised over the years to evaluate microorganisms in a variety of natural environments. Staley and Konopka (1985) present an excellent summary of the most commonly used methods, along with objective considerations of the strengths and weaknesses of each. Technical advances in the last 30 years, both in equipment and most importantly in the ready availability of specific radiolabeled compounds, have inexorably shifted attention to the metabolic (chemical) consequences of the bacteria (Craven and Karl, 1984; Kirchman and Hoch, 1988). There are precautions to be observed here as well. Although these in situ metabolic measurements certainly provide the desired quantitative information on specific chemical processes, one does not know for certain which organisms in the complex natural community were actually carrying them out. When the latter approach is followed to an uncritical extreme, we find papers with titles such as "The Ecology of Sulfate Reduction" or "The Ecology of the Sulfur Cycle", which carry an implication that the process is somehow disconnected from actual organisms. This notion of the ecology of disembodied metabolism is as inappropriate as are the occasional discussions of the evolution of genes as though one could follow their changes without reference to the organisms which contain them.

Many authors address this basic dichotomy (Lynch and Hobbie, 1988;

Bull and Slater, 1982; Burns and Slater, 1982; Ellwood et al., 1980), and the appropriate conclusion seems to be as follows. Neither of these approaches is inherently superior to the other, but they provide fundamentally different information. Consequently, they should be used to answer different questions. I note the terms which have already been used: SRA and sulfate-reducing bacteria are distinctly different terms referring to activity and presence, respectively, and their usage will be chosen carefully. As will be developed in detail, the bacterial production of methane is inseparable from a discussion of sulfate reduction. The counterpart terms are MPA, for methane production activity and methane-producing bacteria. To move from the general dichotomy to the specific topics under discussion here, the majority of my presentation will be concerned with activities (SRA and MPA) and not presence (sulfate-reducing and methane-producing) of the bacteria. Note, however, that there is a recent increase in the study of population sizes of sulfate reducers in nature (at least indirectly) through a variety of new molecular techniques (see Chapter 6 for details).

7.3 Balanced Metabolic and Ecological Processes

It is a common practice in papers about microbially mediated mineralization to present diagramatic representations of the cycles of the elements in question as a framework for details about natural processes (Jorgensen, 1980; Postgate, 1984a; Ferguson, 1988; Hobbie and Fletcher, 1988; Stolp, 1988). An important emphasis of the present chapter is that SRA is only one of several microbiological activities operating in concert in natural anaerobic environments. Therefore, cyclic diagrams of individual elements are less important here than is a more general overview of the metabolic relations of the microbial community. Figure 7.1, which is a combination of figures from Capone and Kiene (1988) and Dicker (1983), summarizes this general picture of interdependent metabolism in an anaerobic community. Several of the conversions depicted in the "terminal metabolism" portion of the figure will be considered in more detail below.

7.4 Metabolic Diversity of Sulfate-Reducing Bacteria

The existence of the present volume is evidence that one cannot summarize all the ecologically significant physiological features of sulfate-reducing bacteria in one chapter. Therefore, choices must be made as to which topics to include. The emphasis in my presentation will be on those natural activities of sulfate reducers that have the greatest direct impact on other members of the microbial community. Before consideration of these primary ecological topics, it is useful to note briefly some of the physiolo-

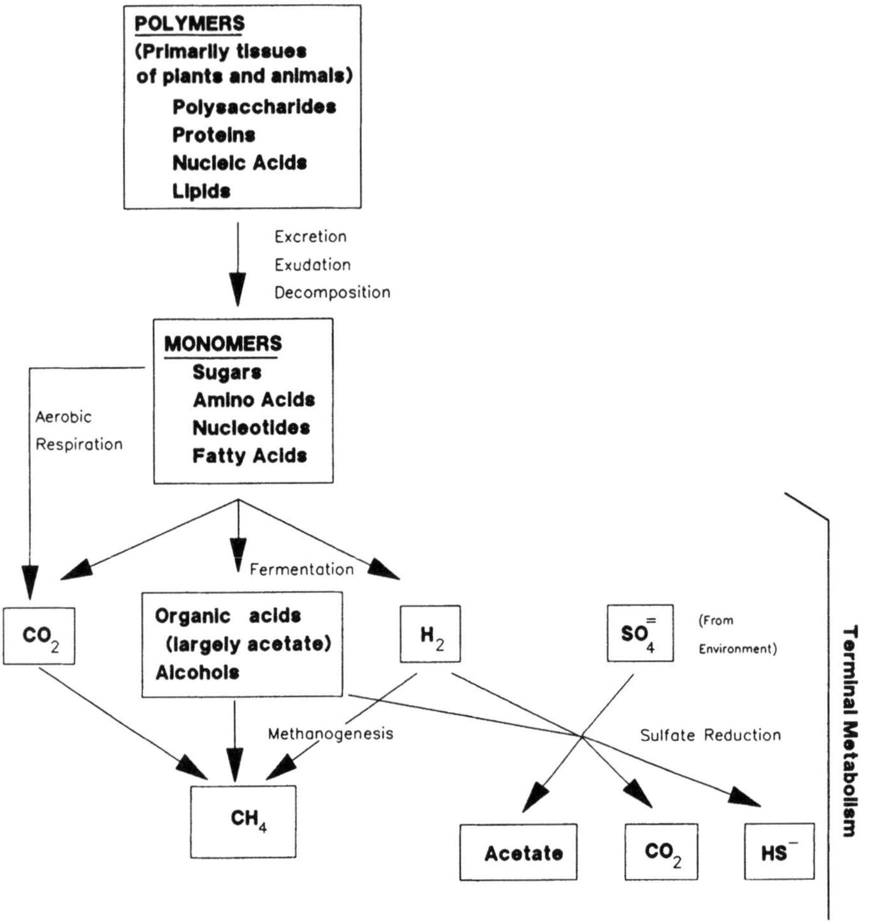

Figure 7.1 Schematic representation of organic transformations by bacteria in an anaerobic environment such as a sediment. Note that all bacterial activities here are associated with the formation of additional bacterial biomass.

gical abilities of sulfate-reducing bacteria that will not receive further detailed attention.

- These bacteria are ubiquitous in terms of environmental adaptation and have been isolated from a wide range of environments, including some classified as extreme (Belkin and Jannasch, 1985; Reysenbach and Deming, 1991).
- They have been shown to be quantitatively significant nitrogen fixers in several areas (Herbert, 1975; Dicker and Smith, 1980b; Lovely and Klug, 1982; Widdel, 1987; Gandy and Yoch, 1988).
- Some strains are apparently capable of carrying out dissimilatory

nitrate reduction in addition to SRA (Seitz and Cypionka, 1986 and references therein).

- After many years of dispute, it is now clear that at least some strains of sulfate-reducing bacteria can grow autotrophically (Brysch et al., 1987; Widdel, 1987; see also Chapter 2 in this volume). At least some of these autotrophs use a reductive TCA cycle, not the Calvin cycle, to incorporate CO_2 (Schauder et al., 1987), although these latter authors do note that acetate is still a preferred carbon source for their strains.
- The metabolism of sulfate reducers has long been associated with environmental metals, mostly by sulfide production, which forms insoluble metal precipitates, making the metal ions unavailable to other organisms (Postgate, 1984a). Among the metals precipitated in this way is iron, which is crucial for synthesis of cytochromes and other electron carriers by sulfate reducers as well as other organisms. Therefore, sulfate-reducing bacteria may actually limit their own metabolism by producing sulfide, which restricts their access to soluble iron (Smith and Oremland, 1982). Oremland et al. (1991) recently reported a new activity for sulfate reducers, namely, the demethylation of the highly toxic methyl mercury.

We therefore see that the list of diverse and significant activities attributable to sulfate-reducing bacteria has become quite impressive in addition to the major emphasis on anaerobic mineralization, which will be presented below.

7.5 Methodology

Measurement of SRA A functional definition of SRA is use of SO_4^{2-} as the terminal electron acceptor in anaerobic respiration (Postgate, 1984a). Even though a variety of compounds, both inorganic and organic, can serve as electron donors for this process, SRA is restricted to a relatively few types of bacteria (Widdel, 1988). Therefore, it is possible to measure the activity in natural samples by focusing on SO_4^{2-} irrespective of what the electron donor may be at any given time. One must remember, however, that the results obtained in situ reflect overall SRA in that sample and are not attributable to individual genera or species of sulfate-reducing bacteria.

The two methodological approaches have been to measure either (1) disappearance of SO_4^{2-} or (2) appearance of sulfide, the reduced product. Of these two, measurement of product formation has become standard, since sulfide may be trapped and analyzed quite specifically, and substrate disappearance always leaves questions about possible alternative fates of the starting SO_4^{2-}. Field investigations have been greatly facilitated by use of $^{35}SO_4^{2-}$, a medium-energy beta emitter which is relatively inexpensive and readily available. The general procedure, which usually has specific modifications for individual studies, involves the following steps:

- collection of sample;
- addition of radiolabeled SO_4^{2-};
- incubation under controlled conditions;
- termination of incubation by poisoning or dilution;
- anaerobic removal and trapping of sulfide;
- radioactivity measurement (usually scintillation counting).

Several research groups have used this general procedure (Skyring et al., 1979; Howarth and Merkel, 1984; Howes et al., 1984; Dicker and Smith, 1985a, 1985b; King, 1988, among others). Recently there has been increased attention to differential extraction procedures to fractionate the various reduced sulfur components from the incubated sediment (Westrich and Berner, 1984; Dicker and Smith, 1985a; Canfield and Des Marais, 1991).

Isotope discrimination effects Use of $^{35}SO_4^{2-}$ presumes that the radiolabeled SO_4^{2-} will be used equally with the isotopically lighter nonradioactive forms of the anion. It has been known for some time that microorganisms metabolize different isotopes of an element at different rates (Kaplan, 1975). What is less often explained is that the degree of fractionation by sulfate-reducing bacteria, in culture at least, is dependent upon the electron donor (Kaplan and Rittenberg, 1964). These authors showed that the magnitude of SO_4^{2-} fractionation depended on the rate at which the reduction was taking place. Since the rates of SRA may vary greatly within the same bacterial community, the potential for deviations due to metabolic fractionation should not be overlooked when making quantitative interpretations about the significance of SRA in any given environment.

Sampling techniques The physical form of the sample examined is a crucial aspect of SRA measurement. As will be developed below, most SRA measurements are made in sediment samples. There are two basic approaches to sampling; either (1) intact cores of sediment are removed and examined or (2) sediment is collected and formed into a slurry with an appropriate solution. Each approach has unique advantages and disadvantages.

Intact core use preserves microenvironments present within the material before sampling. However, it is difficult to add radiolabeled SO_4^{2-} and/or other compounds (alternate nutrients, inhibitors) and obtain uniform distribution throughout the core. The preparation of a slurry solves the distribution problem but destroys all of the native microenvironments originally present. Substantial differences have been noted in parallel core and slurry measurements for SRA and other activities (Jorgensen, 1977b; Smith, 1980; Yavitt and Lang, 1990).

I offer no firm recommendation here as to which is the preferred approach, slurries or cores. My caution is that one should be extremely careful not to make quantitative comparisons between measurements obtained in the two different ways. Additionally, extrapolation from small samples to system-level material budgets and fluxes is seriously compromised by disregard of this precaution (Dicker and Smith, 1980a).

Finally, I note that in the problem of sampling, microbial ecology can benefit greatly from lessons learned in macroecological work. Specifically, Montagna (1982) has put forth statistical considerations in which he stresses that the large variability inherent in sediments must be carefully considered in sampling and experimental design. He describes a hierarchical sampling scheme in which the greatest amount of replication should be done at stages closest to the actual removal of samples from the environment. He notes, however, that in most microbial ecology studies replication is usually concentrated at the other end of the process, with multiple plates or flasks for obtaining pure cultures. These sampling concerns are traditionally dealt with very well in the design of macroecological field studies, and microbial ecologists would be well served to exercise greater care in including them as well.

Use of inhibitors SRA has been measured in samples from many environments to explore the roles of sulfate-reducing bacteria in mineralization. Both quantitative and qualitative questions have been asked, often by using the putatively specific SRA inhibitor, MoO_4^{2-}. Similar questions have been posed about methane-producing bacteria, with bromoethanesulfonic acid (BES) as an inhibitor. Oremland and Capone (1988) have recently reviewed the use of inhibitors of these and other processes in great detail. Their excellent account is recommended for precautions about unexpected side reactions and limitations of interpretation.

7.6 Substrates for SRA in situ

As noted in section 7.2, this chapter will concentrate on metabolic activities of the bacteria in natural environments. An essential part of this discussion is the explicit consideration of both specific electron donors and acceptors that have been found to be important in field studies of SRA. Consequently, following a brief consideration of natural environments where SRA is important, the remaining discussion will be organized around the molecules being metabolized, rather than the environment being studied or the genus and species of organism involved. As a result, the conclusions reached will be more broadly supported and applicable.

Environments used for SRA study There are two broad categories of environment in which SRA has been studied: (1) permanently submerged sediments (either marine or freshwater) or (2) salt marsh sediments, which are intertidal. (Salt marsh materials are variously referred to as soil, sediment, or peat. Sediment will be the preferred term here). These two types of environment do, of course, have significant biological and hydrological differences from each other, but the majority of reports in the last decade or so have not focused on these variations and have instead treated the results as interchangeable. I know of no reports of specific attempts to examine these differences in detail, although it would be welcome if the possibility of systematic variation of SRA rate and character as a function of sediment type were rigorously investigated.

One obvious difference between marsh sediments and permanently submerged sediments is the source of organic material that is degraded. Marsh sediments are extremely productive (Valiella et al., 1976; Schubauer and Hopkinson, 1984) with accumulation of plant material both above and below the marsh surface. Although there are complex organic compounds in this plant tissue, it is degraded essentially where it was formed. On the other hand, submerged sediments generally have very little primary production of their own, although some shallow areas do support rooted vegetation. Therefore, the bulk of organic material initially available to these sediments must be transported there, primarily by settling from the overlying water. During this settling process the material is likely to be ingested, excreted, and otherwise modified by a range of organisms, from fish to invertebrates to fungi and protozoa. Although these metabolic modifications certainly leave substantial amounts of organic material for sediment microorganisms, its character is essentially impossible to determine directly (Yavitt and Lang, 1990; Capone and Kiene, 1988).

Organic electron donors The revolution in perception of sulfate-reducing bacteria was initiated primarily by a series of papers published by Widdel and Pfennig (see Widdel, 1988). These investigators demonstrated that the simple view of two types of sulfate-reducing bacteria limited to a few organic acids or H_2 as possible electron donors was no longer supportable, that in fact a wider range of organic electron donors was usable by the several genera and species they isolated and identified. There are still some metabolic limitations since, for example, polymers such as proteins or carbohydrates or even simple sugars such as glucose do not appear to be generally utilizable. Even though their results were determined with pure cultures, there have been direct consequences to field studies as other research groups have broadened their scope while studying SRA in nature. Recent studies have frequently integrated SRA with other sediment activities in analyses of field results. Five examples of this integration are of special interest:

(1) Westrich and Berner (1984) found biphasic SRA associated with

degradation of carbon from plankton obtained from Long Island Sound. They attributed the biphasic pattern to the presence of two separate categories of organic substrate, one readily degradable and the other more recalcitrant. Interestingly, they found the same biphasic pattern of O_2 respiration as well, indicating that some organic material is being degraded both aerobically and anaerobically.

(2) Crill and Martens (1987) used the marvelous phrase "associative nature of anaerobic organic matter remineralization" to convey the crucial idea that these microbial processes cannot meaningfully be examined in isolation. In a study of sediments off the North Carolina coast, they claim to have demonstrated a bimodal pattern of SRA as measured by $^{35}SO_4^{2-}$ reduction. The first peak occurs at the oxic/anoxic boundary just below the sediment surface (1 cm), and the second is at the interface between high methane concentrations and low sulfate concentrations (10 to 15 cm below the sediment surface). The initial shallow peak was explained as the result of organic material diffusion from the surface and rapid establishment of anaerobic conditions. The lower peak is more problematic, and the authors explain it as a consequence of changes in the identity of the short-chain fatty acids at these depths, changes that could shift the balance between SRA and MPA. Although useful in the future experiments they suggest, these results must be regarded carefully since there is considerable variation in replicate measurements; furthermore, they do not offer rigorous statistical analysis to support the model they use to conclude the existence of the two peaks.

(3) Christensen (1984) measured the accumulation rates of various organic acids in intact sediment cores in the presence of MoO_4^{2-}, the commonly used inhibitor of SRA. The logic of the experiments was that sulfate-reducing bacteria were using fermentation products produced by other members of the community and that inhibition of SRA should cause accumulation of fermentation products, specifically organic acids. The results did indeed show organic accumulation in inhibited samples, with the conclusion that acetate supported an average of 65% of the SRA over all depths, with a high value of 94% in the 0- to 2-cm portion of the core. Similar results were found with slurried sediment samples (Sorensen et al., 1987), and the general correspondence between results with slurries and cores is reassuring (see methodology for comments on the effects of the physical form of samples).

(4) Dicker and Smith (1985a,b) measured the response of SRA to a variety of organic amendments in salt marsh sediment samples. Some putative fermentation products were stimulatory to SRA (lactate, ethanol, methanol, butanol, and formate), others were inhibitory (pyruvate, propionate), and still others (notably acetate) had no effect. This latter result is at first surprising in light of other reports that acetate is the major electron donor for SRA in marine sediments (Christensen, 1984) and that acetate is the major fermentation end product, at least in some freshwater sedi-

ments (Lovely and Klug, 1982). These apparent contradictions may be resolved by noting differences in procedure (Christensen did not directly measure SRA) and that lactate and ethanol were readily metabolized when added to sediments by Lovely and Klug. Furthermore, Dicker and Smith (1985a) conducted substrate competition experiments which indicated that the preferred sequence of oxidation due to SRA is lactate > acetate > ethanol, even though acetate was far more readily oxidized when the three substrates were tested one at a time. Finally, Dicker and Smith (1985a) also found very large and immediate stimulation of SRA by addition of glucose and cellobiose to sediment samples, even though there are no reports that these sugars can be used directly by sulfate-reducing bacteria. The significance of this finding is to reinforce the "associative nature" described by Crill and Martens and to keep in mind that a terminal process such as SRA can show rapid response to organic molecules even if those molecules must be initially metabolized by some intermediate microbial group. In the same vein, the recent report by Haggblom and Young (1990) shows that this tight connection exists to molecules regarded as pollutants of human origin as well, in this case chlorophenol.

(5) Part of the classical definition of sulfate-reducing bacteria was that they performed only partial oxidation, for example, oxidizing lactate to acetate and CO_2 while reducing SO_4^{2-} to sulfide (Postgate, 1984a), of their electron donor. It is, therefore, interesting that Gunnarsson and Ronnow (1982) offer evidence that the in situ metabolism of sulfate-reducing bacteria is determined at least to some extent by ambient SO_4^{2-} concentration. When SO_4^{2-} is abundant, then sulfate reducers oxidize acetate to CO_2; when SO_4^{2-} concentration is low, then they produce acetate from other organic compounds. This observation must be considered in light of the claim by Banat et al. (1981) that there are two different functional groups of sulfate-reducing bacteria in salt marsh sediments. One group preferentially oxidizes acetate, and a second group favors H_2 as an electron donor (see below for more comments about H_2 oxidation). Although it is not possible to distinguish definitively between these two proposals, they remind us of the possible diverse roles for sulfate-reducing bacteria in anaerobic sediments, not only in the reactions they catalyze, but also in the consequences of those reactions. We see here that they may be either the terminal stage in organic oxidation and completely oxidize acetate to CO_2 or serve as intermediates in the mineralization process and release acetate for other bacteria to use.

The previous paragraph referred primarily to studies involving organic electron donors to SRA. There are two categories of inorganic electron donors which are also of great importance for SRA. These are (1) H_2 gas and (2) partially oxidized sulfur compounds.

H_2 as electron donor H_2 has been known as an electron donor for sulfate-reducing bacteria in culture for many years (Postgate, 1984a;

Kaplan and Rittenberg, 1964) and has received increasing attention in situ in the last several years as the focus of competition between sulfate-reducing and methane-producing bacteria. The latter point will be elaborated upon below. A clear conclusion of several studies is that H_2 is metabolized in situ in tightly coupled cycles between producers and consumers and occurs at high rates (Novelli et al., 1987). Consequently, many sediments have very low steady-state H_2 concentrations, and it is imperative for ecological studies that the rate of H_2 turnover be measured, whenever possible, instead of just static H_2 concentration or net H_2 release. It is essential to remember that H_2 in sediments is biogenic and, more specifically, that it arises as a fermentation product during the release of important short-chain fatty acids (Lovely and Klug, 1982, 1986; Oremland, 1988). Also, the amount and nature of organic matter input affects the steady-state concentration of H_2 in sediments (Lovely and Klug, 1982). Conversely, the concentration of H_2 has a strong controlling effect on the proportion of organic acids produced by fermenting organisms (Scranton et al., 1984) owing to thermodynamic constraints associated with H_2 production (Gottschalk, 1986). Therefore, although H_2 is often considered, at least implicitly, as a separate factor in studies of SRA, it is well to remember the functional connections to organic material that enters anaerobic systems.

Sulfite and thiosulfate as electron donors One of the more interesting developments in bacterial physiology in the last decade is the discovery that sulfate-reducing bacteria are capable of novel metabolism of sulfite and thiosulfate (Bak and Cypionka, 1987; Bak and Pfennig, 1987). The net reactions, which are correctly termed disproportionations (see Chapter 1), are as follows:

For sulfite:

$$4SO_3^{2-} + H^+ \rightarrow 3SO_4^{2-} + HS^- \tag{7.1}$$

For thiosulfate:

$$S_2O_4^{2-} + H_2O \rightarrow SO_4^{2-} + HS^- + H^+ \tag{7.2}$$

Many strains of sulfate-reducing bacteria have been found to carry out these reactions (Kramer and Cypionka, 1989). In Chapter 1, Singleton discussed some of the terminology consequences of these unique reactions and agreed with Kelly (1987) that they should both be considered examples of a new category of metabolism, called "inorganic fermentation." These arguments are well founded, and that terminology will be used here.

Two important issues concern the oxidation-reduction nature of these reactions. For sulfite, the disporportionation is clear: three molecules of sulfite are oxidized to sulfate (total of 6 electrons removed) while one molecule of sulfite is reduced to bisulfide (6 electrons added). For thiosul-

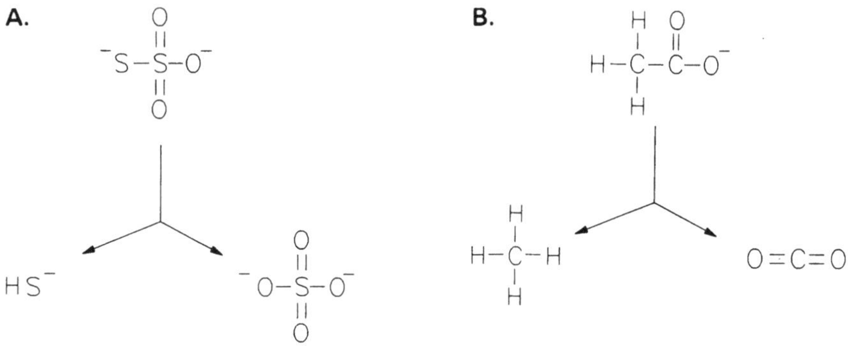

Figure 7.2 Disproportionation of thiosulfate (a) and acetate (b).

fate, the analysis is potentially confusing, owing to a difference in how some investigators view the thiosulfate molecule. I will present two alternative versions and then argue strongly that one of them is clearly to be preferred.

The disagreement arises from oxidation numbers assigned to the two sulfur atoms of thiosulfate (Fig. 7.2a). In the conventional analysis, the outer, or sulfane, sulfur is assigned a value of −1, and the inner, or sulfonate, sulfur is assigned a value of +5 (Ebbing and Wrighton, 1990; Brock and Madigan, 1991). The alternative view is that the sulfane sulfur should be assigned a value of −2 and the sulfonate sulfur a value of +6 (Bak and Cypionka, 1987; Jorgensen, 1990a). The significance of these different assignments is seen when we consider the products of thiosulfate disproportionation, namely, sulfate and bisulfide, which have oxidation numbers of +6 and −2, respectively. In the first view, the disproportionation results in oxidation of the sulfonate sulfur from +5 to +6 (sulfate), coupled to reduction of the sulfane sulfur from −1 to −2 (bisulfide). In the second view, the disproportionation is seen as a simple cleavage of thiosulfate, with no resulting redox change, since the two sulfur atoms of thiosulfate have been assigned oxidation values of the cleavage products.

One important issue that comes from this discrepancy concerns the fundamental notion of biological energy generation. All energy-yielding (ATP-forming) catabolic schemes have at least one redox reaction somewhere in the pathway. By chemical definition (see Chapter 1), disproportionation reactions must be redox reactions. Since there is no dispute about oxidation numbers of the product sulfate and bisulfide, it must therefore be the case that oxidation numbers assigned to thiosulfate should be +5 and −1.

There is an additional argument against assigning oxidation numbers of +6 and −2 to the sulfur atoms of thiosulfate, and that is an argument by analogy. The ability of many methane-producing bacteria to convert acetate to methane and carbon dioxide is a well-known process and is clearly seen as a redox reaction (Thauer et al., 1989). Figure 7.2b presents this reaction in a manner parallel to that of thiosulfate disproportionation (see Thauer et al., 1989 for a summary of the mechanism). Here we see that oxidation of the carboxyl carbon of acetate is coupled to reduction of the methyl carbon, releasing carbon dioxide and methane, respectively. Although there are certainly mechanistic differences between thiosulfate and acetate disproportionation, it is clear that the basic chemical processes in the two reactions are analogous, and bacterial physiologists will be well served to regard them as such.

The objection I have raised is much more than just a narrow interpretation of an arcane point of chemistry. The significance of incorrectly assigning oxidation numbers in this case is that, if +6 and −2 were the correct values, then thiosulfate disproportionation would be placed into a metabolic category all of its own, in which an organism could obtain its energy for growth without using redox reactions. No mechanism by which energy extraction from such a process would be possible was suggested by those who categorize the disproportionation as a non-redox reaction. On the other hand, following the correct chemical convention and assigning values of +5 and −1 allow analysis of thiosulfate disproportionation in the context of other, well-known metabolic processes as an interesting new example of fermentation.

The ecological implications of thiosulfate disproportionation go well beyond these physiological considerations. Jorgensen (1990a,b) has described very nicely a cyclic series of reactions he calls the "thiosulfate shunt" operating in marine and freshwater sediments. The essence of his proposal is that these sediments have high concentrations of sulfide and that the anaerobic oxidation of this sulfide proceeds with thiosulfate as an intermediate. As he reports, thiosulfate is a preferred electron acceptor for sulfate-reducing bacteria (Postgate, 1984a; Widdel, 1988), presumably because, unlike SO_4^{2-}, it does not need activation (as APS) before it can be reduced. Therefore, thiosulfate presence will alter the balance of reactions within the sediment, especially since Jorgensen and Bak (1991) report that all of their sulfate-reducing isolates will reduce thiosulfate. Jorgensen also offers some intriguing arguments that the thiosulfate shunt will account for the different degree of isotope fractionation seen for sulfur compounds in sediments as compared with the pattern observed in pure cultures (Jorgensen, 1990b). As an interesting sidelight, this apparent in situ significance of thiosulfate lends credibility to its long-standing use as an electron donor in the culture of *Thiobacillus* (Smith and Strohl, 1991).

Jorgensen (1990a) notes that at least one of the sulfate-reducing strains that disproportionate thiosulfate is also capable of autotrophic growth. This finding gains significance in light of the proposed role of reduced sulfur compounds in the energy dynamics of highly productive sedimentary bacterial communities (Howarth and Teal, 1980; Peterson et al., 1980). These sulfur compounds can serve as electron donors for lithotrophic autotrophs, such as the aerobic *Thiobacillus*, with the result that there is additional production beyond that due to photosynthetic members of the ecosystem. This production is not accounted for by traditional methods that consider only aerobic respiratory processes. The discovery that secondary autotrophy can also be carried out anaerobically via thiosulfate disproportionation expands the perceived role of sulfur compounds into a larger number of environments. The result is that ecosystem level energy budget calculations must be cognizant of this contribution.

7.7 Interactions with Methane-Producing Bacteria

One of the more lively debates involving sulfate-reducing bacteria over the past 15 years has been their interactions with methane-producing bacteria. There are many physiological and ecological similarities between the two groups of anaerobic bacteria, notably their common presence in a variety of anaerobic environments, particularly sediments. Early reports describe the relation between sulfate-reducing and methane-producing bacteria as mutually exclusive, on the basis of spatial separation between the two groups in sediment (Cappenberg, 1974; Winfrey and Zeikus, 1977; Reeburgh and Heggie, 1977; Abram and Nedwell, 1978a; Hines and Buck, 1982). Possible explanations for this exclusion include metabolic inhibition of MPA by SO_4^{2-} or H_2S (see Figure 7.3) and spatial segregation of methane producers in the center of sulfate-free microenvironments within the sediment. Although there are many reports of zonation within anaerobic environments (Capone and Kiene, 1988), it is now clear that H_2S is not toxic to methane-producing bacteria (Gunnarsson and Ronnow, 1982). Metabolic studies have instead shifted to electron donors each group of bacteria uses, and it has been found that there are two donors which sulfate reducers and methane producers both use readily, namely, acetate and H_2 (Oremland and Policin, 1982). Evidence has been presented to support three general relations between sulfate-reducing and methane-producing bacteria: (1) competition between the two groups for limiting electron donor; (2) coexistence through use of separate resources; or (3) a synergism in which members of one of the two groups supply an electron donor needed by the other.

Competition Most readers will have an intuitive understanding of what is meant by the term "competition". However, closer examination of ecological reports makes it clear that different ecologists vary significantly in their usage of the term and in the types of conclusions they draw about the phenomenon of competition. Microorganisms were used in the classical laboratory demonstrations of competition, but the large majority

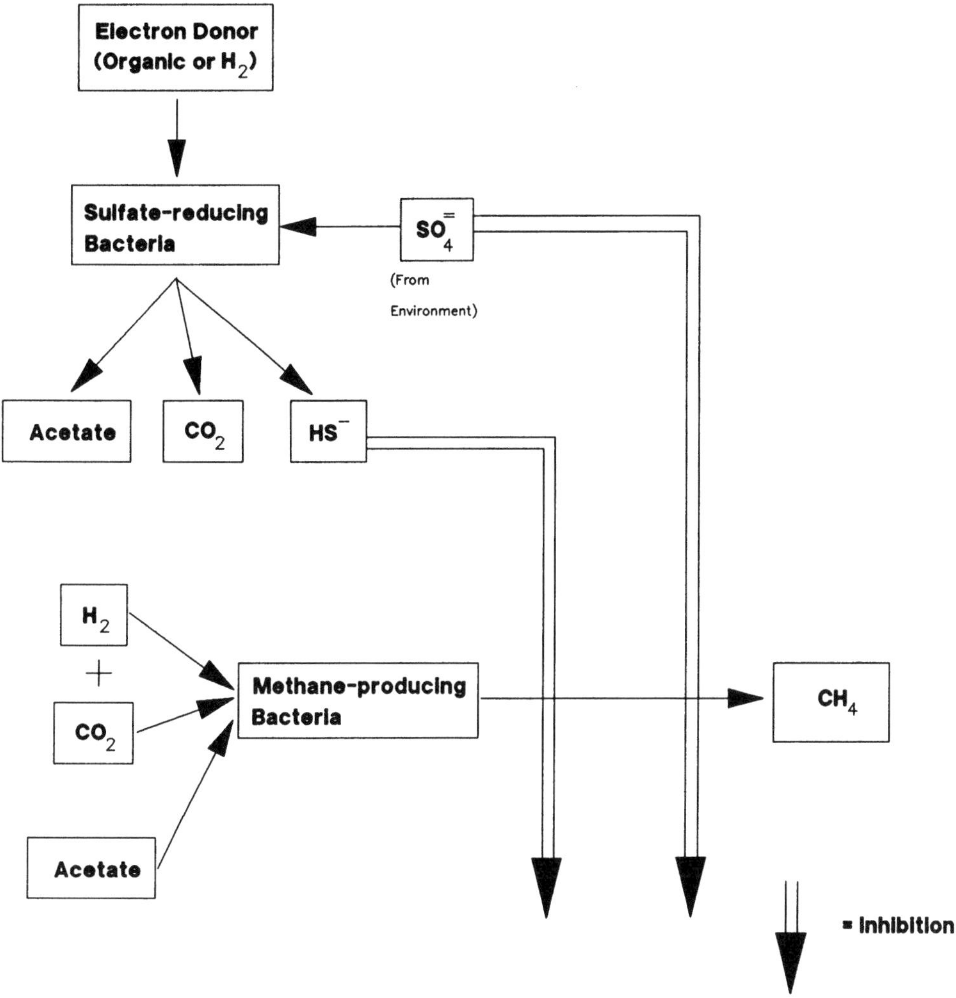

Figure 7.3 Inhibition of methane-producing bacteria by sulfate-reducing bacteria and sulfate.

of field investigations into competition have been done with macroorganisms (see Ricklefs, 1990 for an excellent review of the history of competition as an ecological paradigm). In the present chapter, I will follow the definition presented in Ricklefs' (1990) glossary, with a significant amendment:

Competition [is] the use or defense of a resource by one individual that reduces the availability of that resource to other individuals, whether of the same species (intraspecific competition) or other species (interspecific competition).

The amendment I attach is to make explicit an implicit assumption within Ricklefs' definition, namely, that the resource in question must be growth limiting for the competing organisms. Three important aspects to this definition should be kept in mind as this concept is applied to sulfate-reducing bacteria below.

(1) Several types of resource may be relevant in different locations at different times. Many studies with macroorganisms analyze competition for "food," an imprecise term that encompasses several nutritional components, but which primarily represents an organism's electron donor. The biochemical details of electron processing are almost always too complicated to resolve very well in macroorganisms. However, bacterial metabolism, in general, is much more amenable to precise analysis, especially for the sulfate-reducing bacteria, which have a finite number of biochemical choices for the generation of energy. In keeping with the admonitions about redox balance put forth above, I will explore competition between sulfate-reducing and methane-producing bacteria by explicitly considering both electron donors and electron acceptors separately as resources when possible.

(2) Ricklefs' definition refers to interactions between individual organisms, noting that competitors may be of the same or of different species. Claims about competition involving sulfate-reducing bacteria are exclusively of the interspecific type, primarily with methane-producing bacteria, since, for technical reasons, it would be essentially impossible to detect intraspecific competition on a biochemical level in a natural population of bacteria.

(3) Most importantly, this definition states that the use of a resource by one individual must diminish the availability of that resource to the competitor. Although there is evidence that this criterion is met in some ecosystems, some studies of microbiological competition fail to demonstrate specifically this diminished resource availability.

Competition between macroorganisms is usually measured in terms of the effect on population density (carrying capacity) of competitors at equilibrium. For microbiological studies it is much more common to measure metabolic activities because (1) specific, quantitative enumeration of individual bacterial types in a natural environment is an unrealistic goal (Atlas, 1982; Fry, 1982), and (2) many bacteria may be present in an environment (sediment, for example) but in a metabolically inactive condition, as noted above. The functional groups implied by the designations *sulfate-reducing* and *methane-producing* bacteria fit very well with the traditional physiological approach to bacterial ecology.

Evidence concerning sulfate-reducing/methane-producing competition comes from several reports, most of which rely on using the putatively specific inhibitors discussed above: MoO_4^{2-} to inhibit sulfate reducers, and BES to inhibit methane producers. The most popularly proposed competitive interaction is that sulfate-reducing bacteria are superior

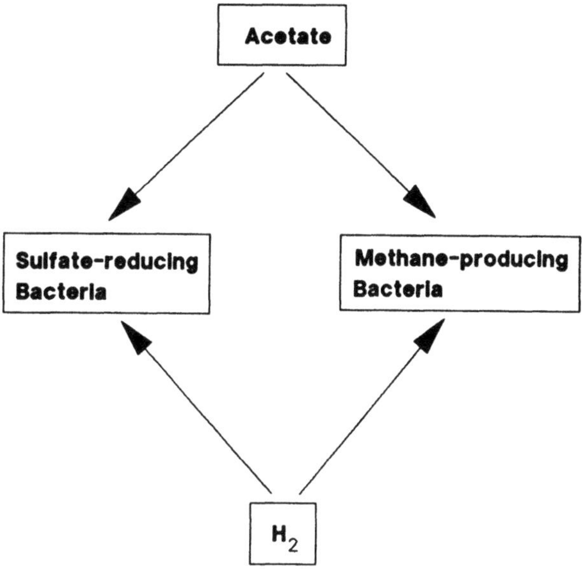

Figure 7.4 Competition between sulfate-reducing and methane-producing bacteria for electron donors.

scavengers of H_2 and acetate (Figure 7.4), which both groups use as an electron donor; this superior resource scavenging limits the opportunities for growth of methane-producing bacteria (Lovely et al., 1982; Robinson and Tiedje, 1984). This assumption has become generally accepted as a starting place for analysis in many ecological studies (Khan and Trottier, 1978; Iversen and Jorgensen, 1985; Yavitt and Lang, 1990, among others). However, it must be remembered that a balanced redox process by sulfate-reducing bacteria, no matter what they use as an electron donor, requires a supply of an electron acceptor, which is classically SO_4^{2-}. The chief competitive result is that sulfate reducers are the primary anaerobic mineralizers in marine and brackish sediments where SO_4^{2-} is abundant, whereas methane producers may account for over 90% of the anaerobic degradation in freshwater sediments that are low in SO_4^{2-} (Abram and Nedwell, 1978b; Senior et al., 1982; Martens and Klump, 1984; Capone and Kiene, 1988; Kuivila et al., 1990).

This general pattern was dissected by Lovely and Klug (1986), who showed that the outcome of competition between sulfate-reducing and methane-producing bacteria for H_2 in lake sediments depends quantitatively on SO_4^{2-} concentration; sulfate reducers get the H_2 when SO_4^{2-} is above 30 μM, and methane producers win at lower concentrations. Working in a different system (peat in freshwater wetlands), Yavitt and Lang (1990) also found a threshold effect of SO_4^{2-} on MPA, but at concentrations that were an order of magnitude higher than those reported by

Lovely and Klug. Specifically, they found no inhibition of MPA by 400 μM SO_4^{2-}, but greater than 90% inhibition by 2000 μM SO_4^{2-} in their slurries, which were incubated for up to 9 days. Inhibition of MPA by SO_4^{2-} in their study was reversible by addition of either acetate or H_2 along with SO_4^{2-}. This relief from inhibition may have come about because electron donors that were previously limiting (acetate and H_2) were added in high enough concentrations that competition ceased. Alternatively, these authors note that reversal of SO_4^{2-} inhibition was usually accompanied by a 2- to 3-day lag period after amendments were made. Their explanation was that added SO_4^{2-} stimulated sulfate-reducing bacteria until SO_4^{2-} was consumed, at which time methane-producing bacteria could oxidize residual acetate or H_2. Unfortunately, SO_4^{2-} concentrations were not measured during the course of the incubations to test this depletion hypothesis.

The final point to be considered about competition is a closer examination of the proposed mechanism for the competitive success of sulfate-reducing bacteria. Capone and Kiene (1988) summarize the results from a number of studies, which show that sulfate reducers have higher substrate affinities (lower K_m) than methane-reducing bacteria for both H_2 and acetate, although differences for H_2 are not pronounced. Lovely et al. (1982) and Lovely and Klug (1986) point out the dangers in extrapolating from such affinity measurements, especially in laboratory cultures, to a presumption of competitive ability in nature. Furthermore, Lovely et al. (1982) point out that the rate at which bacteria actually incorporate a substrate and process it metabolically may not be predicted very well from a simple knowledge of binding efficiencies.

A different line of evidence for a competitive mechanism is to compare theoretical energy yields for oxidation of H_2 and acetate coupled to reduction of either SO_4^{2-} or CO_2 (Capone and Kiene, 1988). The energetic advantage to using SO_4^{2-} as an electron acceptor is 10% for H_2 oxidation

Table 7.1. Calculated energy yields of H_2 and acetate oxidation coupled to either SRA or MPA

	$\Delta G'^{\circ}$ (kJ mole^{-1} of Electron Donor)		
Electron Donor	Sulfate Reduction	Methanogenesis	% Increase
Part A. Reactants under standard conditions (1 M solutes, 1 atm gases)			
H_2	38	34	11.8
Acetate	41	28	46.4
Part B. Reactants at H_2 partial pressure of 10^{-4} atm.			
H_2	151.9	135.6	12.0
Acetate	47.3	31.0	52.6

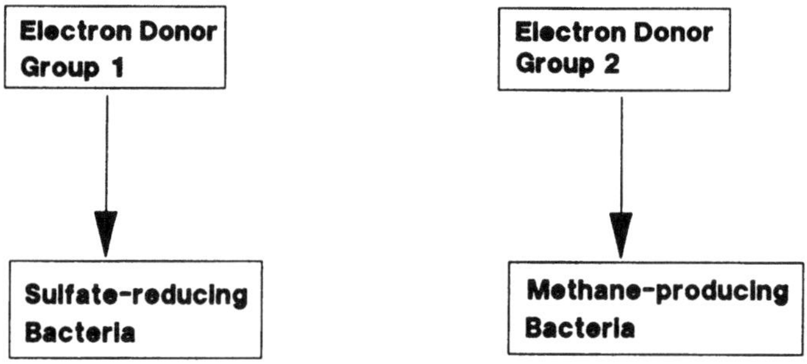

Figure 7.5 Coexistence between sulfate-reducing and methane-producing bacteria through use of separate electron donors.

and 40% for acetate oxidation (Table 1, part A). These authors note the usual precaution, namely, that the calculations were done under assumptions of standard chemical conditions—1 M concentration of soluble substrates and 1 atm partial pressure for gases. These standard conditions are certainly not realized in natural environments, where H_2 concentration is usually on the order of 10^{-4} to 10^{-3} atm (Novelli et al., 1987). Gunnarsson and Ronnow (1982) performed similar calculations, which were adjusted to reflect environmentally realistic substrate concentrations, and found that H_2 oxidation coupled to SO_4^{2-} reduction is still more favorable than CO_2 reduction; the percentages were similar to those obtained under standard conditions (Table 1, part B).

Coexistence An alternative hypothesis to competition is that sulfate-reducing and methane-producing bacteria utilize different electron donors when present in the same microenvironments (Figure 7.5). Oremland and Policin (1982) measured the SO_4^{2-} effect on MPA in response to several amendments. In keeping with the competition results reported above, acetate and H_2 both greatly stimulated MPA in sediment slurries, but the stimulation was substantially less when SO_4^{2-} was added at the same time. On the other hand, MPA in these sediment slurries was also enhanced by addition of methanol, trimethylamine, or methionine, but concomitant SO_4^{2-} addition had no observable effect on this stimulation of MPA. This observation implies that sulfate-reducing bacteria could not use these electron donors. In addition, these authors reported that methanol and trimethylamine had no measurable effect on SRA, further indicating that sulfate-reducing and methane-producing bacteria were coexisting by using separate electron donors. This general proposal of separate substrate use by sulfate reducers and methane producers is partially contradicted by reports that SRA is stimulated by methanol addition (King, 1988; Dicker and Smith, 1985b). A different proposal was put forth by Senior et al. (1982) and Crill and Martens (1984), who reported that low levels of MPA continue even in the presence of active SRA. Apparently, methane-producing bacteria are able to obtain small amounts of H_2, implying that the competitive superiority of sulfate-reducing bacteria is not absolute. Rates of MPA are on the order of 10% of the SRA rates and are not seen as quantitatively significant in most sediments (Capone and Kiene, 1988). Therefore, the majority of MPA in active SRA sediments is attributable to use of substrates such as methylamine, which are unavailable to sulfate-reducing bacteria (Oremland and Policin, 1982; Winfrey and Ward, 1983).

The last, and perhaps simplest, explanation for the coexistence of sulfate-reducing and methane-producing bacteria has been raised by Yavitt and Lang (1990) and Martens and Klump (1984). These investigators noted that in the presence of high concentrations of organic material, either peat or organic sediment, competition between sulfate reducers and methane producers is greatly diminished. This conclusion fits nicely with the amended definition given above, which emphasized that competition can occur only for resources that are limiting for the competitors. The same relief from competition was noted above when the concentration of electron donor, either H_2 or acetate, was experimentally increased; it is reassuringly consistent that the same phenomenon is observed in unamended samples.

Synergism The anaerobic environments of interest to this discussion contain a highly varied mixture of organic compounds. Complete mineralization of this complex starting material requires an association among several different metabolic types of microorganism (Fig. 1). One laboratory observation that has been useful in suggesting possible in situ activities is that some sulfate-reducing bacteria ferment lactate to acetate and H_2 in the absence of SO_4^{2-} (Bryant et al., 1977). Sulfate reducers obtain a small amount of energy from this fermentation, but an additional consequence is that lactate, which cannot be metabolized by methane-producing bacteria, has been transformed into two resources, acetate and H_2, which can be utilized. Gunnarsson and Ronnow (1982) and Mountfort et al. (1980) have observed this metabolic process operating in marine sediments. For a full understanding of this process, it is important to remember that production of H_2 is a thermodynamically unfavorable process unless H_2 is continuously removed. Other organisms, such as a methane producers, serve such a function and remove, via interspecies hydrogen transfer, H_2 as it is produced by sulfate reducers (Bryant et al., 1977; see also Chapter 3). As a result there can be a very close metabolic association between sulfate-reducing and methane-producing bacteria in SO_4^{2-}-depleted zones of a sediment. This association, or consortium, is then capable of significant mineralization that neither member of the group could have carried out alone (see Fig. 7.6).

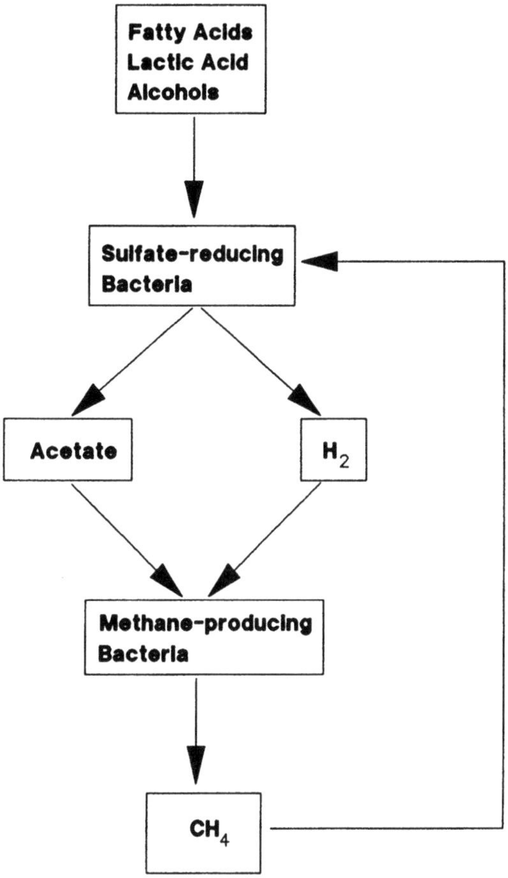

Figure 7.6 Synergistic reactions between sulfate-reducing and methane-producing bacteria.

A similar laboratory observation is the oxidation of ethanol by *Desulfovibrio* in the absence of SO_4^{2-} when cocultured with a methanogen (Kremer et al., 1988a). However, these authors note that this apparent interspecies H_2 transfer did not improve the growth of *Desulfovibrio*, whereas the methanogen was clearly stimulated. In making environmental extrapolations of such observations, however, we should remember to be cautious about using results with laboratory cultures, even cocultures, uncritically as predictive of activities in nature.

A last synergistic exchange between methane-producing and sulfate-reducing bacteria is the proposed methane oxidation by sulfate reducers, which has been reported in certain marine sediments (Reeburgh, 1980; Iversen and Blackburn, 1981; Devol, 1983; Iversen and Jorgensen, 1985). Although the evidence for anaerobic methane oxidation is good, it has not been possible to assign the activity specifically to sulfate-reducing bacteria.

Davis and Yarbrough (1966) reported that sulfate reducers could carry out this oxidation, and others have noted that methane oxidation coupled to SO_4^{2-} reduction is thermodynamically favorable under standard conditions (Howarth, 1984). Nonetheless, it remains true that pure cultures of sulfate-reducing bacteria capable of methane oxidation have not been obtained, despite focused efforts to detect them (Postgate, 1984a; Capone and Kiene, 1988; Smith and Van Ostenbridge, unpublished results). The possibility of direct nutritional transfer between these anaerobes, which are usually thought to be competitors, remains intriguing and is still being pursued (Alperin and Reeburgh, 1985).

7.8 Aerobic SRA

An essential part of the traditional definition of sulfate-reducing bacteria is the unequivocal statement that they are anaerobes (Postgate, 1984a). In the present chapter the ecological considerations have all dealt with anaerobic environments, since there was no reason until fairly recently to suspect that significant amounts of SRA could take place aerobically. Cypionka et al. (1985) reported that eight of their sulfate-reducing bacteria strains could survive aeration for 6 minutes, with four of them viable after 3 hours. Postgate (1984a) has often stressed the need for growth media of sulfate-reducing bacteria to be "poised" with some active oxygen-scavenging agent, often a sulfhydryl compound (see Chapter 1).

Cypionka and coworkers tested several sulfhydryl agents for protection against O_2 exposure and found that these agents actually increased oxygen sensitivity of their cultures; sulfide was the most damaging. These authors proposed that enhanced oxygen sensitivity arose from peroxide and superoxide formation increase by sulfhydryl agents, and that these singlet forms of oxygen caused the actual damage. They also describe O_2-dependent growth in O_2-sulfide gradients, but emphasize that the effect of O_2 was indirect, occurring through a short sulfur cycle in which thiosulfate and elemental sulfur were generated abiologically. These two partially oxidized compounds then served as electron acceptors in place of the usual SO_4^{2-}. The cycle they describe is reminiscent of the thiosulfate shunt proposed by Jorgensen (1990a,b).

Dilling and Cypionka (1990) extended these aerobic experiments and demonstrated that the O_2 consumption they measured immediately upon thiosulfate addition was actually respiration. This conclusion is based on results with several respiratory inhibitors and uncouplers, including a demonstration that addition of CCCP terminates ATP synthesis but stimulates O_2 consumption, as expected for a respiratory uncoupler. These authors note, however, that these bacteria were not able to grow aerobically despite the clear generation of ATP through aerobic respiration of thiosulfate. There must, therefore, be other O_2-sensitive processes within these cells (Dijk et al., 1983). Although the demonstration of aerobic ability in sulfate-reducing bacteria came from laboratory observations, we should be cautious in making ecological inferences based on those observations. Nevertheless, laboratory results do increase the number of possibilities that must be kept in mind when examining activities in nature. One ecological implication of aerobic SRA is that these bacteria might compete for electron donors (thiosulfate, elemental sulfur) with *Thiobacillus* and *Beggiatoa*. One consequence might be diminished secondary autotrophy, since essentially all *Thiobacillus* and *Beggiatoa* species are autotrophs, while autotrophy appears to be narrowly distributed among sulfate-reducing bacteria.

There are two general types of environment in which the O_2 tolerance of SRA has been implicated. The first is at the interface between aerobic and anaerobic (oxic/anoxic) zones in sediments and deep waters. The second is the well-oxygenated portion of cyanobacterial mats.

Aerobic/anaerobic interface (Note: the terms "oxic/anoxic" and "aerobic/anaerobic" are essentially equivalent. There is a tendency to use the former pair to describe the physical environment and the latter pair when speaking of the metabolic characteristics of the organisms present. My usage in each case will reflect that of the authors whose work I am citing).

Hastings and Emerson (1988), studying the permanently anoxic Cariaco Trench basin, measured maximum SRA rates at the oxic/anoxic interface in the water column. They also note that Karl et al. (1977) found maximum values of ATP at this same interface, implying large-scale significance for SRA. The interfacial location of this maximum activity allows sulfate-reducing bacteria access to organic debris from above while avoiding competition from the aerobes, which would almost certainly outcompete the sulfate reducers, in the oxic portion of the water column. These authors also address the traditional notion that sulfate reducers are strict anaerobes and offer two explanations for their survival at the interface. First, they note that it has been established that detritus particles of 100 μm or greater diameter may have reduced microniches inside (Jorgensen, 1977b), thereby allowing sulfate-reducing bacteria to escape from O_2. Such an escape would, of course, have a metabolic price associated with it, namely, that needed nutrients, including SO_4^{2-}, would also be restricted by diffusion into these microniches.

Their second explanation is more conceptual, but quite reasonable. It begins with the observation that the obligately anaerobic character of sulfate reducers is based on laboratory results. It is reasonable to expect that bacterial communities of mixed populations in situ could behave differently. Furthermore, establishment of a pure culture with standard aseptic technique will necessarily select a single genetic type, or clone. If sulfate-reducing bacteria in nature were a mixture of aerobes and anaerobes, then repeated laboratory isolation and transfer always done under anaerobic conditions will select against growth of any aerobic strains that happened to be present at the start.

An important conclusion from these observations is that it is advisable to return frequently to the natural environment to obtain fresh samples of bacteria that have not been separated from their evolutionarily important selective pressures. It is clear that all bacterial cultures undergo genetic change in the laboratory, although perhaps in characteristics not under active observation (Dykhuizen and Hartl, 1983). Therefore, it seems there is a tradeoff between having long-term, well-characterized cultures on the one hand, and recent cultures that probably reflect the natural environment more accurately, on the other hand.

A second important study of the aerobic/anaerobic interface was reported by Jorgensen and Bak (1991). Working in marine sediments, they found that the highest concentrations of sulfate-reducing bacteria (100× higher by an improved most-probable-number technique) were found in the most oxic portions of the sediment. This result is important in supporting the notion that SRA is not a rigidly anaerobic activity. These authors calculate the amount of SO_4^{2-} reduced per sulfate-reducing organism in their samples and concluded that the rates are comparable to those previously measured with pure cultures (Jorgensen, 1978). Given the great variability in bacterial distribution in sediment (see comments from Montagna in Methodology), it is questionable that such a combination of field and laboratory results is very useful. Wiebe (1984) gives similar precautions and stresses the value of the in situ measurements. The fundamental differences between population estimates and activity measurements mentioned earlier in this chapter are also important in interpreting these results.

The well-studied intertidal sediments of salt marshes constitute a third environment with a pronounced aerobic/anaerobic interface. Marshes along the Atlantic Coast of North America are dominated by the cordgrass *Spartina alterniflora* (Loisel) (Howes et al., 1984). There are two growth forms of this grass, termed short form and tall form, which have characteristic distribution patterns. The tall form is found along creek banks in poorly drained sediments, and the short form is found in slightly higher elevations where sediment drainage is substantially greater (Linthurst and Seneca, 1981). One would expect greater O_2 penetration in the better drained short form sediments, and King (1988) measured this expected O_2 profile difference. In the same report King also showed that SRA was higher in the short form sediments. Howarth and Giblin (1983) showed clearly that SRA was highest in sediment samples taken from the surface, in both short and tall form areas. In both of these studies the highest SRA rates were found in samples from the most aerated portions of the sediment. Combining these results with those of Hastings and Emerson (1988) leads to the conclusion that O_2 is probably not the limiting factor in determining the distribution or magnitude of SRA. In all these reports the maximum SRA was associated with areas of greatest availability of electron donor, either by diffusion in the water column or from decomposition of *Spartina* in the salt marsh. These observations lend further weight to my earlier admonition that it is best to consider explicitly the identity and quantity of both the electron donor and acceptor when analyzing terminal mineralization processes.

Cyanobacterial mats These mats are highly diverse miniature ecosystems in which various bacterial populations are in close association with each other. Cohen (1984) reported not only that SRA occurred in the oxic zone of mats, but also that it was in fact stimulated during photosynthesis of cyanobacteria associated with the SRA. Since O_2 concentration must certainly have increased during photosynthesis, it is most reasonable to conclude that stimulation of SRA resulted from an increased supply of electron donor(s) to sulfate-reducing bacteria as release of photosynthate from cyanobacteria. A different mat study (Canfield and Des Marais, 1991) also showed greatest SRA associated with the most oxygenated portion of the mat. These authors state that it was not possible to spatially separate the effects of light and O_2 on SRA. I concur that improved spatial resolution of SRA and O_2 evolution rates would be beneficial, but I note the similarity of these results to those from the marsh and deep sea: availability of electron donor is strongly suggested as the major limiting factor for SRA in several environments.

7.9 Conclusion

It has long been the case that the choice of subject for most microbial ecology studies is determined by the degree of understanding of relevant microbial physiology. During the past 20 years, this connection between field and laboratory has been especially important for studies of sulfate-reducing bacteria and SRA. Two examples stand out. First, there is the laboratory work of Pfennig and Widdel, which revealed that the range of usable electron donors for sulfate reducers was far wider than previously imagined. As a result, field investigations have been modified to take into account the possibility of more complicated metabolic interactions with other organisms. A second example is the work of Cypionka and coworkers, who demonstrated that sulfate-reducing bacteria are not always killed by exposure to oxygen as conventional wisdom has had it. These results have led to studies in oxic marine sediments and cyanobacterial mats which most likely would not have been undertaken without prior laboratory indication that sulfate-reducing bacteria could be expected in such environments.

As Kelly pointed out in 1987 when introducing the articles on thiosulfate disproportionation, sulfur bacteria have on several occasions been at the forefront of studies of new metabolic concepts. As summarized here, our appreciation of the diverse abilities of sulfur bacteria (sulfate reducers in this case) as revealed in the laboratory has led to new ways of approaching field studies as well. Perhaps this successful connection will serve as an effective reminder that ecology and physiology are complementary approaches and that the most useful studies are those which incorporate both explicitly.

[References, see p. 211]

This chapter is contribution number 155 from the Ecology Program at the University of Delaware.

8

Industrial and Environmental Activities of Sulfate-Reducing Bacteria

J.M. Odom

8.1 Introduction

Expansion of the bacterial group loosely termed sulfate reducers from two genera, *Desulfovibrio* and *Desulfotomaculum*, to at least ten new genera (Widdel and Pfennig, 1977, 1981a,b, 1982; Pfennig and Biebl, 1976, 1981; Pfennig et al., 1981) overturned long-established perceptions and practices within the industrial sector concerning the ecology, culture, and taxonomic diversity of this group. Many of the new isolates exibit fastidious nutritional requirements, long generation times, surprising oxygen tolerance, and utilize an expanded list of growth substrates. Consequently, confidence has been seriously eroded in our ability to accurately predict the threat that this diverse group of organisms poses to many industrial processes, and new emphasis has been placed on control and development of methods for determining their presence and type.

Superimposed on the realization that current industrial microbiological practices may deal with only a small aspect of the diversity of sulfate-reducing bacteria is the increasing economic and environmental costs of biogenic sulfide. Public concern with air quality, evidenced by recent stringent regulations governing sulfur emissions from oil refineries and the sulfur content of gaseous fuels, compounds the costs and problems for industry. These trends underscore the need to develop new technologies to measure and control the problem of biogenic sulfide in an environmentally compatible way.

The cost of sulfide pollution is probably greatest for the oil and gas industry and for sewage treatment processes. The spectrum of problems range from contaminated digestor gas, sulfide corrosion of oil drilling equipment, and concrete corrosion due to sulfur-cycle activity, to the plugging of entire rock formations owing to the generation of sulfide precipitates. The two industries employ very different control methods for these problems, reflecting different operational requirements (Chwirka and Satchell, 1990; Dezham et al., 1988; Jarmeel, 1989; Steiner 1983). There is a keen awareness of the problems caused by sulfate-reducing bacteria in these industries, but surprisingly little documentation of the exact extent of the problem. According to the latest National Bureau of Standards study, which is 10 years old, the economic cost of corrosion of metal structures has been estimated at $200 billion a year in the U.S. alone (NBS, 1980). Typically, developed countries are estimated to spend 4% of their gross national product on repair and replacement of corroded metal; however, the fraction of this due to microbially induced or microbially influenced corrosion is impossible to determine.

Over 800,000 miles of concrete sewer conduits are in use in the United States that are subject to sulfide-related corrosion as a result of the concerted activities of sulfate-reducing and sulfide-oxidizing bacteria. Southern coastal cities are particularly prone to these problems because of high average yearly temperatures. Gaseous emissions from anaerobic digestors face more stringent regulations on sulfur content such that secondary treatments are necessary before these emissions can be burned.

The interrelationships between sulfate reducers and heavy metals present in sewers, lakes, and streams are only currently being resolved, but it is already apparent that sulfate reducers play an important role in the mobilization and toxicity of these industrial pollutants.

This chapter will trace the historical association of the oil and sewage treatment industries with the microbiology of sulfate reduction and discuss current perceptions, practices, and future trends dealing with the problem of biogenic sulfide.

8.2 Historical Perspective

Reports appearing in the late 1800s raised questions about whether sulfate reduction in sediments was a biological or chemical process. An early demonstration that sulfate reduction was of biological origin showed that conversion of sulfate to hydrogen sulfide occurred only in the presence of living, as opposed to heat-killed, organic matter (Plauchud, 1877). Hoppe-Seyler (1886) studied the fermentation of a mixture of cellulose and calcium sulfate. He concluded that cellulose breakdown and reduction of sulfate were not independent and that the methane produced would not chemically reduce sulfate to sulfide.

However, the first isolation of a microorganism that could carry out reduction of sulfate to sulfide was reported by Beijerinck (1895). The organism was a strict anaerobe, vibroid in appearance, and originally named *Spirillum desulfuricans*. Subsequently, Van Delden (1903) isolated a sulfate-reducing bacterium from sediments on the Dutch coast. The organism, *Microspira aestuarii*, differed from the Beijerinck isolate because it required a higher salinity for growth. Elion (1924) isolated and characterized the first thermophilic sulfate reducer, *Vibrio thermodesulfuricans*, which had a temperature optimum of 55°C. This observation extended the temperature range and geologic depth at which biological sulfate reduction could occur. These isolates represented the origins of the "classical" sulfate-reducing genera, *Desulfovibrio* and *Desulfotomaculum* (see Chapter 1), which remained the only genera until the 1970s.

The deleterious effects of microbial sulfur-cycle activity on concrete structures is by no means a recent problem. Barr and Buchanan (1912) studied the decomposition of a concrete sewage disposal plant at the Inebriate Hospital in Knoxville, Tennessee. In this work they demonstrated that biological reduction of sulfate to sulfide, rather than release of sulfide from thiol-containing organics, was responsible for high sulfide levels observed in the plant. They isolated a sulfate-reducing organism which was shown to be very similar to Beijerinck's isolate. Olmstead and Hamlin (1900) described in great detail the effects of microbial sulfur-cycle activity on a concrete sewer that received water from a nearby oil field in the Los Angeles county area.

Early observations indicated that water in proximity to oil-bearing formations exhibited a decreased sulfate content and a proportionately higher sulfide level (Meyer,1864; Rogers, 1919; Mills and Wells, 1919; Bastin, 1926). Two processes might have explained this relationship: (1) a chemical process, in which sulfate was reduced by methane or the oil itself; (2) a microbial process which held that sulfate-reducing bacteria were present in the oil formations. The problem for the microbial theory was to explain the presence and metabolic activity of bacteria under hundreds or thousands of feet of sediment.

A number of workers conducted important work in oil field microbiology to validate the theory of microbial action in oil-bearing formations. This included work by Bastin (1926), Gahl and Anderson (1928), Ginter (1930), von Wolzogen Kuhr (1922), and of course ZoBell (1958), who did extensive studies on oil well microbiology. These workers established that sulfate reducers were present in high numbers in produced water fluids from a large percentage of the oil wells examined.

Bastin (1926) explored the chemical and biological theories in a remarkable paper on the subject. Employing a lactate-sulfate medium with high ferrous iron concentrations and rigorous microbiological technique, Bastin sampled the effluent from 67 oil wells in Illinois and California that ranged in depth from 400 to 4000 feet. Sulfate-reducing bacteria were detected in 39 of the samples, evidenced by iron sulfide formation and blackening of the media. [The same basic culture method is recommended today by the American Petroleum Institute (API) for monitoring populations of sulfate-reducing bacteria (API, Recommended Practice #38).] From these data and the lack of convincing data for chemical reduction of sulfates, Bastin conlcuded that sulfate-reducing bacteria were primarily responsible for sulfide production in oil reservoirs and the associated problems of sour oil and down-hole corrosion.

Ginter (1930) suggested the possibility of subterranean bacterial populations in relation to the oil industry and concluded, as had Bastin (1926), that sulfate-reducing bacteria were present in the deep subsurface of the earth. How the organisms came to be there, and what they grew on, was a major stumbling block for proponents of biological sulfate reduction. Beckman (1926) collected soil specimens adjacent to a wrecked oil tanker in San Francisco Bay and showed that microorganisms from these samples could effect changes in density of the oil. This experiment provided evidence that bacteria, generally, could use oil as a source of carbon and energy and may also have been the first report demonstrating the role of bacteria in degrading oil spills. Beckman's report was also indirect evidence that oil could provide a carbon source for growth of sulfate reducers.

ZoBell and Sisler personally supervised an attempt to aseptically drill and sample an oil well to ensure a minimum possibilty of surface contamination. They collected numerous samples that contained sulfate-reducing bacteria, implying that these organisms did indeed exist in deep subsurface environments (Sneddon, 1951; Doig and Wachten, 1951). While there is little doubt today that sulfate reducers can gain access to and flourish in aquifers and oil-bearing sediments, knowledge of the ecological distribution and the nutritional diversity of these organisms has developed only within the last 15 years (Jones et al., 1989). This ecological and nutritional diversity is demonstrated by the isolation of new genera *Desulfobacterium, Desulfobacter, Desulfococcus, Desulfosarcina, Desulfonema, Desulfuromonas,* in large part owing to the work of Pfennig et al. (1981). It has also become apparent that the question of chemical versus biological sulfide generation in subterranean environments is ongoing, with new chemical mechanisms being proposed, as will discussed below.

The influence of sulfate-reducing bacteria on metal corrosion is another controversial area. Gaines (1910) investigated the relatively high sulfur content of metal corrosion products obtained from underground steel conduits. He observed that, whereas the sulfur content of uncorroded metal was only a small fraction of a percent, the tubercles obtained from around corrosion pits contained from 1% to 6% sulfur and a substantial amount of organic matter. Gaines discussed the relative contributions of the oxidative and reductive sides of the microbial sulfur cycle including the role of Beijerinck's *Spirillum desulfuricans* in formation of these high-sulfur tubercles.

To explain microbially induced metal corrosion, Von Wolzogen Kuhr and Van Der Vlugt (1934) proposed what has become a classic scheme termed the cathodic depolarization mechanism. This proposal (discussed in detail below) integrates the hydrogen metabolism of sulfate reducers with the distinctive characteristics of anaerobic corrosion to explain the corrosive effect of these organisms on steel. Hydrogen metabolism and hydrogenase are key elements of this proposal; this observation has, in part, driven research on the physiology and enzymology of hydrogenase to the present day. The possible involvement of other hydrogen-dependent anaerobic respirations, including methanogenesis, in cathodic depolarization reactions was predicted by Bunker (1939) and Hadley (1940) and later demonstrated by Daniels et al. (1987).

In the nearly 100 years since the first isolation of a sulfate-reducing organism, enormous advances have been made in understanding the physiology, hydrogen metabolism, and, more recently, the molecular biology of these organisms. However, the ecology of the different taxonomic types, their involvement in anaerobic corrosion, and our understanding of anaerobic corrosion has, until recently, advanced very little. This situation is due in large part to the inadequacy of classical microbiological methods for in situ identification and monitoring of these organisms. However, current research into molecular biological and immunological approaches to these problems is rapidly changing this picture (see Devereaux and Stahl, Chapter 6, and Voordouw, Chapter 5, this volume).

8.3 The Oil and Gas Industry

For this industry, hydrogen sulfide has traditionally been of concern for three reasons: its role in metal corrosion; the economics of oil recovery and refining; and as a health hazard for oil field personnel. Because of the volumes of oil and gas currently being refined, the ability of refineries to meet sulfur emission regulations is now also a key concern. More stringent regulations by the State of California (Rule 431.1) on the sulfur content of gaseous fuels—including natural gas, landfill and sewage digestor gas, and production and vehicular gaseous emissions—are indicative of future trends in the rest of the United States. These regulations state that after May, 1993 natural gas containing more than 16 ppm of hydrogen sulfide may not be sold in California. However, until that time, sale of natural gas with 80 ppm hydrogen sulfide is acceptable. As concerns about air quality increase, these restrictions may well become limiting factors in the processing of sulfur-containing fuels. Currently, sulfide-laden refinery emissions are treated by the Claus process, which can remove 90% to 95%

of the hydrogen sulfide from a waste gas stream (Dearson, 1973). However, residual sulfide emissions from the Claus process may still exceed regulations. Biological reactors of photosynthetic sulfur bacteria have been described that could function to remove this residual sulfide (Cork, 1982).

Petroleum exploration is the specific area which has historically experienced the most severe problems with sulfide pollution. Metal sulfide precipitates and bacterial biomass generated within oil reservoir formations cause plugging of rock pores within the reservoir and eventual cessation of oil recovery. Sulfide-related stress cracking of metal components involved in surface and downhole drilling operations can lead to equipment failures and costly downtime. The oil industry refers to sulfide-laden and sulfide-free oil as "sour" and "sweet" respectively, and the former is generally worth less than the latter. All of these factors add to the cost of oil in the form of corrosion prevention and maintenance, biocide treatment, and refining as originally sweet reserves go sour (Hamilton, 1985).

Oil souring is usually associated with a practice known as waterflooding, a technique used to maintain pressure within a reservoir. Water from a nearby aquifer, or seawater (in off-shore operations), is injected into the rock formation with surface injection pumps. The resulting water pressure forces an emulsion of oil, water, and gas to the surface at the production well. Surface operations then separate the three fractions, and the recovered water, and sometimes the gas, are reinjected into the ground to sustain pressure. This loop is not completely closed; thus, additional water is required to sustain the pressure. The volume of water involved in the process may be on the order of millions of gallons, and it may take years to make a complete circuit of the reservoir.

Waterflood is a way to collect a large fraction of remaining oil, or oil adsorbed to the rock itself, that would otherwise be unobtainable after inherent loss of reservoir pressure. From the microbiological perspective the danger in this practice is that oil-water emulsions create conditions favorable for growth of hydrocarbon-degrading bacteria, which in turn generate substrates for other bacteria including sulfate reducers. If seawater is injected, then 28 mM sulfate is also available for sulfate respiration by sulfate-reducing bacteria that are certainly present as an inoculum in the seawater. In very deep wells, injected water may provide regional cooling at the injection well site, thereby allowing growth of mesophilic bacteria. Bacterial growth is restricted by both temperature and pressure. When these effects are combined, a "temperature-pressure growth envelope" can be defined. This envelope for sulfate reduction in very deep wells indicates that the process can occur at up to 640 atmospheres at 45°C (Stott and Herbert, 1986; Herbert et al., 1985).

Subsequent to waterflooding, colonization of the rock formation at the injection well site could be followed by secondary growth of thermophilic sulfate reducers in a warmer zone of 60° to 80°C near the production well. The role of thermophiles may be important, since it is unlikely that the mesophilic growth volume of a reservoir (i.e., the capacity of a reservoir to support growth of mesophilic organisms) is sufficient to account for levels of hydrogen sulfide observed in many cases. Case studies have documented the souring of initially sweet oil reservoirs by the practice of waterflooding (Dewar, 1986; Galbraith et al., 1985; Hill et al., 1987; Antloga and Griffen, 1985; Kuznetsova et al., 1962, 1963).

A recent report indicates the existence of a sulfate-reducing organism capable of direct utilization of long-chain hydrocarbons (Aeckersberg et al., 1990). Thus, there may be unrecognized and as yet uncharacterized bacterial populations capable of sulfate reduction and direct utilization of oil. Sulfate reducers may also generate free hydrogen sulfide from the principal organic thiol, dibenzothiophene, present in crude oil, by reductive transformation of this compound to biphenyl and hydrogen sulfide (Kim et al., 1990).

There are alternate hypotheses to the biological mechanism for well souring. One hypothesis involves chemical reactions of oil, gas, sulfate, and the sediments themselves. Thermochemical reduction of sulfate, an early proposal, derives from observations of methane depletion in boundary areas where sulfate-containing waters were in contact with oil formations (Gahl and Anderson, 1928). The supposition here was that natural gas, or methane, and sulfate react chemically to form sulfide and carbon dioxide. However, this process, at least in the laboratory, requires temperatures in excess of 700°C, and even at the substantially increased pressures present at depths of up to 10,000 feet the reaction should not occur below 250°C (Ginter, 1930; Bastin, 1926).

Thermal breakdown of organic sulfur compounds in oil is another possible source of hydrogen sulfide in petroleum. The organic thiol component of petroleum is normally removed via thermal breakdown during refining. However, the possibility of thermal breakdown at ambient oil formation temperatures is again unlikely, and secondly, the amount of organic thiol present in the oil is often insufficient to account for the free sulfide observed (Orr, 1978).

Formation of free sulfide during waterflooding by dissolution of pyrites or metal sulfide minerals present in rock formations is another mechanism that requires some triggering mechanism. Acidification arising from microbial metabolism of hydrocarbons could induce dissolution of mineral sulfides, and a correlation between the mineral sulfide content of a formation and occurrence of souring has been reported (Bourgeois et al., 1979). Alternatively there may be uncharacterized chemistry between some of the myriad injection water additives (biocides, corrosion inhibitors, etc.) and the mineral sulfides resulting in release of hydrogen sulfide (Marsland et al., 1989).

In summary, the early days of oil exploration were marked by interest in and controversy over the origin and mechanism of reservoir souring. No single mechanism or theory is adequate to explain all cases of reservoir contamination. The biological mechanism involving sulfate-reducing bacteria is probably a reasonable one to account for the gradual onset of souring in fields under waterflood, particularly waterflood with seawater. It should also be noted that many fields under waterflood operate for years without souring or antimicrobial treatment of any kind. A lack of good correlative data which show a direct cause-and-effect relationship and the difficulty in simulating the phenomenon under controlled conditions contribute to the continuing controversy (Marsland et al., 1989).

8.4 Detection and Identification of Sulfate-Reducing Bacteria

Standard industrial practices As noted in Chapter 1, a clear conclusion of recent studies of sulfate-reducing bacteria is that they are an extremely diverse group of organisms with sulfate respiration as the only common trait among them. Despite this diversity, *Desulfovibrio* species are the most frequently isolated organisms in most lactate-sulfate enrichments. Consequently, lactate-sulfate-based culture detection methods have become suspect, and a renewed interest in developing better detection methods has grown. The traditional method, issued by the American Petroleum Institute, defines media, media preparation, and culturing of oil field samples for detection of sulfate-reducing bacteria (API, 1965).

This procedure (known as "RP-38 practice") has been the industry standard for many years and is heavily relied upon by oil field operators today. The method employs a serial extinction dilution technique, with a standardized medium, and essentially provides an order of magnitude estimate of the numbers of bacteria present. The medium contains sodium lactate as carbon and electron source for sulfate reduction. A small iron nail is added as an iron source and to react with sulfide, thereby producing black iron sulfide as a visual growth indicator. The number of blackened bottles in a serial dilution assay is indicative of the order of magnitude of bacteria present.

Although "RP-38 practice" is established standard industrial procedure, the method suffers from a number of disadvantages. Not the least of these are (1) long incubation periods, on the order of weeks, and (2) the standard lactate-based medium may not be nutritionally suited to many species of sulfate reducers. These drawbacks limit the usefulness of culture methods as dose indicators for real-time determinations of biocide effectiveness in many applications. In response to these deficiencies, industrial and academic groups have explored alternative methods for direct assessment of the types and numbers of sulfate-reducing bacteria potentially present in a sample. New strategies have centered on non-culture-dependent assays for intracellular or extracellular biological markers specific for sulfate-reducing bacteria.

Taxonomic markers Identification of the taxonomic types of sulfate reducers present at a particular site is important to develop an appropriate detection method. However, the existing basis for classification of sulfate reducers, based on morphological and nutritional characteristics, is tentative at best (Postgate, 1984a). More recently the phylogeny of this group has been investigated on the basis of 16S and 23S rRNA homologies, wherein diversity was found to be greater than expected, suggesting a reassignment of a number of strains formally classified as *Desulfovibrio* (Devereaux et al., 1990; also, this volume, Chapter 6). This approach also suggests that sulfate reducers can be viewed as two large, distantly related, groups: *Desulfovibrio*-like organisms are one group, and the remaining sulfate-reducing genera the other. Thus, there are inherent deep ancestral divisions in the group known as sulfate-reducing bacteria which make it difficult to identify a universal marker for these organisms. Historically, the presence of the pigment desulfoviridin was considered a taxonomic trait of *Desulfovibrio* sp. (Postgate, 1984a). However, it is now clear that desulfoviridin may not be present in many sulfate reducers including some *Desulfovibrio* species (see Chapter 3). The search for marker proteins of taxonomic significance has revealed a number of species-specific or genus-specific markers, but discovery of a reliable universal marker protein for sulfate-reducing bacteria, as a whole, has been elusive.

Genetic approaches and chemical markers The majority of sulfate-reducing isolates are known to contain a very active hydrogenase. *Desulfovibrio* are now known to contain one or more of three different hydrogenases depending on the particular species (see Chapters 3 and 5), and thus this variation in gene pattern may be a good target for identifying the type of sulfate reducer present. Voordouw et al. (1990a) have explored this approach in *Desulfovibrio* and shown that specific probes to these hydrogenase genes could be used to identify subgroups of *Desulfovibrio* species in environmental samples. However, these probes also failed to react with a number of environmental samples that contained sulfate-reducing bacteria.

Lipid fatty acid composition profiles, as species-specific biomarkers, have been investigated in enrichments and pure cultures. Taylor and Parkes (1985) were able to discern acetate-utilizing sulfate reducers from hydrogen utlizers based on fatty acid profiles. Similarly, Edlund et al. (1985) detected characteristic fatty acid profiles for hydrogen-utilizing *Desulfovibrio* sp. This approach can be useful, particularly when a strain may be detected by a unique or distinctive fatty acid. However, fatty acid compositions may also be significantly affected by nutritional factors, thereby making it difficult to extrapolate identification of environmental samples from laboratory cultures profiles (Edlund et al., 1985).

Immunological approaches to detection Immunoassay of cell-surface antigens or of intracellular proteins of taxonomic significance has been studied by several groups. Norqvist and Roffey (1985) examined cell surface antigens in *Desulfovibrio* and *Desulfotomaculum* and found genus-specific antigens to *Desulfovibrio*, but species-specific antigens to *Desulfotomaculum*. Similarly, Smith (1982) used an indirect fluorescent antibody approach to screen 44 *Desulfovibrio* and *Desulfotomaculum* and found a high degree of species specificity. Singleton et al. (1984) examined immunolgical cross-reactivities of antibodies to cytochrome c_3 and found a considerable variation in cross-reactivitiy even within the genus *Desulfovibrio*, although there was some degree of clustering of cross-reactivities. In general, the problem of species specificity has limited the usefulness of extracellular antigenic markers.

APS reductase is a cytoplasmic protein (Odom and Peck, 1981b) obligatory for respiratory sulfate reduction, but which should not be found in other bacterial types with the exception of sulfide-oxidizing bacteria. Odom et al. (1991) investigated the immunological diversity of adenosine 5′-phosphosulfate (APS) reductases in *Desulfovibrio*, *Desulfotomaculum*, *Desulfobulbus*, and *Desulfosarcina* as well as colorless and photosynthetic sulfide-oxidizers. APS reductases from sulfate reducers were found to be immunologically distinct from sulfide oxidizers, with cross-reactivities strongest within the genus *Desulfovibrio*. Cross-reactivities of *Desulfobulbus*, *Desulfosarcina*, and *Desulfotomaculum* were much lower than those of *Desulfovibrio* sp. but were clearly distinguishable from non-sulfate-reducing bacteria. Interestingly, monoclonal antibodies were found which cross-reacted strongly with all APS reductases including the enzyme from *Thiobacillus denitrificans*, suggesting that some common epitopes are present in both *Desulfovibrio* and *Thiobacillus*.

The work on extracellular markers demonstrates that immunological approaches, as general detection methods, are limited by species specificity. The lack of strong cross-reactivity across all the known genera of sulfate reducers observed with APS reductase probably reflects the deep ancestral division between *Desulfovibrio* and other sulfate-reducing genera (Devereaux et al., 1990).

8.5 Corrosion of Iron and Steel: Mechanisms

General considerations Anaerobic corrosion of iron or steel is a physsically distinctive, poorly understood process that may result from concerted chemical, biological, and electrochemical activities (Iverson, 1981; Tiller, 1982). Our poor understanding of anaerobic corrosion is evident from the general use of terms such as "microbially associated" or "microbially influenced" corrosion to describe the microbiological component. These terms express the possibility that some microbial activity observed at corroded sites on metal surfaces may not result from bacterial growth on metal, but rather that chemical or electrochemical attack on the metal may provide a favorable niche for bacteria to grow. The phenomenon is difficult to duplicate in the laboratory, and direct cause-and-effect relationships of field studies are tenuous. Rates of corrosion observed in laboratory studies can rarely account for rates observed in many field studies where highly corrosive oxidation products of hydrogen sulfide, such as elemental sulfur or sulfuric acid, may also contribute to corrosion (Schaschl, 1980). A broader view of microbially associated corrosion has now evolved to include biofilms of many other physiological types such as organic acid-producing bacteria and the concerted effect of oxygen depletion zones (Hamilton, 1985; Costerton and Lappin-Scott, 1989).

Sulfate-reducing bacteria, however, have a long-standing association with metal corrosion dating back to the report by Gaines (1910) and remain primary suspects in most types of anaerobic corrosion. It is unequivocally established that exposure of mild or stainless steels to pure cultures of sulfate-reducing bacteria under strictly anaerobic conditions induces a characteristic corrosion dependent upon the type of metal (Figure 8.1). Whereas stainless steels are subject to a rapid and unpredictable pitting type of corrosion, which often leads to complete and localized metal penetration, mild steels exhibit a more generalized type of corrosion characterized by end-grain graphitization, leaving only a carbon skeleton (Weimer et al., 1988; Figure 8.2).

Elemental analysis of corrosion deposits from mild or stainless steels reveals the presence of characteristic ferrous sulfide minerals such as mackinawite (Hardy and Brown, 1984; Weimer et al., 1988). However, the organic content of these deposits may be quite low, indicating they are not masses of bacterial cells but rather are primarily mineral in composition (Weimer et al., 1988).

Mechanisms involving sulfate-reducing bacteria The mechanism whereby sulfate-reducing bacteria induce corrosion has been controversial. The controversy has centered on the question of whether active bacterial metabolism or their subsequent metabolites and oxidation products are the principal causative agents. The most inspired idea on the mechanism of sulfate-reducing bacteria-induced corrosion is termed the cathodic depolarization hypothesis and was first put forth in 1934 by Von Wolzogen Kuhr and Van der Vlugt. This mechanism, essentially a microbially driven electrochemical process, is schematically shown in Figure 8.3. On a clean metal surface, atomic hydrogen spontaneously forms on the iron surface. In the absence of hydrogen removal, an equilibrium is established, which results in polarization of anodic and cathodic sites on the metal surface. Cathodic depolarization entails removal of this polarizing

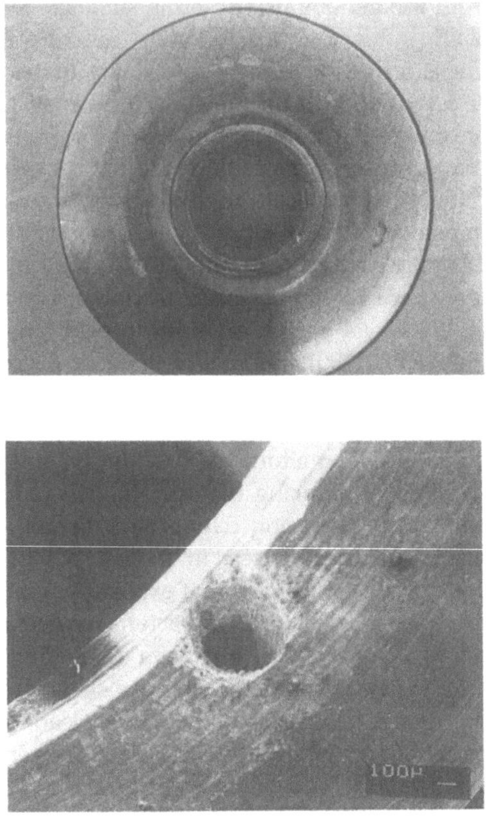

Figure 8.1 Pitting corrosion of stainless steel flange exposed to a pure chemostat culture of *Desulfovibrio desulfuricans* G100A for a period of several weeks. **Top**, macroscopic view of the flange showing pits around the inner ring and shallower pits elsewhere on the surface; **bottom**, closeup of one of the deep pits. (Courtesy of Paul Weimer.)

atomic hydrogen layer. Speculation evolves around the role of sulfate-reducing bacterial action in removing this layer.

The hydrogenase system of the sulfate reducer can utilize hydrogen to reduce sulfate to sulfide. This action depolarizes the cathodic sites and establishes a microbially driven electrolytic cell. In this scheme, electrochemical activity would be facilitated at both the cathode and the anode, i.e., cathodic hydrogen consumption and chemical reaction of sulfide with iron liberated at the anode. This mechanism requires an electron acceptor, perhaps contact between the sulfate reducer and the metal surface, and spontaneous conversion of atomic hydrogen to molecular hydrogen at the metal surface (Booth and Tiller, 1960; Costello, 1974; Hardy, 1983).

It has been recognized that ferrous sulfide itself can chemically depolarize iron or steel in the absence of bacteria (Costello, 1974; King and Miller, 1971). Ferrous sulfide does not behave like a catalytic cathode; rather, depolarization is proportional to the amount of ferrous sulfide added to the metal. Thus, there are two independent components to be

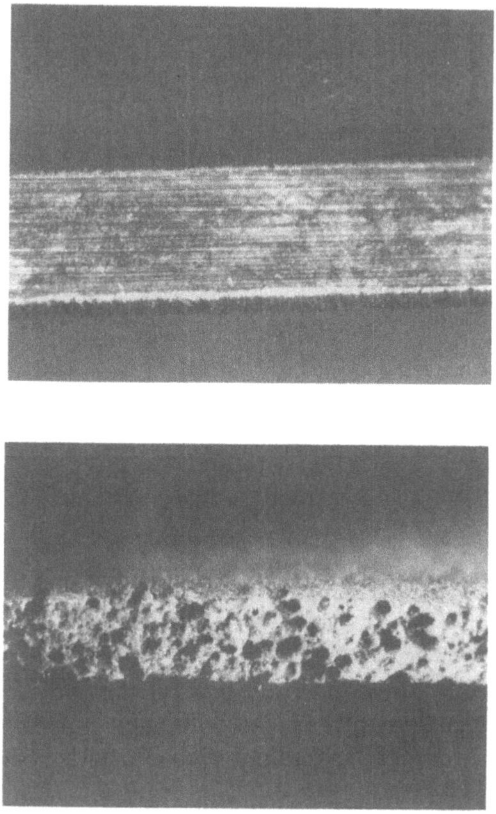

Figure 8.2 End-grain corrosion of mild 1020 carbon steel plates subjected to a pure culture of *Desulfovibrio desulfuricans* G100A for a period of several weeks. **Top**, edge of the plate before exposure; **Bottom**, after exposure. (Courtesy of Paul Weimer.)

considered in polarization studies involving sulfide production by sulfate-reducing bacteria: (1) classical cathodic depolarization involving biological hydrogen removal by bacterial hydrogenase, and (2) chemical depolarization due to formation of a ferrous sulfide cathode.

Indirect evidence for the cathodic depolarization mechanism would be to demonstrate sulfate respiration with iron as sole electron donor. Evidence of this kind has come from experiments showing the generation of small amounts of ^{35}S-sulfide from ^{35}S-sulfate with steel as electron donor and cathodic depolarization by cultures of sulfate-reducing bacteria (Hardy, 1983). Hardy also concluded, on the basis of the very low level of sulfate respiration observed, that growth and corrosion due to removal of cathodic hydrogen are a minor contributor to the corrosion mechanism. Hydrogenase positive *Desulfovibrio* strains have been shown to produce more sulfide from steel wool than did the hydrogenase-negative strain *Dv. sapovorans* (Cord-Ruwisch and Widdel, 1986). However, all strains required lactate as a supplemental electron donor in order to observe the

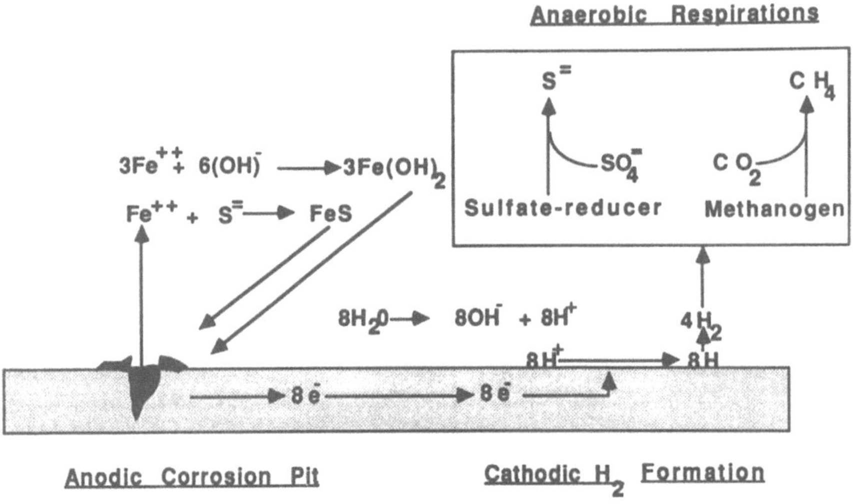

Figure 8.3 Generalized scheme for cathodic depolarization by methanogenic or sulfate-reducing bacteria.

effect. These results, on the whole, show that iron or steel can provide some reducing equivalents in the form of cathodic hydrogen, but that spontaneously formed cathodic hydrogen from iron is probably not a very favorable substrate for sulfate respiration.

Experiments with cultures of *Dv. desulfuricans* and *Dtm. orientis* (Booth and Tiller, 1960) provided direct evidence for the cathodic depolarization mechanism. This work showed that either organism could depolarize anodic sites via chemical reaction of iron and sulfide. Only *Dv. desulfuricans*, however, induced cathodic depolarization; this observation could be rationalized by the greater hydrogenase activity of the *Desulfovibrio* strain. However, this interpretation is complicated by the more recent finding of hydrogen-dependent growth in *Dtm. orientis* (Klemps et al., 1985). Additional support for the role of hydrogenase in cathodic depolarization reaction comes from experiments with cell-free hydrogenase preparations which couple reduction of viologen dyes to oxidation of iron (Bryant and Laishley, 1990; Chatelus et al., 1987).

Many sulfate reducers can also produce hydrogen in the absence of an electron acceptor, although accumulation of high partial pressures is unlikely (Pankhania et al., 1988). Metal embrittlement due to bacterial production of hydrogen and metallic absorption of hydrogen is another biological corrosion mechanism. In this scheme, partial pressures of hydrogen generated within microbial biofilms may be sufficient to weaken the metal (Ford and Mitchell, 1990b; Parker et al., 1988; Walch et al., 1989).

Phosphate was first implicated in corrosion of mild steel by pure cultures of sulfate-reducing bacteria by Iverson (1968, 1981; Iverson and Olson, 1984). The corrosive species was postulated to be a reduced phosphorous compound, possibly a phosphide; however, definitive evidence for a corrosive phosphorus-containing metabolite has been elusive. In a rigorous study of factors affecting corrosion in two-stage chemostat studies with *Dv. desulfuricans*, phosphate content in the medium clearly increased corrosion. Furthermore, actively growing cells were more corrosive than were resting cells (Weimer et al., 1988). The corrosion products contained mackinawite and up to 7% by weight of phosphorus in an unidentified form. These results tend to support a role for phosphorus in the corrosion mechanism. Bryant and Laishley et al. (1990) also demonstrated a strong enhancement by phosphate on cathodic depolarization of steel by cell-free hydrogenase.

Cathodic depolarization and methanogenic bacteria Thermodynamic considerations suggest that sulfate is not the only electron acceptor that could function in a cathodic depolarization scheme. In fact, a role for methanogenic bacteria in anaerobic corrosion was predicted by Bunker (1939). Methanogenic bacteria have recently been shown to couple oxidation of cathodic hydrogen from iron with reduction of carbon dioxide as electron acceptor to form methane (Daniels et al., 1987; Deckena and Blotevogel, 1990; Figure 8.3). This observation confirms the cathodic depolarization mechanism, extends the model to microorganisms other than sulfate-reducing bacteria, and raises the possibility that methanogens could be important in anaerobic corrosion. Bak et al. (1990) reported the frequent occurrence of methanogenic and homoacetogenic bacteria in heavily corroded pipes used for transport of crude oil and concluded that other microbial types besides sulfate-reducing bacteria may be important in anaerobic corrosion.

The available data suggest that biological cathodic depolarization reactions can clearly play a role in anaerobic corrosion. Furthermore, carbon dioxide, as well as sulfate, may be important electron acceptors. However, the quantitative contribution of this mechanism may be minor in comparison with the purely chemical reactions involving sulfide and sulfide oxidation products.

8.6 Anaerobic Corrosion and Sulfate Reducers: Commercial Cures

Irrespective of the role of sulfate-reducing bacteria in metal corrosion, the practice known as "cathodic protection" is probably the best insurance for metal structures submerged in an anaerobic environment. This practice consists of placing a negative potential of around −850 mV (relative to the copper/copper sulfate half cell) on the metal structure. The amount of current required for protection depends on the environment (corrosion rate) and may be quite substantial on offshore drilling platforms (Massad and Holmes, 1958). Maxwell (1986) showed that steel coupons subjected to −950 mV were less prone to colonization by sulfate reducers, although the effect was thought to be indirect rather than a direct electrochemical inhibition. Regardless of the precise mechanism of action of cathodic protection, where applicable, it can be used with great success against anaerobic electrochemical attack.

Cathodic protection is not always feasible to use; thus, there are two industrial areas where biocide treatments against sulfate reducers are appropriate. Biocides are generally toxic, broad-spectrum antimicrobials used in the range of ten to hundreds of parts per million. They are suited for use in situations where equipment and processes handle large volumes of water subject to extensive microbial growth. The first occurs primarily in oil drilling operations (i.e., waterfloods) or industrial cooling towers where cost effectiveness, rather than specificity, absolute effectiveness, or environmental compatibility, is the important factor. The market for biocides in this sector is at least $200 million in the United States alone (Steiner, 1983).

A second necessity for biocide use occurs in the sewage treatment industry. Here serious sulfide pollution problems within sewage conduits and anaerobic digestors require treatments that are specific and compatible with environmental regulations. At present, however, none of the currently available commercial biocides are truly specific for sulfate reducers, but many have demonstrated effectiveness against these organisms. Complexing agents, such as ferric chloride, are currently used to physically trap sulfide; however, large quantities of this highly corrosive chemical are required for stoichiometric sulfide removal.

Biocides may be organic or inorganic and fall generally into the following categories: halogenated organics, quarternary detergents, fatty acids, aldehydes, inorganic oxidizers, and organosulfur compounds. Most are sold as corrosion inhibitors, rather than microbial inhibitors, though data for their effectiveness in either role may be scant. Nitrogen- or sulfur-containing organics are some of the more effective industrial biocides against sulfate reducers. These compounds include the isothiazolines, thiocyanates, and 2,2-dibromo-3-nitrilopropionamide. All are effective against sulfate reducers at concentrations below 30 ppm. Glutaraldehyde, formaldehyde, and acrolein are widely used at concentrations of several hundred ppm to treat the aqueous phase in waterflood projects for sulfide generation; they are inexpensive and good general microbiocides. Quarternary amines are a diverse group of compounds that contain a quarternary nitrogen and a long-chain alkyl or aromatic substituent. These compounds have detergent properties and are generally less hazardous than many industrial microbiocides. Halogenated compounds such as chlorhexidine (a biguanide also known as Hibitane®) are widely used antimicrobial compounds that appear to be particularly effective against sulfate-reducing bacteria. Chlorhexidine is nonspecific and appears to function by disrupting the cell membrane (Davies et al., 1954; Quesnel et al., 1978). Inorganic oxidizing agents, such as liquid chlorine, chlorine dioxide, hypochlorites, and chloroisocyanurates, are all extremely hazardous, but are cost effective and have found widespread use in combating down-hole sulfide generation. Where heavy sulfide contamination is a problem, however, oxidizing agents may be inactivated by sulfide, thereby necessitating excessively high doses.

Classical inhibitors of sulfate reduction, such as molybdate and fluorophosphate, are analogs of sulfate and have been shown to interfere with the primary enzymatic step in activation of sulfate, i.e., the ATP sulfurylase reaction (Taylor and Oremland, 1979). These compounds are impractical as commercial biocides but have found use as research tools because of their reputed specificity for sulfate reduction.

8.7 Environmental Activities: The Sulfur Cycle

The microbial sulfur cycle is most active in intertidal saltmarsh zones where a combination of sulfate, light, the anaerobic-aerobic interface, and abundant organic matter drive cyclical conversions of sulfur from sulfide (−2) to sulfate (+6) with thiosulfate and elemental sulfur as ecologically significant intermediates (Jorgensen, 1977a, 1982). Physiologically, sulfate reduction is a strictly anaerobic process, since most of the redox proteins of sulfate-reducing bacteria are readily oxidized by molecular oxygen. For example, cytochrome c_3, which is thought to be the first electron carrier after hydrogenase, is autooxidized by molecular oxygen. Autooxidation of this protein would divert reducing equivalents away from phosphorylative electron transfer (see Chapter 3).

On the macroscopic scale, sulfate reduction can occur in anaerobic microniches within partially aerated environments (Jorgensen, 1977b). The ability of sulfate reducers to flourish under apparently adverse conditions may be ascribed to their general ability to create anaerobic microniches suitable for growth via the chemical reaction of sulfide and oxygen (Jorgensen 1977b). These situations can occur in soils, biofilms, or small particles where aerobic microbial growth and sulfide production create conditions suitable for establishment of a sulfur cycle sustaining the energy requirements of sulfate reducers as well as sulfide-oxidizing bacteria.

Oxygen tolerance by sulfate-reducing bacteria may be due to the chemically protective effect of sulfide, as well as an innate sensitivity of a particular strain to oxygen. However, these organisms can also take advantage of partially oxidized sulfur compounds present at aerobic-anaerobic interfaces. Cypionka et al. (1985) found that all species of *Desulfovibrio*, *Desulfotomaculum*, *Desulfococcus*, and *Desulfobacter* tested withstood oxygen exposure for 6 minutes in the absence of sulfide without loss of viability. Four species of sulfate-reducing bacteria were found to withstand aeration for up to 3 hours, in the absence of sulfide, without impaired viability. An oxygen-dependent growth mode was also described. Growth resulted from the ability to use reaction products of oxygen and sulfide, such as sulfur and thiosulfate as electron acceptors, in the absence of sulfate. Spore formation is another mechanism by which certain sulfate reducers such as *Desulfotomaculum* sp. may survive air exposure (Widdel, 1986). Many species of *Enterobacteria* can reduce thiosulfate or sulfur to hydrogen sulfide and thus may also participate in an anaerobic-aerobic sulfur cycle (Barrett and Clark, 1987).

Water-logged rice paddies are a prime example of an aerobic/anaerobic interface where limited sulfur cycling occurs with significant economic consequences for rice cultivation. Sulfate reducers, at 10,000 to 100,000 cells per gram of dry soil, have been reported, and these organisms may contribute to the nitrogen-fixing capacity of the soil (Durbin and Watanabe, 1980). Excessive sulfide accumulation, however, is injurious to the rice plant and can cause significant crop loss (Vamos, 1958; Wakao and Furusaka, 1976). Beneficial effects of sulfate reducers on rice cultivation have also been noted and arise from sulfide toxicity toward nematodes which infest the rice plant roots. Nematodes are apparently more sensitive to sulfide than the plant itself and thus may be selectively killed (Jacq and Fortuner, 1979). One testimony to the ability of sulfate reducers to render an aerobic environment anaerobic is the observation that application of sulfur-containing fertilizers to certain turfgrasses results in soil blackening, death of the grass, and an expanding zone of turf decline (even the putting greens are vulnerable! Berndt and Vargas, 1987).

8.8 Sulfur Cycling and Sewage Treatment

The esthetic and human health problem of obnoxious and potentially dangerous gases emanating from sewage conduits is one aspect of the sulfur cycle that most of us have noticed on occasion. Hydrogen sulfide is perceptible to human beings at 0.5 to 2 ppm; the National Institute for Occupational Safety and Health (NIOSH) sets 10 ppm (14 mg/m^3) as a maximum exposure level for humans, and 300 ppm as a deadly concentration (NIOSH Pocket Guide to Chemical Hazards, U.S. Department of Health and Human Services, June 1990). High levels of sulfate respiration can also adversely affect microbial activities, including methane formation, in anaerobic digestors. However, the principal economic cause for concern from biogenic sulfide is the substantial damage inflicted to the sewage treatment infrastructure of municipalities in the form of concrete corrosion.

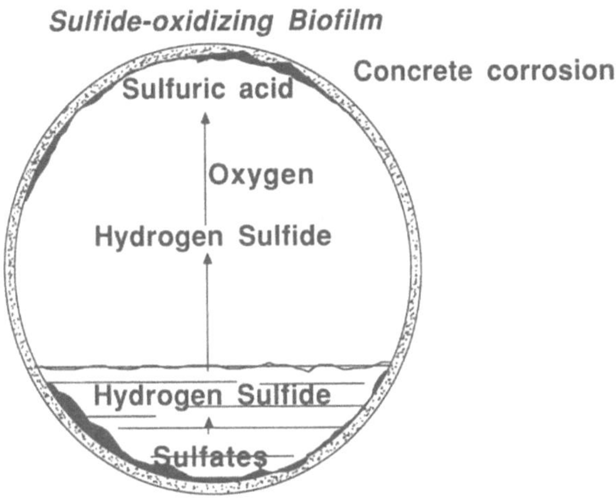

Figure 8.4 Sulfur cycle activity and sulfur compounds present within sewage conduits.

Concrete corrosion The chemical effects of oxidized sulfur species on concrete are well known and were vividly described in the account by Olmstead and Hamlin (1900) nearly 100 years ago. Sulfide, sulfur, and thiosulfate are quite corrosive to steel and concrete, yet these materials continue to be used in construction of sewage conduits, holding tanks, and digestors. Sand and Bock (1984) examined the microbiology of the Hamburg sewer system, which is noted for exhibiting extensive sulfuric acid-induced concrete corrosion. Their results are consistent with a model (Figure 8.4) that involves a chemical-biological sulfur cycle; sulfate reduction occurs in the biofilm, and sulfide oxidation to sulfur occurs on the aerobic inner surface of the conduit. In this scheme, sulfide and its chemical oxidation products are oxidized by *Thiobacilli* to sulfuric acid, which reacts with concrete to form calcium sulfate or gypsum. Sand and Bock (1984) observed a correlation between the extent of concrete corrosion and cell numbers of *T. thiooxidans*, although *T. neapolitanus*, *T. intermedius*, and *T. novellus* were also found. Methylmercaptan, dimethylsulfide, and dithiabutane have been detected in higher concentrations than hydrogen sulfide in the Hamburg sewer system. These compounds appear to originate from methionine sulfur and not from sulfate respiration, but it is unclear what contribution volatile organic thiols make to concrete corrosion (Pohl et al., 1984). Sivela and Sundman (1972), however, demonstrated utilization of methyl sulfides by *Thiobacilli*.

Biogenic sulfide can be a particularly serious problem for sewer systems in coastal sunbelt cities where flat topography and high temperatures translate into long sewage transit times and high levels of microbial activity. Mesa, Arizona, is one of the few cities to establish upper limits for the content of dissolved sulfide in sewage. Since 1964 Mesa has controlled dissolved sulfides to a maximum of 0.5 ppm largely by use of ferrous chloride to precipitate sulfide (Jarmeel, 1989). While ferrous chloride may effectively remove the sulfide odor, sulfide remains as the ferrous mineral and may be utilized by *Thiobacilli*. Cost effectiveness, rather than absolute effectiveness, is the key factor in treatments for hydrogen sulfide pollution. Furthermore, use of broad-spectrum biocides is out of the question owing to interference with other microbial processes involved in the waste treatment process. Thus, simple precipitation of dissolved sulfide with ferrous chloride has become a common practice in many sewage treatment programs. Alternatively, gas-scrubbing systems utilizing activated charcoal or chemical oxidizing agents can be employed to purify and recirculate trapped sewer gases (Chwirka and Satchell, 1990).

Substantial populations of sulfate-reducing bacteria can be sustained in anaerobic digestors, in the absence of sulfate, owing to growth by interspecies hydrogen transfer. Sulfate reducers can be routinely isolated from these environments (Bryant et al., 1977; Kaspar and Wuhrman, 1978), and cell densities as high as 10,000 cells/ml have been reported (Toerien et al., 1968). Sulfate influx into a digestor is rapidly converted to hydrogen sulfide, with concomitant loss in methane production and contamination of the fuel gas produced. The San Francisco Bay Area Air Quality Management District has successfully used iron salts to lower hydrogen sulfide concentration in digestor gas from an untreated level of 2000 to 3000 ppm to 430 ppm with treatment (Dezham et al.,1988).

8.9 Anaerobic Transformations of Mercury

The environmental fate of mercury has attracted considerable attention in recent years for two reasons: its occurrence as an industrial pollutant of air, soil, and water; and its bioconversion to the volatile neurotoxin, methylmercury (Jeffries, 1982; Robinson and Tuovinen, 1984). The ability of aerobic bacteria to methylate mercury is widespread; however, the role of anaerobes in mercury methylation is poorly understood. In sulfate-reducing sediments the situation is complicated by the fact that hydrogen sulfide will chemically precipitate Hg^{+2} as insoluble HgS and thus essentially remove the element from the food chain (Craig and Bartlett, 1978). Consequently, in marine situations, where high rates of sulfate reduction occur, mercury may be immobilized via precipitation as the insoluble sulfide. Additionally, HgS has been shown to be a poor substrate for methylation reactions until chemical oxidation (air oxidation) releases the Hg^{+2} ion, whereupon methylation by a variety of aerobic bacteria may commence (Fagerstrom and Jernelov, 1971). However, this picture has been modified with the report that a pure culture of *Dv. desulfuricans* exhibited high mercury methylation activity, which was fully expressed under sulfate limitation in the presence of fermentable organic substrates (Compeau and Bartha, 1985). These workers propose that not only are sulfate reducers capable of this activity, but that they are the principal mercury methylators in anaerobic sediments. Demethylation of methylmercury was also demonstrated with a culture of *Dv. desulfuricans* (Shariat et al., 1979). In studies with sediments, indirect evidence, with molybdate as an inhibitor of sulfate-reducing bacteria (Compeau and Bartha, 1985; Oremland et al., 1990), supported the role of sulfate reducers as both methylators and demethylators of mercury. Methylation by *Dv. desulfuricans* was shown to be inhibited by propyl iodide, and the inhibition was light reversible (Berman et al., 1990).

8.10 Transformation of Other Industrial Pollutants

The pulp and paper industry produces a large volume of spent sulfite liquor as a waste product of its operations. A component (50% to 65%) of this waste is lignosulfonate generated by the reaction of lignin with bisulfite used during the pulping process. A process has been described in which a sulfate-reducing enrichment was used to treat this waste with the result that 30% of the sulfur was removed as hydrogen sulfide (Jurgensen and Patton, 1979). Tyagi et al. (1988) studied a mixed culture fermentation of lignosulfonate by *Dv. desulfuricans* and *Lactobacillus bulgaricus* and observed a 37% decline in sulfur content of the spent sulfite liquor. However, it is unclear in these studies whether the desulfurization could be accounted for solely through the desulfonation of lignin. Ziomek and Williams (1989) found only high-molecular-weight decreases in defined lignosulfonate preparations incubated with *Dv. desulfuricans* and an actual increase in the sulfur content of the modified lignosulfonate.

As discussed in Chapter 2, sulfate reducers have been found to modify or degrade an array of aromatic and aliphatic organics. Growth on aniline and catechol as sole electron and carbon sources by *Desulfobacterium catecholicum* sp. nov. and *Desulfobacterium anilini* respectively have been described (Szewzyk and Pfennig, 1987; Schnell et al., 1989). Benzaldehydes, such as vanillin, were shown to be utilized as electron donor for sulfate reduction by *Desulfovibrio* sp. (Zellner et al.,1990). Dechlorinations of perchloroethene, tetrachloromethane, dichloroethane, and chlorobenzoic acid have been described for *Desulfobacterium* sp. and *Desulfomonile tiedje* (Fathepure et al., 1987; Egli et al., 1987b; Dolfing and Tiedje, 1991). Dv. *desulfuricans* has also been shown to reduce dibenzothiophene, a major sulfur contaminant of crude oil, to biphenyl and hydrogen sulfide (Kim et al., 1990).

8.11 Conclusions

The industrial and environmental impact of biogenic sulfide is changing substantially in scale as our tolerance of air and water pollution by sulfide decreases and the cost of this biological process to industry increases. We are currently seeing significant advances in our understanding of the sulfate-reducing bacteria, particularly at the molecular level. However, application of this understanding for development of new technologies to control biogenic sulfide in sensitive environmental and industrial situations is only beginning to emerge. Development of truly specific inhibitors and rapid methods for in situ detection and identification of sulfate-reducing bacteria are the tools necessary to reliably assess and control the industrial and environmental threat of biogenic sulfide.

[References, see p. 211]

References

Abdollahi, H. and Nedwell, D.B. 1980. Serological characteristics within the genus *Desulfovibrio. Antonie van Leeuwenhoek* 46:73–83. (#1, 6)*

Abelson, P.H. 1966. Chemical events on the primitive earth. *Proceedings of the National Academy of Sciences of the United States of America* 55:1365–1372. (#3)

Abram, J.W. and Nedwell, D.B. 1978. Inhibition of methanogenesis by sulphate reducing bacteria competing for transferred hydrogen. *Archives of Microbiology* 117:8–92. (#7)

Achenbach-Richter, L., Stetter, K.O., and Woese, C.R. 1987. A possible biochemical missing link among archaebacteria. *Nature* 327:348–349. (#6)

Adams, M.W.W. 1990. The structure and mechanism of iron hydrogenases. *Biochimica et Biophysica Acta* 1020:115–145. (#3)

Adams, M.W.W. and Mortenson, L.E. 1984. The physical and catalytic properties of hydrogenase II of *Clostridium pasteurianum. Journal of Biological Chemistry* 259:7045–7055. (#5)

Adman, E.T., Sieker, L.C., and Jensen, L.H. 1973. The structure of a bacterial ferredoxin. *Journal of Biological Chemistry* 259:7045–7055. (#5)

Adman, E.T., Sieker, L.C., Jensen, L.H., Bruschi, M., and LeGall, J. 1977. A structural model of rubredoxin from *Desulfovibrio vulgaris* at 2Å resolution. *Journal of Molecular Biology* 112:113–120. (#5)

Aeckersberg, F., Bak, F., and Widdel, F. 1991. Anaerobic oxidation of saturated hydrocarbons by a new type of sulfate-reducing bacterium. *Archives of Microbiology* 156:5–14. (#2, 8)

Akagi, J.M. 1967. Electron carriers for the phosphoclastic reaction of *Desulfovibrio desulfuricans. Journal of Biological Chemistry* 242:2478–2483. (#3)

Akagi, J.M. 1981. Dissimilatory sulphate reduction, mechanistic aspects, pp. 178–187. In Bothe, H. and Trebst, A. (editors) *Biology of Inorganic Nitrogen and Sulfur*. Springer-Verlag, New York. (#1)

Aketagawa, J., Kobayaski, K., and Ishimoto, M. 1985a. Purification and properties of thiosulfate reductase from *Desulfovibrio vulgaris*, Miyazaki. *Journal of Biochemistry* (Tokyo) 97:1025–1032. (#3)

Aketagawa, J., Kojo, K., Kobayashi, K., and Ishimoto, M. 1985b. Comparative immunological studies of sulfite reductases, hydrogenases, and somatic antigens from strains of the genus *Desulfovibrio. Journal of General and Applied Microbiology* 31:347–357. (#1)

Alperin, M.J. and Reeburgh, W.S. 1985. Inhibition experiments on anaerobic methane oxidation. *Applied and Environmental Microbiology* 50:940–945. (#2, 7)

*Numbers in parentheses refer to chapter citation.

Amann, R.I., Krumholz, L., and Stahl, D.A. 1990a. Fluorescent-oligonucleotide probing of whole cells for determinative, phylogenetic, and environmental studies in microbiology. *Journal of Bacteriology* 172:762–770. (#6)
Amann, R.I., Binder, B.J., Olson, R.J., Chisholm, S.W., Devereux, R., and Stahl, D.A. 1990b. Combination of 16S rRNA-targeted oligonucleotide probes with flow cytometry for analyzing mixed microbial populations. *Applied and Environmental Microbiology* 56:1619–1625. (#6)
Amann, R.I., Stromley, J., Devereux, R., and Stahl D.A. 1992. Molecular and microscopic identification of sulfate-reducing bacteria in multispecies biofilms. *Applied and Environmental Microbiology* 58:614–623.
Amber, R.P. 1968. The amino acid sequence of cytochrome c_3 from *Desulfovibrio vulgaris* (NCIB 8303). *Biochemical Journal* 109:47. (#5)
Ambler, R.P., Bruschi, M., and LeGall, J. 1969. The structure of cytochrome c_3 from *Desulfovibrio gigas* (NCIB 9332). *FEBS Letters* 5:115–117. (#5)
Ambler, R.P., Bruschi, M., and LeGall, J. 1971. The amino acid sequence of cytochrome c_3 from *Desulfovibrio desulfuricans* (strain El Agheila Z, NCIB 8380). *FEBS Letters* 5:347–350. (#5)
Antloga, K.M. and Griffen, W.M. 1985. Characterization of sulfate-reducing bacteria isolated from oil field waters. *Developments in Industrial Microbiology* 26:597–610. (#8)
API 1965 (American Petroleum Institute) 1965. *Recommended Practice for Biological Analysis of Subsurface Injection Waters*. 2nd edition, RP 38. Dallas Division. (#8)
Aranki, A. and Freter, R. 1972. Use of anaerobic glove boxes for cultivation of strictly anaerobic bacteria. *American Journal Clinical Nutrition* 25:1329–1334. (#1)
Aranki, A., Syed, S.A., Kenney, E.B., and Freter, R. 1969. Isolation of anaerobic bacteria from human gingiva and mouse cecum by means of a simplified glove box procedure. *Applied Microbiology* 17:568–576. (#1)
Argyle, J.L., Rapp-Giles, B.J., and Wall, J.D. 1992. Plasmid transfer by conjugation in *Desulfovibio desulfuricans*. *FEMS Microbiology Letters*. In press. (#4)
Atlas, R.M. 1982. Enumeration and estimation of microbial biomass, pp. 84–102. In Burns, R.G. and Slater, J.H. (editors) *Experimental Microbial Ecology*. Blackwell Scientific Publications, Oxford. (#7)
Baars, J.K. 1930. Over sulfaatreductie door bacteriën. Ph.D. thesis, University of Delft. (#2)
Badziong, W. and Thauer, R.K. 1978. Growth yields and growth rates of *Desulfovibrio vulgaris* (Marburg) growing on hydrogen and sulfate and hydrogen and thiosulfate as sole energy sources. *Archives of Microbiology* 117:209–214. (#1, 3)
Badziong, W. and Thauer, R.K. 1980. Vectorial electron transport in *Desulfovibrio vulgaris* (Marburg) growing on hydrogen plus sulfate as sole energy source. *Archives of Microbiology* 125:167–184. (#1, 3)
Badziong, W., Thauer, R.K., and Zeikus, J.G. 1978. Isolation and characterization of *Desulfovibrio* growing on hydrogen plus sulfate as the sole energy source. *Archives of Microbiology* 116:41–49. (#1, 2, 3)
Badziong, W., Ditter, B., and Thauer, R.K. 1979. Acetate and carbon dioxide assimilation by *Desulfovibrio vulgaris* (Marburg), growing on hydrogen and sulfate as sole energy source. *Archives of Microbiology* 123:301–305. (#2)
Bagdasarian, M., Lurz, M.R., Ruckert, B., Franklin, F.C.H., Bagdasarian, M.M., Frey, J., and Timmis, K.N. 1981. Specific purpose cloning vectors II. Broad host range, high copy number, RSF1010-derived vectors, and a host-vector system for gene cloning in Pseudomonas. *Gene* 16:237–247. (#4)
Bak, F. and Cypionka, H. 1987. A novel type of energy metabolism involving

the fermentation of inorganic sulphur compounds. *Nature* 326:891–892. (#1, 3, 7)

Bak, F. and Pfennig, N. 1987. Chemolithotrophic growth of *Desulfovibrio sulfodismutans* sp. nov. by disproportionation of inorganic sulfur compounds. *Archives of Microbiology* 147:184–189. (#1, 2, 3, 7)

Bak, F. and Widdel, F. 1986a. Anaerobic degradation of indolic compounds by sulfate-reducing enrichment cultures, and description of *Desulfobacterium indolicum* gen. nov., sp. nov. *Archives of Microbiology* 146:170–176. (#6)

Bak, F. and Widdel, F. 1986b. Anaerobic degradation of phenol and phenol derivatives by *Desulfobacterium phenolicum* sp. nov. *Archives of Microbiology* 146:177–180. (#6)

Bak, F., Behrendt, H.G., and Kleinitz, W. 1990. Pipeline corrosion caused by bacteria. *Erdoel, Erdgas, Kohle*. 106:109–113. (#8)

Baker, F.D., Paiska, H.R., and Campbell, L.L. 1962. Choline fermentation by *Desulfovibrio desulfuricans*. *Journal of Bacteriology* 84:973–978. (#2)

Balch, W.E. and Wolfe, R.S. 1976. New approaches to the cultivation of methanogenic bacteria: 2-mercaptoethane sulfonic acid (HS-CoM)-dependent growth of *Methanobacterium ruminantium* in a pressurized atmosphere. *Applied and Environmental Microbiology* 32:781–791. (#1)

Balch, W.E., Fox, G.E., Magrum, L.J., Woese, C.R., and Wolfe, R.S. 1979. Methanogens: reevaluation of a unique biological group. *Microbiological Reviews* 43:260–296. (#1, 3)

Baltscheffsky, M. and Nyrén, P. 1987. PP_i in the energy conservation system of *Rhodospirillum rubrum*, pp. 260–263 In: Torriani-Gorini, A., Rothman, F.G., Silver, S., Wright, A. and Yagil, E. (editors) *Phosphate Metabolism and Cellular Regulation in Microorganisms*. American Society of Microbiology, Washington. (#1)

Banat, I.M., Lindstron, E.B., Nedwell, D.B., and Balba, M.T. 1981. Evidence for coexistence of two distinct functional groups of sulfate-reducing bacteria in salt marsh sediments. *Applied and Environmental Microbiology* 42:985–992. (#7)

Barnes, E.M. and Mead, G.C. 1986. *Anaerobic Bacteria in Habitats Other Than Man*. 444 pp. Blackwell Scientific Publications, Oxford. (#1)

Barr, W.M. and Buchanan, R.E. 1912. The production of excessive hydrogen sulfide in sewage disposal plants and consequent disintegration of the concrete. *Bulletin of Iowa State College of Agriculture and Mechanical Arts*. Vol X, No. 26, pp. 1–15. (#8)

Barrett, E.L. and Clark, M.A. 1987. Tetrathionate reduction and production of hydrogen sulfide from thiosulfate. *Microbiological Reviews* 51:192–205. (#8)

Barton, L.L., LeGall, J., and Peck, H.D. 1970. Phosphorylation coupled to oxidation of hydrogen with fumarate in extracts of the sulfate-reducing bacterium *Desulfovibrio gigas*. *Biochemical and Biophysical Research Communications* 41:1036–1042. (#1)

Barton, L.L., LeGall, J., Odom, J.M., and Peck, H.D. Jr. 1983. Energy coupling in nitrite respiration in the sulfate-reducing bacterium *Desulfovibrio gigas*. *Journal of Bacteriology* 154:867–871. (#3)

Bastin, E.S. 1926. The problem of the natural reduction of sulfates. *Bulletin of the American Association Petroleum Geologists* 10:1270–1299. (#8)

Beckman, J.W. 1926. Action of bacteria on mineral oil. *Journal of Industrial Engineering Chemistry* 4:21–22. (#8)

Beerstecher, E. 1954. *Petroleum Microbiology: An Introduction to Microbiological Petroleum*. 375 pp. Elsevier, Amsterdam. (#1)

Behrens, M.I. and DeMeis, L. 1985. Synthesis of pyrophosphate by chromatophores of *Rhodospirillum rubrum* in the light and by soluble yeast inorganic pyrophosphatase in water-organic solvent mixtures. *European Journal of Biochemistry* 152:221–227. (#3)

Beijerinck, M.W. (1895) Über *Spirillum desulfuricans* als Ursache von Sulfatreduction. *Zentralblatt fuer Bakteriologie, Parasitenkunde und Infereionskrankheiten* (Abt. 2). 1: 1–9, 49–59, 104–14. (Preface, #8)

Belkin, S. and Jannasch, H.W. 1985. A new extremely thermophilic, sulfur-reducing heterotrophic, marine bacterium. *Archives of Microbiology* 141:181–186. (#7)

Bell, G.R., LeGall, J., and Peck, H.D. Jr. 1974. Evidence for the periplasmic location of hydrogenase in *Desulfovibrio gigas*. *Journal of Bacteriology* 120:994–997. (#3)

Benson, S.A., Hall, M.N., and Silhavy, T.J. 1985. Genetic analysis of protein export in *Escherichia coli* K12. *Annual Review of Biochemistry* 54:101–134. (#5)

Berg, W.A.M. van den, Dongen, W., M.A.M. van, Veeger, C. 1991. Reduction of the amount of periplasmic hydrogenase in *Desulfovibrio vulgaris* (Hildenborough) with antisense RNA: direct evidence for an important role of this hydrogenase in lactate metabolism. *Journal of Bacteriology* 173:3688–3694. (#3)

Berman, M., Chase, T. Jr., and Bartha, R. 1990. Carbon flow in mercury biomethylation by *Desulfovibrio desulfuricans*. *Applied and Environmental Microbiology* 56:298–300. (#8)

Berndt, W.L. and Vargas, J.M. Jr. 1987. Etiology and impact of dissimilatory sulfate reduction in highly maintained turfgrass soils. Abstract, *Annual Meeting of Phytopathological Society*, North Central Division, August 2–6, 1987, Cincinnati, Ohio. (#8)

Betzl, D., Ludwig, W., and Schleifer, K.H. 1990. Identification of Lactococci and Enterococci by colony hybridization with 23S rRNA-targeted oligonucleotide probes. *Applied and Environmental Microbiology* 56:2927–2929. (#6)

Booth, G.H. and Tiller, A.K. 1960. Polarization studies of mild steel in cultures of sulfate-reducing bacteria. *Transactions of the Faraday Society* 56:1689–1696. (#8)

Bourgeois, J.P., Bloise, T., Millet, J.L., and Aypaix, M. 1979. Suggested explanation of hydrogen sulfide in natural gas underground storage structures by reduction of mineral sulfides contained in reservoir rock. *Revue de l'Institute Francais du Petrole*. 34:371–373. (#8)

Brandis, A. and Thauer, R.K. 1981. Growth of *Desulfovibrio* species on hydrogen and sulphate as sole energy source. *Journal of General Microbiology* 126:249–252. (#2)

Brandis-Heep, A., Gebhardt, N.A., Thauer, R.K., Widdel, F., and Pfennig, N. 1983. Anaerobic acetate oxidation to CO_2 by *Desulfobacter postgatei*. 1. Demonstration of all enzymes required for the operation of the citric acid cycle. *Archives of Microbiology* 136:222–229. (#2)

Bremer, H. and Dennis, P.P. 1987. Modulation of chemical composition and other parameters of the cell by growth rate, pp. 1527–1542. In Neidhardt, F.C. (editor) *Escherichia coli and Salmonella typhimurium*. American Society for Microbiology, Washington, D.C. (#6)

Brock, T.D. and Madigan, M.T. 1991. *Biology of Microorganisms*, 835 pp. 6th edition. Prentice-Hall, Englewood Cliffs, NJ. (#1, 7)

Brumlik, M.J. and Voordouw, G. 1989. Analysis of the transcriptional unit encoding the genes for rubredoxin (*rub*) and a putative rubredoxin oxidoreductase (*rbo*) in *Desulfovibrio vulgaris* Hildenborough. *Journal of Bacteriology* 171:4996–5004. (#5)

Brumlik, M.J., LeRoy, G., Bruschi, M., and Voordouw, G. 1990. The nucleotide sequence of the *Desulfovibrio gigas* desulforedoxin gene indicates that the *Desulfovibrio vulgaris* rbo-gene originated from a gene fusion event. *Journal of Bacteriology* 172:7289–7292. (#5)

Brune, G., Schoberth, S.M., and Sahm, H. 1983. Growth of a strictly anaerobic bacterium on furfural. *Applied and Environmental Microbiology* 46:1187–1192. (#6)

Bruschi, M. 1976. Non-heme iron proteins: the amino acid sequence of rubredoxin from *Desulfovibrio vulgaris. Biochimica et Biophysica Acta* 434:4–17. (#5)

Bruschi, M. and Guerlesquin, F. 1988. Structure, function and evolution of bacterial ferredoxins. *FEMS Microbiology Reviews* 54:155–176. (#5)

Bruschi, M. and LeGall, J. 1972. C-type cytochromes of *Dv. vulgaris*: The primary structure of cytochrome c_{553}. *Biochimica et Biophysica Acta* 271:48–60. (#3, 5)

Bruschi, M. Moura, I, LeGall, J., Xavier, A.V., and Sieker, L.C. 1979. The amino acid sequence of desulforedoxin, a new type of non heme iron protein from *Desulfovibrio gigas. Biochemical and Biophysical Research Communications* 90:596–605. (#5)

Bryant, M.P. 1979. Microbial methane production—theoretical aspects. *Journal of Animal Science* 48:193–201. (#3)

Bryant, R.D. and Laishley, E.J. 1990. The role of hydrogenase in anaerobic biocorrosion. *Canadian Journal of Microbiology* 36:259–264. (#8)

Bryant, M.P., Campbell, L.L., Reddy, C.A., and Crabill, M.R. 1977. Growth of *Desulfovibrio* in lactate and ethanol media low in sulfate in association with H_2-utilizing methanogenic bacteria. *Applied and Environmental Microbiology* 33:1162–1169. (#1, 2, 3, 7, 8)

Brysch, K., Schneider, C., Fuchs, G., and Widdel, F. 1987. Lithoautotrophic growth of sulfate-reducing bacteria and description of *Desulfobacterium autotrophicum* gen. nov., sp. nov. *Archives of Microbiology* 148:264–274. (#1, 2, 6)

Bull, A.T. and Slater, J.H. (editors) 1982. *Microbial Interactions and Communities,* 567pp. volume 1. Academic Press, London. (#7)

Bunker, H.J. 1939. Microbiological experiments in anaerobic corrosion. *Journal of the Society of Chemical Industry,* LVIII: 93. (#8)

Burggraf, S., Jannasch, H.W., Nicolaus, B., and Stetter, K.O. 1990. *Archaeoglobus profundus* sp. nov. represents a new species within the sulfate-reducing archaebacteria. *Systematic and Applied Microbiology* 13:24–28. (#6)

Burns, R.G. and Slater, J.H. (editors) 1982. *Experimental Microbial Ecology,* 683 pp. Blackwell Scientific Publications, Oxford. (#7)

Butlin, K.R. 1949. Some malodorous activities of sulphate-reducing bacteria. *Proceedings of the Society for Applied Bacteriology* 2:39–42. (#1)

Cammack, R., Fauque, G., Moura, J.J.G., and LeGall, J. 1984. ESR studies of cytochrome c_3 from *Desulfovibrio desulfuricans* Norway 4. Midpoint potentials of the four haems and interactions with ferredoxin and colloidal sulphur. *Biochimica et Biophysica Acta* 784:68–74. (#1)

Cammack, R., Patil, D.S., Hatchikian, E.C., and Fernandez, V. M. 1987. Nickel and iron-sulphur centres in *Desulfovibrio gigas* hydrogenase: ESR spectra, redox properties and interactions. *Biochimica et Biophysica Acta* 912:98–109. (#5)

Campbell, L.L. and Postgate, J.R. 1965. Classification of the spore-forming sulfate-reducing bacteria. *Bacteriological Reviews* 29:359–363. (#1, 6)

Canfield, D.E. and Des Marais, D.J. 1991. Aerobic sulfate reduction in microbial mats. *Science* 251:1471–1473. (#1, 7)

Cannack, V., Caffrey, M.S., Voordouw, G., and Cusanovich, M.A. 1991. Expression of the gene encoding cytochrome c_3 from the sulfate-reducing bacterium *Desulfovibrio vulgaris* in the purple photosynthetic bacterium *Rhodobacter sphaeroides. Archives of Biochemistry and Biophysics* 286:629–632. (#5)

Capone, D.G. and Kiene, R.P. 1988. Comparison of microbial dynamics in marine and freshwater sediments: contrasts, in anaerobic carbon catabolism. *Limnology and Oceanography* 33:725–749. (#7)

Cappenberg, T.E. 1974. Interrelations between sulfate-reducing and methane-producing bacteria in bottom deposits of a fresh-water lake. 1. Field observation. *Antonie van Leeuwenhoek* 40:285–295. (#7)

Carr, M.C., Curley, G.P., Mayhew, S.G., and Voordouw, G. 1990. Effects of substituting asparagine for glycine-61 in flavodoxin from *Desulfovibrio vulgaris* (Hildenborough). *Biochemistry International* 20:1025–1032. (#5)

Chambers, I., Frampton, J., Goldfarb, P., Affara, N., McBain, W., and Harrison, P.R. 1986. The structure of the mouse glutathione peroxidase gene: the selenocysteine in the active site is encoded by the "termination" codon, TGA. *EMBO Journal* 5:1221–1227. (#5)

Chambers, L.A. and Trudinger, P.A. 1979. Microbiological fractionation of stable sulfur isotopes: a review and critique. *Geomicrobiology Journal* 1:249–293. (#3)

Chartrain, M. and Zeikus, J.G. 1986. Microbial ecophysiology of whey biomethanation: characterization of bacterial trophic populations and prevalent species in continuous culture. *Applied and Environmental Microbiology* 51:188–196. (#2)

Chatelus, C., Carrier, P., Saignes, P., Libert, M.F., Berlier, Y., Lespinat, P.A., Fauque, G., and LeGall, J. 1987. Hydrogenase activity in aged, nonviable *Desulfovibrio vulgaris* cultures and its significance to anaerobic biocorrosion. *Applied and Environmental Microbiology* 53:1708–1710. (#8)

Chen, J.-S. and Mortenson, L.E. 1974. Purification and properties of hydrogenase from *Clostridium pasteurianum* W5. *Biochimica et Biophysica Acta* 371:283–298. (#3)

Christensen, D. 1984. Determination of substrates oxidized by sulfate reduction in intact cores of marine sediments. *Limnology and Oceanography* 29:189–192. (#7)

Chwirka, J.D. and Satchell, T.T. 1990. A 1990 guide for treating hydrogen sulfide in sewers. *Water Engineering Management*, January 1990, pp 32–35. (#8)

Clark, W.M. 1972. *Oxidation-Reduction Potentials of Organic Systems*, 584 pp. Robert E. Krieger Publishing Co., Huntington, NY. (#1)

Cohen, Y. 1984. Micro-sulfate reduction measurements at the H_2S-O_2 interface in organic-rich sediments. *Eos* 65:905. (#7)

Collins, M.D. and Widdel, F. 1986. Respiratory quinones of sulphate-reducing and sulphur-reducing bacteria: a systematic investigation. *Systematic and Applied Microbiology* 8:8–18. (#6)

Colwell, R.R., MacDonell, M.T., and Swartz, D. 1989. Identification of an Antarctic endolithic microorganism by 5S rRNA sequence analysis. *Systematic and Applied Microbiology* 11:182–186. (#6)

Compeau, G.C. and Bartha, R. 1985. Sulfate-reducing bacteria: principal methylators of mercury in anoxic estuarine sediments. *Applied and Environmental Microbiology* 50:498–502. (#8)

Conner, B.J., Reyes, A.A., Morin, C., Itakura, K., Teplitz, R.L., and Wallace, R.B. 1983. Detection of sickle cell β^s-globin allele by hybridization with synthetic oligonucleotides. *Proceedings of the National Academy of Sciences of the United States of America* 80:278–282. (#6)

Conrad, R., Phelps, T.J., and Zeikus, J.G. 1985. Gas metabolism evidence in support of the juxtaposition of hydrogen-producing and methanogenic bacteria in sewage sludge and lake sediments. *Applied and Environmental Microbiology* 50:595–601. (#7)

Cord-Ruwisch, R. and Widdel, F. 1986. Corroding iron as a hydrogen source for sulfate-reduction in growing cultures of sulfate-reducing bacteria. *Applied and Microbial Biotechnology* 25:169–174. (#8)

Cord-Ruwisch, R., Ollivier, B., and Garcia, J.-L. 1986. Fructose degradation by *Desulfovibrio* sp. in pure culture and coculture with *Methanospirillum hungatei*. *Current Microbiology* 13:285–289. (#5)
Cord-Ruwisch, R., Seitz, H.-J., and Conrad, R. 1988. The capacity of hydrogenotrophic anaerobic bacteria to compete for traces of hydrogen depends on the redox potential of the terminal electron acceptor. *Archives of Microbiology* 149:350–357. (#3)
Cork, D.J. 1982. Acid waste gas bioconversion—an alternative to the Claus desulfurization process, pp. 379–387. In Underkofler L.A. (editor), *Developments in Industrial Microbiology*, Volume 23, Chapter 34. Society for Industrial Microbiology, Arlington, Virginia. (#8)
Corliss, J.B., Baross, J.A., and Hoffman, S.E. 1981. An hypothesis concerning the relationship between submarine hot springs and the origin of life on earth, pp. 59–69. *Oceanologica Acta* Proceedings, 26th International Geological Congress, Geology of Oceans Symposium. (#3)
Costa, C., Moura, J.J.G., Moura, I., Liu, M.Y., Peck, H.D. Jr., LeGall, J., Wang, Y., and Huynh, B.H. 1990. Hexaheme nitrite reductase from *Desulfovibrio desulfuricans* Mössbauer and EPR characterization of the heme groups. *Journal of Biological Chemistry* 265:14382–14387. (#3)
Costello, J.A. 1974. Cathodic depolarization by sulfate-reducing bacteria. *Science* 70:202–204. (#8)
Costerton, J.W. and Lappin-Scott, H.M. 1989. Behavior of bacteria in biofilms. *ASM News* 55:650–654. (#8)
Cragnolino, C. and Tuovinen, O.H. 1984. The role of sulphate-reducing and sulphur-oxidizing bacteria in the localized corrosion of iron-based alloys—A review. *International Biodeterioration* 20:9–26 (#1)
Craig, P.J. and Bartlett, P.D. 1978. The role of hydrogen sulfide in environmental transport of mercury. *Nature* 275:635–637. (#8)
Craven, D.B. and Karl, D.M. 1984. Microbial RNA and DNA synthesis in marine sediments. *Marine Biology* 83:129–139. (#7)
Crill, P.M. and Martens, C.S. 1984. Spatial segregation of methane production from ^{4}C-labeled acetate and bicarbonate in Cape Lookout Bight sediments. *Eos* 65:960. (#7)
Crill, P.M. and Martens, C.S. 1987. Biogeochemical cycling in an organic-rich coastal marine basin. 6. Temporal and spatial variation in sulfate reduction rates. *Geochimica et Cosmochimica Acta* 51:1175–1186. (#7)
Crombie, D.J., Moody, G.J., and Thomas, J.D.R. 1980. Corrosion of iron by sulphate-reducing bacteria. *Chemistry and Industry*, 21 June 1980: 500–504. (#1)
Curley, G.P. and Voordouw, G. 1988. Cloning and sequencing of the gene encoding flavodoxin from *Desulfovibrio vulgaris* Hildenborough. *FEMS Microbiology Letters* 49:295–299. (#5)
Cypionka, H. and Pfennig, N. 1986. Growth yields of *Desulfotomaculum* with hydrogen in chemostat culture. *Archives of Microbiology* 143:396–399. (#2, 3)
Cypionka, H., Widdel, F., and Pfennig, N. 1985. Survival of sulfate-reducing bacteria after oxygen stress and growth in sulfate-free oxygen sulfide gradients. *FEMS Microbiology Ecology* 85:31–42. (#1, 3, 7, 8)
Czechowski, M.H. and Rossmore, H.W. 1980. Factors affecting *Desulfovibrio desulfuricans* lactate dehydrogenase, pp. 349–356. In Underkoffler, L.A. and Wulf, M.L. (editors) *Developments in Industrial Microbiology*, vol. 21. Society for Industrial Microbiology, Arlington, Virginia. (#2)
Daniels, L., Belay, N., Rajagopal, B.S., and Weimer, P.J. 1987. Bacterial methanogenesis and growth from CO_2 with elemental iron as the sole source of electrons. *Science* 237:509–511. (#8)
Daumas, S., Cord-Ruwisch, R., and Garcia, J.L. 1988. *Desulfotomaculum geoth-*

ermicum sp. nov., a thermophilic, fatty acid-degrading, sulfate-reducing bacterium isolated with H_2 from geothermal ground water. *Antonie van Leeuwenhoek* 54:165–178. (#2)

Davies, G.E., Francis, J., Martin, A.R., Rose, F.L., and Swain, G. 1954. 1:6 Di-4′-chlorophenyldiguanidohexane ("Hibitane"). Laboratory investigation of a new antibacterial agent of high potency. *British Journal of Pharmacology* 9:192–196. (#8)

Davis, J.B. 1967. *Petroleum Microbiology*, 604 pp. Elsevier, Amsterdam. (#1)

Davis, J.B. and Yarbrough, J.F. 1966. Anaerobic oxidation of hydrocarbons by *Desulfovibrio desulfuricans*. *Chemical Geology* 1:137–144. (#7)

Davison, J., Heusterpreute, M., Chevalier, N., Vinh, H.T., and Brunel, F. 1987. Vectors with restriction site banks V. pJRD215, a wide-host range cosmid vector with multiple cloning sites. *Gene* 51:275–280. (#4, 5)

Dearson, J.J. 1973. Developments in Claus catalysis. *Hydrocarbon Proceedings*, February, pp. 81–85. (#8)

Deckena, S. and Blotevogel, K.H. 1990. Growth of methanogenic and sulfate-reducing bacteria with cathodic hydrogen. *Biotechnology Letters* 12:615–620. (#8)

Deckers, H.M., Wilson, F.R., and Voordouw, G. 1990. Cloning and sequencing of a [NiFe] hydrogenase operon from *Desulfovibrio vulgaris* Miyazaki F. *Journal of General Microbiology* 136:2021–2028. (#3, 5)

DeLong, E.F., Wickham, G.S., and Pace, N.R. 1989. Phylogenetic stains: ribosomal RNA-based probes for the identification of single cells. *Science* 243:1360–1363. (#6)

Devereux, R., Delaney, M., Widdel, F., and Stahl, D.A. 1989. Natural relationships among sulfate-reducing eubacteria. *Journal of Bacteriology* 171:6689–6695. (#1, 2, 4, 5, 6)

Devereux, R., He, S.-H., Doyle, C.L., Orkland, S., Stahl, D.A., LeGall, J., and Whitman, W.B. 1990. Diversity and origin of *Desulfovibrio* species: phylogenetic definition of a family. *Journal of Bacteriology* 172:3609–3619. (#1, 2, 4, 6, 8)

Devol, A.H. 1983. Methane oxidation rates in anaerobic sediments of Sannich inlet. *Limnology and Oceanography* 28:783–742. (#7)

Dewar, E.J. 1986. Control of microbiologically induced corrosion and accumulation of solids in a seawater flood system. *Materials Performance* July:39–47. (#8)

DeWeerd, K., Mandelco, L., Tanner, R.S., Woese, C.R., and Suflita, J.M. 1990. *Desulfomonile tiedjei* gen. nov. and sp. nov., a novel anaerobic dehalogenating, sulfate-reducing bacterium. *Archives of Microbiology* 154:23–30. (#2, 6)

DeWeerd, K.A., Suflita, J.M., Linkfield, T., Tiedje, J.M., and Pritchard, P.H. 1986. The relationship between reductive dehalogenation and other aryl substituent removal reactions catalyzed by anaerobes. *FEMS Microbiology Ecology* 38:331–339. (#6)

Dezham, P., Rosenblum, D., and Jenkins, E. 1988. Digestor gas H_2S control using iron salts. *Journal of Water Pollution Control* 60:514–517. (#8)

Dicker, H.J. 1983. Metabolism of low molecular weight organic compounds and H_2 by sulfate reducing bacteria in a Delaware salt marsh. Ph.D. Dissertation, University of Delaware, 186 pp. (#7)

Dicker, H.J. and Smith, D.W. 1980a. Acetylene reduction (nitrogen fixation) in a Delaware, USA salt marsh. *Marine Biology* 57:241–250. (#7)

Dicker, H.J. and Smith, D.W. 1980b. Enumeration and relative importance of acetylene-reducing (nitrogen-fixing) bacteria in a Delaware salt marsh. *Applied and Environmental Microbiology* 39:1019–1025. (#7)

Dicker, H.J. and Smith, D.W. 1985a. Effects of organic amendments on sulfate reduction activity, H_2 consumption, and H_2 production in salt marsh sediments. *Microbial Ecology* 11:299–315. (#1, 7)
Dicker, H.J. and Smith, D.W. 1985b. Metabolism of low molecular weight compounds by sulfate-reducing bacteria in a Delaware salt marsh. *Microbial Ecology* 11:317–335. (#1, 7)
Dijk, C.V., van Berkel-Arts, A., and Veeger, C. 1983. The effect of re-oxidation on the reduced hydrogenase of *Desulfovibrio vulgaris* strain Hildenborough and its oxygen stability. *FEBS Letters* 156:340–344. (#7)
Dilling, W. and Cypionka, H. 1990. Aerobic respiration in the sulfate-reducing bacteria. *FEMS Microbiological Letters* 71:123–128. (#1, 3, 7)
Dixon, R.O.D. 1972. Hydrogenase in legume root nodule bacterioids: occurrence and properties. *Archives of Microbiology* 85:193–201. (#3)
Dockins, W.S., Olson, G.J., McFeters, G.A., and Turbak, S.C. 1980. Dissimilatory bacterial sulfate reduction in Montana ground water. *Geomicrobiology Journal* 2:83–98. (#1)
Doetsch, R.N. (editor) (1960) *Microbiology. Historical Contributions from 1776 to 1908 etc.*, pp. 179–205. Rutgers Univ. Press, New Brunswick, N.J. (Preface)
Doetsch, R.N. and Cook, T.M. 1973. *Introduction to Bacteria and Their Ecobiology*, 371 pp. University Park Press, Baltimore. (#7)
Doig, K. and Wachten, A. 1951. Bacterial casing corrosion in the Ventura Field. *Corrosion* 7:212–216. (#8)
Dolfing, J. 1988. Acetogenesis, pp. 417–468. In Zehnder, A.J.B. (editor) *Biology of Anaerobic Organisms*. John Wiley, New York. (#3)
Dolfing, J. and Tiedje, J.M. 1991. Influence of substituents on reductive dehalogenation of 3 chlorobenzoate analogs. *Applied and Environmental Microbiology* 157:800–824. (#8)
Dolla, A., Fu, R., Brumlik, M.J., and Voordouw, G. 1992. Nucleotide sequence of dcrA, a *Desulfovibrio vulgaris* Hildenborough chemoreceptor gene and its expression in *Escherichia coli*. *Journal of Bacteriology* 174:1726–1733. (#5)
Donelly, L.S. and Busta, F.F. 1980. Heat resistance of *Desulfotomaculum nigrificans*. *Applied and Environmental Microbiology* 40:721–725. (#1)
Drobner, E., Huber, H., Wächlershäuser, G., Rose, D., and Stetter, K.O. 1990. Pyrite formation linked with hydrogen evolution under anaerobic conditions. *Nature* (London) 346:742–744. (#3)
Dubourdieu, M. and Fox, J.L. 1977. Amino acid sequence of *Desulfovibrio vulgaris* flavodoxin. *Journal of Biological Chemistry* 252:1453–1462. (#5)
Dumont, M.E., Ernst, J.F., Hampsey, D.M., and Sherman, F. 1987. Identification and sequence of the gene encoding cytochrome *c* heme lyase in the yeast *Saccharomyces cerevisiae*. *EMBO Journal* 6:235–241. (#5)
Durbin, K.J. and Watanabe, I. 1980. Sulfate-reducing bacteria and nitrogen fixation in flooded rice soil. *Soil Biology and Biochemistry* 12:11–14. (#8)
Dwyer, D.F. and Tiedje, J.M. 1986. Metabolism of polyethylene glycol by two anaerobic bacteria, *Desulfovibrio desulfuricans* and a *Bacteroides* sp. *Applied and Environmental Microbiology* 52:852–856. (#2)
Dykhuizen, D.E. and Hartl, D.L. 1983. Selection in chemostats. *Microbiological Reviews* 47:150–168. (#7)
Ebbing, D.D. and Wrighton, M.S. 1990. *General Chemistry*, 3rd edition, 1035 pp. Houghton Mifflin, Boston. (#7)
Edlund, A., Nichols, P.D., Roffey, R., and White, D.C. 1985. Extractable and lipopolysaccharide fatty acid and hydroxy acid profiles from *Desulfovibrio* species. *Journal of Lipid Research* 26:982–987. (#8)
Egli, C., Scholtz, R., Cook, A.M., and Leisinger, T. 1987. Anaerobic dechlorination of tetrachloromethane and 1,2 dichloroethene to degradable pro-

ducts by pure cultures of *Desulfobacterium* sp. and *Methanobacterium* sp. *FEMS Microbiology Letters* 43:257–261. (#8)

Egli, C., Stromeyer, S., Cook, A.M., and Leisinger, T. 1990. Transformation of tetra- and trichloromethane to CO_2 by anaerobic bacteria is a non-enzymic process. *FEMS Microbiology Letters* 68:207–212. (#2)

Eidsness, M.K., Scott, R.A., Prickril, B., DerVartanian, D.V., LeGall, J., Moura, I., Moura, J.J.G., and Peck Jr., H.D. 1989. Evidence for selenocysteine coordination to the active site nickel in the [NiFeSe] hydrogenase from *Desulfovibrio baculatus*. *Proceedings of the National Academy of Science of the United States of America* 86:147–151. (#5)

Elion, L. 1924. A thermophilic sulfate-reducing bacterium. *Zentralblatt für Bakteriologie* (Abt. 2) 63:58–67. (#8)

Elion, L. 1927. Formation of hydrogen sulfide by the natural reduction of sulfates. *Journal of Industrial Engineering Chemistry* 19:1368–1370. (#8)

Ellwood, D.C., J.N. Hedger, M.J. Latham, J.M. Lynch, and J.H. Slater (editors). 1980. *Contemporary Microbial Ecology*, 438 pp. Academic Press, London. (#7)

Esnault, G., Caumette, P., and Garcia, J.L. 1988. Characterization of *Desulfovibrio giganteus* sp. nov., a sulfate-reducing bacterium isolated from a brackish coastal lagoon. *Systematic and Applied Microbiology* 10:147–151. (#2)

Evans, H.J., Harker, A.R., Papen, H., Russell, S.A., Hanus, F.J., and Zuber, M. 1987. Physiology, biochemistry and genetics of the uptake hydrogenase in Rhizobium. *Annual Review of Microbiology* 41:335–361. (#5)

Fagerstrom, T. and Jernelov, A. 1971. Formation of methyl mercury from pure mercuric sulfide in aerobic organic sediments. *Water Research* 5:121–122. (#8)

Fathepure, B.Z., Nengu, J.P., and Boyd, S.A. 1987. Anaerobic bacteria that dechlorinate perchloroethene. *Applied and Environmental Microbiology* 53:2671–2674. (#2, 8)

Fauque, G., Herve, D., and LeGall, J. 1979. Structure-function relationships in hemoproteins: the role of cytochrome c_3 in the reduction of colloidal sulfur by sulfate reducing bacteria. *Archives of Microbiology* 121:261–264. (#3)

Fauque, G.D., Barton, L.L., and LeGall, J. 1980. Oxidative phosphorylation linked to the dissimilatory reduction of elemental sulphur by *Desulfovibrio*, pp. 71–86. In *Sulphur in Biology, Ciba Foundation Symposium 72*. Elsevier, Amsterdam. (#3)

Fauque, G., Peck, H.D. Jr., Moura, J.J.G., Huynh, B.H., Berlier, Y., Teixeira, M., Przybyla, A.E., Lespinat, P.A., Moura, I. and LeGall, J. 1989. The three classes of hydrogenases from sulfate-reducing bacteria of the genus *Desulfovibrio*. *FEMS Microbiology Reviews* 54:299–344. (#3)

Fauque, G., LeGall, J., and Barton, L.L. 1991. Sulfate-reducing and sulfur-reducing bacteria, pp. 271–337. In Shively, J.M. and Barton, L.L. (editors) *Variations in Autotrophic Life*. Academic Press, New York. (#3)

Felsenstein, J. 1984. The statistical approach to inferring evolutionary trees and what it tells us about parsimony and compatibility, pp. 169–191. In Duncan, T. and Stuessy, T.F. (editors) *Cladistics: Perspectives in the Reconstruction of Evolutionary History*. Columbia University Press, New York. (#6)

Ferguson, S.J. 1988. The redox reactions of the nitrogen and sulphur cycles, pp. 1–29. In Cole, J.A. and Ferguson, S.J. *The Nitrogen and Sulphur Cycles, Symposium 42 of the Society for General Microbiology*, Cambridge University Press, Cambridge. (#7)

Fernandez, V.M., Hatchikian, E.C., Patil, D.S., and R. Cammack. 1986. ESR detectable nickel and iron-sulphur centers in relation to the reversible activation of *Desulfovibrio gigas* hydrogenase. *Biochimica et Biophysica Acta* 883:145–154. (#5)

Festl, H., Ludwig, W., and Schleifer, K.H. 1986. DNA hybridization probe for the *Pseudomonas fluorescens* group. *Applied and Environmental Microbiology* 52:1190–1194. (#6)

Fitz, R.M. and Cypionka, H. 1989. A study on electron transport-driven proton translocation in *Desulfovibrio desulfuricans*. *Archives of Microbiology* 152:369–376. (#1, 3)

Folkerts, M., Ney, U., Kneifel, H., Stackebrandt, E., Witte, E.G., Förstel, H., Schoberth, S.M., and H. Sahm, 1989. *Desulfovibrio furfuralis* sp. nov., a furfural degrading strictly anaerobic bacterium. *Systematic and Applied Microbiology* 11:161–169. (#2, 6)

Ford, C.M., Garg, N., Garg, R.P., Tibelius, K.H., Yates, M.G., Arp, D.J., and Seefeldt, L.C. 1990. The identification, characterization, sequencing and mutagenesis of the genes (*hupS,L*) encoding the small and large subunits of the H_2-uptake hydrogenase of *Azotobacter chroococcum*. *Molecular Microbiology* 4:999–1009. (#5)

Ford, T. and Mitchell, R. 1990a. The ecology of microbial corrosion. *Advances in Microbial Ecology* 11:231–262. Plenum Press, New York. (#7)

Ford, T. and Mitchell, R. 1990b. Metal embrittlement by bacterial hydrogen—an overview. *MTS Journal* 24:29–35. (#8)

Forsberg, C.W. 1980. Sulfide production from cysteine by *Desulfovibrio desulfuricans*. *Applied and Environmental Microbiology* 39:453–455. (#2)

Fowler, V.J., Widdel, F., Pfennig, N., Woese, C.R., and Stackebrandt, E. 1986. Phylogenetic relationships of sulfate- and sulfur-reducing eubacteria. *Systematic and Applied Microbiology* 8:32–41. (#6)

Fox, G.E., Pechman, K.R., and Woese, C.R. 1977. Comparative cataloging of 16S ribosomal ribonucleic acid: molecular approach to prokaryotic systematics. *International Journal of Systematic Bacteriology* 27:44–57. (#6)

Fox, G.E., Stackebrandt, E., Hespell, R.B., Gibson, J., Maniloff, J., Dyer, T.A., Wolfe, R.S., Balch, W.E., Tanner, R.S., Magrum, L.J., Zablen, L.B., Blakemore, R., Gupta, R., Bonen, L., Lewis, B.J., Stahl, D.A., Luehrsen, K.R., Chen, K.N., and Woese, C.R. 1980. The phylogeny of the prokaryotes. *Science* 209:457–463. (#6)

Frey, M., Sieker, L., Payan, F., Haser, R., Bruschi, M., Pepe, G., and LeGall, J. 1987. Rubredoxin from *Desulfovibrio vulgaris*. A molecular model of the oxidized form at 1.4Å resolution. *Journal of Molecular Biology* 197:525–541. (#5)

Fry, J.C. 1982. The analysis of microbial interactions and communities in situ, pp. 103–152. In Bull, A.T. and Slater, J.H. (editors) *Microbial Interactions and Communities*, volume 1. Academic Press, London. (#7)

Gahl, R. and Anderson, B. 1928. Sulfate-reducing bacteria in California oil well waters. *Zentralblatt für Bakteriologie* (Abt. 2) 73:331–340. (#8)

Gaines, R.H. 1910. Bacterial activity as a corrosive influence in the soil. *Journal of Industrial Engineering Chemistry* 2:128–130. (#8)

Galbraith, N.J., Mabile, J., and VanBuskirk, K.A. 1985. Monitoring sulfate-reducing bacteria at Prudhoe Bay, Alaska with corrosion coupons. *Symposium, National Association of Corrosion Engineers*, Boston, Mass. (#8)

Gandy, E.L. and Yoch, D.C. 1988. Relationship between nitrogen-fixing sulfate reducers and fermenters in salt marsh sediments and roots of *Spontina alterniflora*. *Applied and Environmental Microbiology* 54:2031–2036. (#7)

Gebhardt, N.A., Linder, D., and Thauer, R.K. 1983. Anaerobic acetate oxidation to CO_2 by *Desulfobacter postgatei*. 2. Evidence from ^{14}C-labelling studies for the operation of the citric acid cycle. *Archives of Microbiology* 136:230–233. (#2, 6)

Gibson, G.R., Parkes, R.J., and Herbert, R.A. 1987. Evaluation of viable counting procedures for the enumeration of sulfate-reducing bacteria in estuarine sediments. *Journal of Microbiological Methods* 7:201–210. (#6)

Ginter, R.L. 1930. Causative agents of sulfate-reduction in oil well waters. *Bulletin of American Association of Petroleum Geologists* 14:139–152. (#8)
Giovannoni, S.J., DeLong, E.F., Olsen, G.J., and Pace, N.R. 1988. Phylogenetic group-specific oligodeoxynucleotide probes for identification of single microbial cells. *Journal of Bacteriology* 170:720–726. (#6)
Giovannoni, S.J., Britschgi, T.B., Moyer, C.L., and Field, K.G. 1990a. Genetic diversity in Sargasso Sea bacterioplankton. *Nature* 345:60–63. (#6)
Giovannoni, S.J., DeLong, E.F., Schmidt, T.M., and Pace, N.R. 1990b. Tangential flow filtration and preliminary phylogenetic analysis of marine picoplankton. *Applied and Environmental Microbiology* 56:2572–2575. (#6)
Gottschalk, G. 1986. *Bacterial Metabolism*, 2nd edition, 359 pp. New York, Springer-Verlag. (#7)
Greathouse, G.H. and Wessel, C.J. 1954. *Deterioration of Materials*, 835 pp. Reinhold, New York. (#1)
Guarraia, L.J. and Peck, H.D Jr. 1971. Dinitrophenol-stimulated adenosine triphosphatase activity in extracts of *Desulfovibrio gigas*. *Journal of Bacteriology* 106:890–895. (#3)
Gunnarsson, L.A.H. and Ronnow, P.H. 1982. Interrelationships between sulfate reducing and methane producing bacteria in coastal sediments with intense sulfide production. *Marine Biology* 69:121–128. (#7)
Hadley, R.F. 1940. Studies in microbiological anaerobic corrosion. *American Gas Association Proceedings* 22:764–788. (#8)
Hagen, W.R., van Berkel-Arts, A., Kruse-Wolters, K.M., Voordouw, G., and Veeger, C. 1986. The iron-sulfur composition of the active site of hydrogenase from *Desulfovibrio vulgaris* (Hildenborough) deduced from its subunit structure and total iron-sulfur content. *FEBS Letters* 203:59–63. (#5)
Haggblom, M.M. and Young, L.Y. 1990. Chlorophenol degradation coupled to sulfate reduction. *Applied and Environmental Microbiology* 56:3255–3260. (#7)
Hahn, D., Kester, R., Starrenburg, M.J.C., and Akkermans, A.D.L. 1990. Extraction of ribosomal RNA from soil for detection of *Frankia* with oligonucleotide probes. *Archives of Microbiology* 154:329–335. (#6)
Hahn, D., Starrenburg, M.J.C., and Akkermans, A.D.L. 1990. Oligonucleotide probes that hybridize with rRNA as a tool to study *Frankia* strains in root nodules. *Applied and Environmental Microbiology* 56:1342–1346. (#6)
Hamilton, W.A. 1983. Sulphate-reducing bacteria and the offshore oil industry. *Trends in Biotechnology*. 1:36–40. (#1)
Hamilton, W.A. 1985. Sulphate-reducing bacteria and anaerobic corrosion. *Annual Review of Microbiology* 39:195–217. (#7, 8)
Handley, J., Adams, V., and Akagi, J.M. 1973. Morphology of bacteriophage-like particles from *Desulfovibrio vulgaris*. *Journal of Bacteriology* 115:1205–1207. (#4)
Hansen, T.A. 1988. Physiology of sulphate-reducing bacteria. *Microbiological Sciences* 5:81–84. (#2)
Hansen, T.A. and Kremer, D.R. 1990. Dissimilation of ethanol and related compounds by *Desulfovibrio* strains, pp. 185–190. In Bélaich, J.P, Bruschi, M., and Garcia, J.L. (editors), *Microbiology and Biochemistry of Strict Anaerobes Involved in Interspecies Hydrogen Transfer*. Plenum, New York. (#2)
Hardie, J.M. 1986. Methods for the isolation and identification of anaerobes, pp. 397–410. In Barnes, E.M. and Mead, G.C. 1986. *Anaerobic Bacteria in Habitats Other Than Man*. Blackwell Scientific Publications, Oxford. (#1)
Hardy, J.A. 1983. Utilization of cathodic hyudrogen by sulfate-reducing bacteria. *British Corrosion Journal* 18:190–193. (#8)
Hardy, J.A. and Brown, J.L. 1984. The corrosion of mild steel by biogenic sulfide films exposed to air. *Corrosion* 40:650–654. (#8)
Hardy, J.A. and Hamilton, W.A. 1981. The oxygen tolerance of sulfate-

reducing bacteria isolated from North Sea waters. *Current Microbiology* 6:259–262. (#3)

Haschke, R.H. and Campbell, L.L. 1971. Purification and properties of a hydrogenase from *Desulfovibrio vulgaris*. *Journal of Bacteriology* 105:249–258. (#3, 5)

Haser, R., Pierrot, M., Frey, M., Payan, F., Astier, J.P., Bruschi, M., and LeGall, J. 1979. Structure and sequence of the multihaem cytochrome c_3. *Nature* 282:806–810. (#5)

Hastings, D. and Emerson, S. 1988. Sulfate reduction in the presence of low oxygen levels in the water column of the Cariaco Trench. *Limnology and Oceanography* 33:391–396. (#7)

Hatchikian, E.C. 1975. Purification and properties of thiosulfate reductase from *Desulfovibrio gigas*. *Archives of Microbiology* 105;249–256. (#3)

Hatchikian, E.C., LeGall, J., Bruschi, M., and Dubourdieu, M. 1972. Regulation of the reduction of sulfite and thiosulfate by ferredoxin, flavodoxin and cytochrome c_3 in extracts of the sulfate reducer *Desulfovibrio gigas*. *Biochimica et Biophysica Acta* 258:701–708. (#3)

Hatchikian, E.C., Chaigneau, M., and LeGall, J. 1976. Analysis of gas production by growing cultures of three species of sulfate reducing bacteria, pp. 109–118. In Schlegel, H.G., Gottschalk, G., and Pfenning, N. (editors) *Microbial Production and Utilization of Gases*. E. Goltze KG, Göttingen. (#3)

Hatchikian, E.C., LeGall, J., and Bell, G.R. 1977. Significance of superoxide dismutase and catalase activities in the strict anaerobes, sulfate reducing bacteria, pp. 159–172. In Michelson, A.M., McCord, J.M., and Fridovich, I. (editors) *Superoxide and Superoxide Dismutases*. Academic Press, London. (#3)

Hatchikian, E.C., Bruschi, M., and LeGall, J. 1978. Characterization of the periplasmic hydrogenase from *Desulfovibrio gigas*. *Biochemical and Biophysical Research Communications* 82:451–461. (#5)

Hatchikian, E.C., Fernandez, V.M., and Cammack, R. 1990. The hydrogenases of sulfate-reducing bacteria: Physiological, biochemical and catalytic aspects, pp. 53–74. In Belaich, J.P., Bruschi, M., and Garcia, J.L. (editors), *Microbiology and Biochemistry of Strict Anaerobes Involved in Interspecies Hydrogen Transfer*. Plenum Press, New York. (#5)

Hayward, H.R. 1960. Anaerobic degradation of choline. III. Acetaldehyde as an intermediate in the fermentation of choline by extracts of *Vibrio cholinicus*. *Journal of Biological Chemistry* 235:3592–3596. (#2)

He, S.-H., Texeira, M., LeGall, J., Patil, D.S., DerVartanian, D.V., Huyn, B.H., and Peck Jr., H.D. 1989. EPR studies with ^{77}Se enriched [NiFeSe] hydrogenase of *Desulfovibrio baculatus*. Evidence for a selenium ligand to the active-site nickel. *Journal of Biological Chemistry* 264:2678–2682. (#5)

Heijthuijsen, J.H.F.G. and Hansen, T.A. 1989. Anaerobic degradation of betaine by marine *Desulfobacterium strains*. *Archives of Microbiology* 152:393–396. (#2)

Herbert, B.N., Gillert, P.D., Stockdale, H., and Watkinson, R.J. 1985. Factors controlling the activity of sulfate-reducing bacteria in reservoirs during water injection. *Society of Petroleum Engineers* paper # 13978/1. (#8)

Herbert, R.A. 1975. Heterotrophic nitrogen fixation in shallow estuarine sediments. *Journal of Experimental Marine Biology* 18:215–225. (#7)

Herriott, J.R., Sieker, L.C., Jensen, L.H., and Lovenberg, W. 1970. Structure of rubredoxin: an X-ray study to 2.5Å resolution. *Journal of Molecular Biology* 50:391–406. (#5)

Hespell, R.B., Paster, B.J., Macke, T.J., and Woese, C.R. 1984. The origin and phylogeny of the bdellovibrios. *Systematic and Applied Microbiology* 5:196–203. (#6)

Higuchi, Y., Kusunoki, M., Matsuura, Y., Yasuoka, W., and Kakudo, M. 1984.

Refined structure of cytochrome c_3 at 1.8Å resolution. *Journal of Molecular Biology* 172:109–139. (#5)
Higuchi, Y., Inaka, K., Yasuoka, N., and Yagi, T. 1987a. Isolation and crystallization of high molecular weight cytochrome from *Desulfovibrio vulgaris* Hildenborough. *Biochimica et Biophysica Acta* 911:341–348. (#3, 5)
Higuchi, Y., Yasuoka, N., Kakudo, M., Katsube, Y., Yagi, T., and Inokuchi, H. 1987b. Single crystals of hydrogenase from *Desulfovibrio vulgaris* Miyazaki F. *Journal of Biological Chemistry* 262:2823–2825. (#5)
Hill, D.E., Remus, B.J., and Undesser, M.J. 1987. Control of microbiological related problems in the Kuparuk River unit water flood. *National Association of Corrosion Engineers Corrosion 87*, Paper # 378. San Francisco, Calif. (#8)
Hines, M.E. and Buck, J.D. 1982. Distribution of methanogenic and sulfate-reducing bacteria in nearshore marine sediments. *Applied and Environmental Microbiology* 43:447–453. (#7)
Hines, M.E., Knollmeyer, S.L., and Tugel, J.T. 1989. Sulfate reduction and other sedimentary biogeochemistry in a northern New England salt marsh. *Limnology and Oceanography* 34:578–590. (#1)
Hobbie, J.E. and Fletcher, M.M. 1988. The aquatic environment, pp. 132–262. In Lynch, J.M. and Hobbie, J.E. (editors) *Micro-Organisms in Action: Concepts and Applications in Microbial Ecology*. Blackwell Scientific Publications, Oxford. (#7)
Hooper, A.B. and DiSpirito, A.A. 1985. In bacteria which grow on simple reductants, generation of a proton gradient involves extra-cytoplasmic oxidation of substrates. *Microbiological Reviews* 49:140–157. (#3)
Hoppe-Seyler, F. 1886. Uber die Gährung der Cellulose mit Bildung von Methane und Kohlensäure. *Zeitschrift fuer Physiologische Chemie* X:401–440. (#8)
Hori, H., and Osawa, S. 1979. Evolutionary change in 5S RNA secondary structure and a phylogenetic tree of 54 5S RNA species. *Proceedings of the National Academy of Sciences of the United States of America* 16:381–385. (#6)
Hormel, S., Walsh, K.A., Prickril, B.C.,Titani, K., LeGall, J., and Sieker, L.C. 1986. Amino acid sequence of rubredoxin from *Desulfovibrio desulfuricans* strain 27774. *FEBS Microbiology Letters* 201:147–150. (#5)
Howarth, R.W. 1979. Pyrite: its rapid formation in a salt marsh and its importance in ecosystem metabolism. *Science* 203:49–57. (#1)
Howarth, R.W. 1984. The ecological significance of sulfur in the energy dynamics of salt-marsh and coastal marine sediments. *Biogeochemistry* 1:5–27. (#1, 7)
Howarth, R.W. and Hobbie, J.E. 1982. The regulation of decomposition and heterotrophic microbial activity in salt marsh soils: A review, pp. 103–127. In Kennedy, V.S. (editor) *Estuarine Comparisons*. Academic Press, New York. (#1)
Howarth, R.W. and Giblin, A. 1983. Sulfate reduction in the salt marshes at Sapelo Island, Georgia. *Limnology and Oceanography* 28:70–82. (#7)
Howarth, R.W. and Marino, R. 1984. Sulfate reduction in salt marshes with some comparison to sulfate reduction in microbial mats, pp. 245–263. In Cohen, Y., Castenholz, R., and Halvorson, H.O. (editors) *Microbial Mats: Stromatolites*. A. R. Liss, New York. (#1)
Howarth, R.W. and Merkel, S. 1984. Pyrite formation and the measurement of sulfate reduction in salt marsh sediments. *Limnology and Oceanography* 29:598–608. (#7)
Howarth, R.W. and Teal, J.M. 1980. Energy flow in a salt marsh ecosystem: the role of reduced inorganic sulfur compounds. *The American Naturalist* 116:862–872. (#7)

Howes, B.L., Dacey, J.W., and King, G.M. 1984. Carbon flow through oxygen and sulfate reduction pathways in salt marsh sediments. *Limnology and Oceanography* 29:1037–1051. (#7)

Huang, C.J. and Barrett, E.L. 1991. Sequence analysis and expression of the *Salmonella typhimurium asr* operon encoding production of hydrogen sulfide from sulfite. *Journal of Bacteriology* 173:1544–1553. (#3)

Huber, R., Stoffers, P., Cheminee, J.L., Richnow, H.H., and Stetter, K.O. 1990. Hyperthermophilic archaebacteria within the crater and open-sea plume of erupting Macdonald Seamount. *Nature* 345:179–182. (#6)

Hungate, R.E. 1950. The anaerobic mesophilic bacteria. *Bacteriological Reviews* 14:1–63. (#1)

Hungate, R.E., Bryant, M.P., and Mah, R.A. 1964. The rumen bacteria and protozoa. *Annual Review of Microbiology* 18:131–166. (#1)

Imhoff-Stuckle, D. and Pfennig, N. 1983. Isolation and characterization of a nicotinic acid-degrading sulfate-reducing bacterium, *Desulfococcus niacini* sp. nov. *Archives of Microbiology* 136:194–198. (#2, 6)

Irie, K., Kobayashi, K., Kobayashi, M., and Ishimoto, M. 1973. Biochemical studies on sulfate-reducing bacteria XII. Some properties of flavodoxin from *Desulfovibrio vulgaris*. *Journal of Biochemistry* (Tokyo) 73:353–366. (#3)

Ishimoto, M. 1959. Sulfate reduction in cell-free extracts of *Desulfovibrio*. *Journal of Biochemistry* (Tokyo) 46:105–106. (#3)

Ishimoto, M., Koyama, J., and Nagai, Y. 1954a. Biochemical studies on the sulfate reducing bacteria: IV. The cytochrome system of sulfate-reducing bacteria. *Journal of Biochemistry* (Tokyo) 41:763–770. (Dedication, #3)

Ishimoto, M., Koyama, J., and Nagai, Y. 1954b. Role of a cytochrome in thiosulfate reduction by a sulfate reducing bacterium. *Seikagaku zasshi* 26:303. (#5)

Iversen, N. and Blackburn, T.H. 1981. Seasonal rates of methane oxidation in anaerobic marine sediments. *Applied and Environmental Microbiology* 41:1295–1300. (#7)

Iversen, N. and Jorgensen, B.B. 1985. Anaerobic methane oxidation rates at the sulfate-methane transition in marine sediments from Kattegat and Skagerrak (Denmark). *Limnology and Oceanography* 30:944–955. (#7)

Iverson, W.P. 1968. Corrosion of iron and formation of iron phosphide by *Desulfovibrio desulfuricans*. *Nature* 217:1265–1267. (#8)

Iverson, W.P. 1974. Microbial corrosion of iron, pp. 475–513. In Neilands, J.B. (editor) *Microbial Iron Metabolism*. Academic Press, New York. (#1)

Iverson, W.P. 1981. An overview of the anaerobic corrosion of underground metal structures: evidence for a new mechanism. In Escalante, E. (editor) *Underground Corrosion, Technical Publication #741*. American Society for Testing and Materials, Philadelphia. (#8)

Iverson, W.P., and Olson G.J. 1984. Anaerobic corrosion of iron and steel: a novel mechanism, pp. 623–627. In Klug, M.J. and Reddy, C.A. (editors), *Current Perspectives in Microbial Ecology*. American Society for Microbiology, Washington D.C. (#8)

Iwabe, N., Kuma, K.-I., Hasegawa, M., Osawa, S., and Miyata, T. 1989. Evolutionary relationship of archaebacteria, eubacteria and eukaryotes inferred from phylogenetic trees of duplicated genes. *Proceedings of the National Academy of Sciences of the United States of America* 86:9355–9359. (#3)

Jablonski, E., Moomaw, E.W., Tullis, R.H., and Ruth, J.L. 1986. Preparation of oligodeoxynucleotide–alkaline phosphatase conjugates and their use as hybridization probes. *Nucleic Acids Research* 14:6115–6128. (#6)

Jacq, V.A. and Fortuner, R. 1979. Biological control of rice nematodes using sulfate-reducing bacteria. *Revue Nematologie* 2:41–50. (#8)

Jansen, K., Thauer, R.K., Widdel, F., and Fuchs, G. 1984. Carbon assimilation pathways in sulfate-reducing bacteria. Formate, carbon dioxide, carbon monoxide, and acetate assimilation by *Desulfovibrio baarsii*. *Archives of Microbiology* 138:257–262. (#2)

Jarmeel, P. 1989. The use of ferrous chloride to control dissolved sulfides in interceptor sewers. *Journal of Water Pollution Control* 61:230–236. (#8)

Jeffries, T.W. 1982. The microbiology of mercury, pp. 21–75. In Bull, M.J. (editor) *Progress in Industrial Microbiology*. Elsevier, Amsterdam. (#8)

Jones, R.E., Beeman, R.E., Suflita, J.M. 1989. Anaerobic metabolic processes in the deep terrestrial subsurface. *Geomicrobiology Journal* 7:117–130. (#8)

Jorgensen, B.B. 1977a. The sulfur cycle of a coastal marine sediment (Limfjordan, Denmark). *Limnology and Oceanography* 22:814–832. (#1, 8)

Jorgensen, B.B. 1977b. Bacterial sulfate reduction within reduced microniches of oxidized marine sediments. *Marine Biology* 41:7–17. (#7, 8)

Jørgensen, B.B. 1978. A comparison of methods for the quantification of sulfate reduction in coastal marine sediments. II. Estimation from chemical and bacteriological field data. *Geomicrobiology Journal* 1:49–64. (#6, 7)

Jorgensen, B.B. 1980. Mineralization and the bacterial cycling of carbon, nitrogen and sulphur in marine sediments, pp. 239–251. In Ellwood, D.C., Hedger, J.N., Latham, M.J., Lynch, J.M., and Slater, J.H. (editors) *Contemporary Microbial Ecology*. Academic Press, London. (#7)

Jorgensen, B.B. 1982. Ecology of the bacteria of the sulfur cycle with special reference to anoxic-oxic interface environments. *Philosphical Transactions of the Royal Society*, Series B 298:543–561. (#8)

Jorgensen, B.B. 1990a. The sulfur cycle of freshwater sediments: role of thiosulfate. *Limnology and Oceanography* 35:1329–1342. (#7)

Jorgensen, B.B. 1990b. A thiosulfate shunt in the sulfur cycle of marine sediments. *Science* 249:152–154. (#7)

Jorgensen, B.B. and Bak, F. 1991. Pathways and microbiology of thiosulfate transformations and sulfate reduction in a marine sediment (Kattegat, Denmark). *Applied and Environmental Microbiology* 57:846–856. (#1, 7)

Joubert, W.A. and Britz, T.J. 1987. Isolation of saccharolytic dissimilatory sulfate-reducing bacteria. *FEMS Microbiology Letters* 48:35–40. (#2)

Jurgensen, M.F. and Patton, J.T. 1979. Bioremoval of lignosulfonates from sulfite pulp mill effluents. *Process Biochemistry* 14:2–4. (#8)

Kamimura, K. and Araki, M. 1989. Isolation and characterization of a bacteriophage lytic for *Desulfovibrio salexigens*, a salt requiring, sulfate-reducing bacterium. *Applied and Environmental Microbiology* 55:645–648. (#4)

Kaplan, I.R. 1975. Stable isotopes as a guide to biogeochemical processes. *Proceedings of the Royal Society of London* B189:183–211. (#7)

Kaplan, I.R. and Rittenberg, S.C. 1962. The microbiological fractionation of sulfur isotopes, pp. 80–93. In Jensen, M.L. (editor) *Biogeochemistry of Sulfur Isotopes*. Yale University Press, New Haven. (#1, 3)

Kaplan, I.R. and Rittenberg, S.C. 1964. Microbial fractionation of sulphur isotopes. *Journal of General Microbiology* 34:195–212. (#1, 3, 7)

Karl, D.M., LaRock, P.A., and Schultz, D.J. 1977. Adenosine triphosphate and organic carbon in the Cariaco Trench. *Deep-Sea Research* 24:105–113. (#7)

Karn, J., Matthes, H.W.D., Gait, M.J., and Brenner, S. 1984. A new selective phage cloning vector, l2001, with sites for *Xba*I, *Bam*HI, *Hind*III, *Eco*RI, *Sst*I and *Xho*I. *Gene* 32:217–224. (#5)

Kaspar, H.F. and Wuhrmann, K. 1978. Kinetic parameters and relative turnovers of some important catabolic reactions in digesting sludge. *Applied and Environmental Microbiology* 36:1–7. (#8)

Keen, N.T., Tamaki, S., Kobayashi, D., and Trollinger, D. 1988. Improved broad-host-range plasmids for DNA cloning in Gram-negative bacteria. *Gene* 70:191–197. (#4)

Keith, S.M. and Herbert, R.A. 1983. Dissimilatory nitrate reduction by a strain of *Desulfovibrio desulfuricans*. *FEMS Microbiology Letters* 18:55–59. (#3)

Kelly, D.P. 1987. Sulphur bacteria first again. *Nature* 326:830. (#1, 7)

Kelly, D.P. 1989. Physiology and biochemistry of uicellular sulfur bacteria, pp. 193–217. In Schlegel, H.G. and Bowien, B. (editors) *Autotrophic Bacteria*. Springer Verlag, New York. (#3)

Kent, H.M., Buck, M., and Evans, D.J. 1989. Cloning and sequencing of the *nifH* gene of *Desulfovibrio gigas*. *FEMS Microbiology Letters* 61:73–89. (#5)

Khan, A.W. and Trottier, T.M. 1978. Effect of sulfur-containing compounds on anaerobic degradation of cellulose to methane by mixed cultures obtained from sewage sludge. *Applied and Environmental Microbiology* 35:1027–1034. (#7)

Kim, J.-H., and Akagi, J.M. 1985. Characterisation of a trithionate reductase system from *Desulfovibrio vulgaris*. *Journal of Bacteriology* 163:472–475. (#5)

Kim, T.S., Kim, H.Y., and Kim, B.H. 1990. Petroleum desulfurization by *Desulfovibrio desulfuricans* M6 using electrochemically supplied reducing equivalents. *Biotechnology Letters* 12:757–760. (#8)

King, G.M. 1983. Sulfate reduction in Georgia salt marsh soils: an evaluation of pyrite formation by use of ^{35}S and ^{55}Fe tracers. *Limnology and Oceanography* 28:987–995. (#7)

King, G.M. 1988. Patterns of sulfate reduction and the sulfur cycle in a South Carolina salt marsh. *Limnology and Oceanography* 33:376–390. (#7)

King, R.A. and Miller, J.D.A. 1971. Corrosion by sulphate-reducing bacteria. *Nature* 233:491–492. (#8)

Kirchman, D.L. and Hoch, M. 1988. Bacterial production in the Delaware Bay estuary: estimation from thymidine and leucine incorporation rates. *Marine Ecology Progress Series* 45:169–178. (#7)

Kissinger, C.R., Adman, E.T., Sieker, L.C., Jensen, L.H.,and LeGall, J. 1989. The crystal structure of the three-iron ferredoxin II from *D. gigas*. *FEBS Letters* 242:477–450. (#5)

Klemps, R., Cypionka, H., Widdel, F., and Pfennig, N. 1985. Growth with hydrogen, and further physiological characteristics of *Desulfotomaculum* species. *Archives of Microbiology* 143:203–208. (#2, 8)

Kobayashi, K., Hasegawa, H., Takagi, M., and Ishimoto, M. 1982. Proton translocation associated with sulfite reduction in a sulfate-reducing bacterium, *Desulfovibrio vulgaris*. *FEMS Microbiology Letters* 142:235–237. (#3)

Kobayashi, K., Takahashi, E., and Ishimoto, M. 1972. Biochemical studies on sulfate-reducing bacteria, XI. Purification and some properties of sulfite reductase, desulfoviridin, *Journal of Biochemistry, Tokyo* 72:879–887. (#3)

Konings, W.N. 1985. Generation of metabolic energy by end-product efflux. *Trends in Biochemical Sciences* 10:317–319. (#1)

Krämer, M. and Cypionka, H. 1989. Sulfate formation via ATP sulfurylase in thiosulfate- and sulfite-disproportionating bacteria. *Archives of Microbiology* 151:232–237. (#3, 7)

Kremer, D.R. and Hansen, T.A. 1987. Glycerol and dihydroxyacetone dissimilation in *Desulfovibrio* strains. *Archives of Microbiology* 147:249–256. (#2)

Kremer, D.R. and Hansen, T.A. 1988. Pathway of propionate degradation in *Desulfobulbus propionicus*. *FEMS Microbiology Letters* 49:273–277. (#2)

Kremer, D.R. and Hansen, T.A. 1989. Demonstration of HOQNO and antimycin A sensitive coupling of NADH oxidation and APS and sulfite reduction in a marine *Desulfovibrio* strain. *FEMS Microbiology Letters* 59:43–48. (#3)

Kremer, D.R., Nienhuis-Kuiper, H.E., and Hansen, T.A. 1988a. Ethanol dis-

similation in *Desulfovibrio*. *Archives of Microbiology* 150:552–557. (#2)
Kremer, D.R., Veenhuis, M., Fauque, G., Peck, H.D. Jr., Le Gall, J., Moura, J.J.G., and Hansen, T.A. 1988b. Immunocytochemical localization of APS reductase and bisulfite reductase in three *Desulfovibrio* strains. *Archives of Microbiology* 150:296–301. (#2, 3)
Kremer, D.R., Nienhuis-Kuiper, H.E., Timmer, C.J., and Hansen, T.A. 1989. Catabolism of malate and related dicarboxylic acids in various *Desulfovibrio* strains and the involvement of an oxygen-labile NADP dehydrogenase. *Archives of Microbiology* 151:34–39. (#2)
Krey, G.D., Vanin, E.F., and Swenson, R.P. 1988. Cloning, nucleotide sequence and expression of the flavodoxin gene from *Desulfovibrio vulgaris* (Hildenborough). *Journal of Biological Chemistry* 263:15436–15443. (#5)
Kröger, A. 1977. Phosphorylative electron transport with fumarate and nitrate as terminal hydrogen acceptors, pp. 61–93. In Haddock, B.A. and Hamilton, W.A. (editors) *Microbial Energetics*. Cambridge University Press, Cambridge, U.K. (#3)
Kuivila, K.M., Murray, J.W., and Devol, A.H. 1990. Methane production in the sulfate-depleted sediments of two marine basins. *Geochimica et Cosmochimica Acta* 53:403–411. (#7)
Kumar, A., Tchen, P., Roullet, F., and Cohen, J. 1988. Nonradioactive labeling of synthetic oligonucleotide probe with terminal deoxynucleotidyl transferase. *Analytical Biochemistry* 169:376–382. (#6)
Kuznetsova, V.A. and Pantskava, E.S. 1962. Effect of freshening of stratal waters on the development of halophilic sulfate-reducing bacteria. *Mikrobiologiya* 31:129–134. (#8)
Kuznetsova, V.A., Li, A.D., and Tiforova, N.N. 1963. A determination of source of contamination of oil-bearing strata of Romashkino field by sulfate-reducing bacteria. *Mikrobiologiya* 32:683–688. (#8)
Laanbroek, H.J., Abee, T., and Voogd, I.L. 1982. Alcohol conversions by *Desulfobulbus propionicus* Lindhorst in the presence and absence of sulfate and hydrogen. *Archives of Microbiology* 133:178–184. (#2)
Lane, D.J. 1991. 16S/23S rRNA Sequencing, pp. 115–176. In Stackebrandt, E., and Goodfellow, M., (editors) *Nucleic Acid Techniques in Bacterial Systematics*. John Wiley & Sons Ltd., Chichester, UK. (#6)
Lane, D.J., Pace, B., Olsen, G.J., Stahl, D.A., Sogin, M.L., and Pace, N.R. 1985a. Rapid determination of 16S ribosomal RNA sequences for phylogenetic analysis. *Proceedings of the National Academy of Sciences of the United States of America* 82:6955–6959. (#6)
Lane, D.J., Stahl, D.A., Olsen, G.J., Heller, D.J., and Pace, N.R. 1985b. Phylogenetic analysis of the genera *Thiobacillus* and *Thiomicrobium* by 5S rRNA sequences. *Journal of Bacteriology* 163:75–81. (#6)
Länge, S., Scholtz, R., and Fuchs, G. 1989. Oxidative and reductive acetyl CoA/carbon monoxide dehydrogenase pathway in *Desulfobacterium autotrophicum* 1. Characterization and metabolic function of the cellular tetrahydropterin. *Archives of Microbiology* 151:77–83. (#2)
Laudenbach, D.E., Reither, M.E., and Straus, N.A. 1988. Isolation, sequence analysis and transcriptional studies of the flavodoxin gene from *Anacystis nidulans* R2. *Journal of Bacteriology* 170:258–265. (#5)
Leclerc, M., Colbeau, A., Cauvin, B., and Vignais, P.M. 1988. Cloning and sequencing of the genes encoding the large and the small subunits of the H2 uptake hydrogenase (*hup*) of *Rhodobacter capsulatus*. *Molecular and General Genetics* 214:97–108. (#5)
Lee, J.P., LeGall, J., and Peck, H.D. Jr. 1973a. Isolation of assimilatory and dis-

similatory type sulfite reductases from *Desulfovibrio vulgaris. Journal of Bacteriology* 115:529–542. (#3)

Lee, J.P., Yi, C.-S., LeGall, J., and Peck, H.D., Jr. 1973b. Isolation of a new pigment, desulforubidin, from *Desulfovibrio desulfuricans* (Norway strain) and its role in sulfite reduction. *Journal of Bacteriology* 115:453–455. (#6)

LeGall, J. and Fauque, G. 1988. Dissimilatory reduction of sulfur compounds, pp. 587–639. In Zehnder, A.J.B. (editor) *Biology of Anaerobic Organisms*. John Wiley, New York. (Intro, #1, 3, 5, 6)

LeGall, J. and Peck, H.D., Jr. 1987. Amino-terminal amino acid sequences of electron transfer proteins from Gram-negative bacteria as indicators of their cellular localization: the sulfate-reducing bacteria. *FEMS Microbiological Reviews* 46:35–40. (#5)

LeGall, J., DerVartanian, D.V., and Peck, H.D., Jr. 1979. Flavoproteins, iron proteins and hemoproteins as electron-transfer components of the sulfate-reducing bacteria. *Current Topics in Bioenergetics* 9:237–265. (#1)

Li, C., Peck, H.D., Jr. and Przybyla, A.E. 1986. Complementation of an *Escherichia coli* pyrF mutant with DNA from *Desulfovibrio vulgaris. Journal of Bacteriology* 165:644–646. (#5)

Li, C., Peck, H.D., Jr., LeGall, J., and Przybyla, A.E. 1987. Cloning, characterization and sequencing of the genes encoding the large and small subunits of the periplasmic [NiFe] hydrogenase of *Desulfovibrio gigas. DNA* 6:539–551. (#5)

Lin, C.C. and Lin, K.C. 1970. Spoilage bacteria in canned foods. II. Sulfide spoilage bacteria in canned mushrooms and a versatile medium for the enumeration of *Clostridium nigrificans. Applied Microbiology* 19:283–286. (#1)

Linthurst, R.A. and Seneca, E.D. 1981. The effects of standing water and drainage potential on the *Spartina alterniflora*-substrate complex in a North Carolina salt marsh. *Estuarine and Coastal Marine Science* 2:41–52. (#7)

Lissolo, T., Choi, E.S., LeGall, J., and Peck Jr, H.D. 1986. The presence of multiple intrinsic membrane nickel containing hydrogenase in *Desulfovibrio vulgaris* (Hildenborough). *Biochemical and Biophysical Research Communications* 139:701–708. (#5)

Liu, M.-C. and Peck, H.D., Jr. 1988. Ammonia-forming, dissimilatory nitrite reductases as a homologous group of hexaheme-*c*-type cytochromes in metabolically diverse bacteria, pp. 685–691. In Kauf, K., von Dohren, K. and Jaenicke, Peck, H.D., Jr. (editors) *The Roots of Modern Biochemistry* (Lipmann Memorial Symposium). Walter de Gruyter and Co., Berlin. (#3)

Liu, M.-C., Costa, C., Coutinho, I.B., Moura, J.J.G., Moura, I., Xavier, A.V., and LeGall, J. 1988. Cytochrome components of nitrate- and sulfate-respiring *Desulfovibrio desulfuricans* ATCC 27774. *Journal of Bacteriology*. 170:5545–5551. (#5)

Liu, C.L., Hart, N., and Peck, H.D. Jr. 1982. Inorganic pyrophosphate energy source for sulfate-reducing bacteria of the genus *Desulfotomaculum. Science* 217:363–364. (#1)

Ljungdahl, L.G. 1986. The autotrophic pathway of acetate synthesis in acetogenic bacteria. *Annual Review of Microbiology* 40:415–450. (#3)

Ljungdahl, L.G., Hugenholtz, J., and Wiegel, J. 1989. Acetogenic and acid-producing clostridia, pp. 145–191. In Minton, N.P. and Clarke, D.J. (editors) *Clostridia*. Plenum Publishing Corp, New York. (#3)

Loutfi, M., Guerlesquin, F., Bianco, P., Haladjian, J., and Bruschi, M. 1989. Comparative studies of polyhemic cytochromes *c* isolated from *Desulfovibrio vulgaris* (Hildenborough) and *Desulfovibrio desulfuricans* (Norway). *Biochemical and Biophysical Research Communications*. 159:670–676. (#5)

Lovely, D.R. and Klug, M.J. 1982. Intermediary metabolism of organic matter in the sediments of a eutrophic lake. *Applied and Environmental Microbiology* 43:552–560. (#7)
Lovely, D.R. and Klug, M.J. 1986. Model for the distribution of sulfate reduction and methanogenesis in freshwater sediments. *Geochimica et Cosmochimica Acta* 50:11–18. (#7)
Lovely, D.R., Dwyer, D.F., and Klug, M.J. 1982. Kinetic analysis of competition between sulfate reducers and methanogens for hydrogen in sediments. *Applied and Environmental Microbiology* 43:1373–1379. (#7)
Lupton, F.S., Conrad, R., and Zeikus, J.G. 1984a. CO metabolism of *Desulfovibrio vulgaris* strain Madison: physiological function in the absence and presence of exogenous substrates. *FEMS Microbiology Letters* 23:263–268. (#3)
Lupton, F.S., Conrad, R., and Zeikus, J.G. 1984b. Physiological function of hydrogen metabolism during growth of sulfidogenic bacteria on organic substrates. *Journal of Bacteriology* 159:843–849. (Intro, #3)
Lynch, J.M. and Hobbie, J.E. (editors) 1988. *Micro-Organisms in Action: Concepts and Applications in Microbial Ecology*, 363 pp. Blackwell Scientific Publications, Oxford. (#7)
Maaloe, O. 1979. Regulation of the protein synthesizing machinery—ribosomes, tRNA, factors, and so on, pp. 487–542. In Goldberger, R. (editor) *Biological Regulation and Development*. Plenum Publishing Corp., New York. (#6)
Maaloe, O. and Kjellgard, N.O. 1966. *Control of Macromolecular Synthesis*, 284 pp. W.A. Benjamin, Inc., New York. (#6)
Maeshima, M. 1991. H^+-Translocating inorganic pyrophosphatase of plant vacuoles-inhibition by Ca^{2+}, stabilization by Mg^{2+} and immunological comparison with other inorganic pyrophosphatases. *European Journal of Biochemistry* 196:11–18. (#3)
Maloney, P.G. 1983. Relationship between phosphorylation potential and electrochemical H^+ gradient during glycolysis in *Streptococcus lactis*. *Journal of Bacteriology* 153:1461–1470. (#3)
Maniatis, T., Fritsch, E.F., and Sambrook, J. 1982, p. 271. In *Molecular Cloning: A Laboratory Manual*. Cold Spring Harbor Laboratory, Cold Spring Harbor, New York. (#5)
Maroney, M.J., Colpas, G.J., Bagyinka, C., Baidya, N., and Mascharak, P.K. 1991. EXAFS investigations of the Ni Site in *Thiocapsa roseopersinica* hydrogenase: Evidence for a novel Ni, Fe, S cluster. *Journal of the American Chemical Society* 113:3962–3972. (#5)
Marsland, S.D., Dowe, R.A., and Kelsall, G.H. 1989. Inorganic chemical souring of oil reservoirs. *Society of Petroleum Engineers* Paper #18480. International Symposium Oilfield Chemistry, Houston, Texas. (#8)
Martens, C.S. and Klump, J.V. 1984. Biogeochemical cycling in an organic-rich coastal marine basin. 4. An organic carbon budget for sediments dominated by sulfate reduction and methanogenesis. *Geochimica et Cosmochimica Acta* 48:1987–2004. (#7)
Massad, A.H. and Holmes, B.G. 1958. Cathodic protection of offshore structures and submerged piping systems in the Gulf of Mexico, pp. 44. *Drilling and Production Practice*, American Petroleum Institute, New York. (#8)
Matthews, J.A. and Kricka, L.J. 1988. Analytical strategies for the use of DNA probes. *Analytical Biochemistry* 169:1–25. (#6)
Maxwell, S. 1986. Effect of cathodic protection on the activity of microbial biofilms. *Materials Performance*, November, pp. 53–56. (#8)
Mayhew, S.G. and O'Connor, M.E. 1982. Structure and mechanism of bacterial hydrogenase. *Trends in Biochemical Sciences* 7:18–21. (#5)

Mayhew, S.G. and Tollin,G. 1991. General properties of flavodoxins, chapter 14. In Mueller, F. (editor) *Chemistry and Biochemistry of Flavoenzymes*, Vol. III. CRC Press, Boca Raton, FL, in press. (#5)

Mayr, E. 1982. *The Growth of Biological Thought: Diversity, Evolution, and Inheritance*. Belknap Press, Cambridge, Mass. (#6)

Medlin, L., Elwood, H.J., Stickel, S., and Sogin, M.L. 1988. The characterization of enzymatically amplified eukaryotic 16S-like rRNA-coding regions. *Gene* 71:491–499. (#6)

Menon, N.K., Peck, H.D., Jr., LeGall, J., and Przybyla, A.E. 1987. Cloning and sequencing of the genes encoding the large and small subunits of the periplasmic (NiFeSe) hydrogenase of *Desulfovibrio baculatus*. *Journal of Bacteriology* 169:5401–5407. (#5)

Menon, N.K., Robbins, J., Peck, H.D., Jr., Chatelus, C.Y., Choi, E.-S., and Przybyla., A.E. 1990. Cloning and sequencing of a putative *Escherichia coli* [NiFe] hydrogenase-1 operon containing six open reading frames. *Journal of Bacteriology* 172:1969–1977. (#5)

Meyer, J. and Gagon, J. 1991. Primary structure of hydrogenase I from *Clostridium pasteurianum*. *Proceedings of the 3rd International Symposium on the Molecular Biology of Hydrogenase*, Troia, Port. *Biochemistry* 30:9697–9704 (#3, 5)

Meyer, L. 1864. Chemische Untersuchung der Thermen zu Landeck in der Grafschaft. *Journal fuer Praktische Chemie*. 91:5–6. (#8)

Meyer, T.E. and Kamen, M.D. 1982. New perspectives on *c*-type cytochromes. *Advances in Protein Chemistry* 35:105–212. (#1)

Mills, R.V.N. and Wells, R.C. 1919. The evaporation and concentration of waters associated with petroleum and natural gas. *U.S. Geological Survey Bulletin No. 693*, p. 70. (#8)

Min, H. and Zinder, S.H. 1990. Isolation and characterization of a thermophilic sulfate-reducing bacterium *Desulfotomaculum thermoacetoxidans* sp. nov. *Archives of Microbiology* 153:399–404. (#2)

Miyada, C.G. and Wallace, R.B. 1987. Oligonucleotide hybridization techniques, *Methods in Enzymology* 154:94–107 (#6)

Mohn, W.W., Linkfield, T.G., Pankratz, H.S., and Tiedje, J.M. 1990. Involvement of a collar structure in polar growth and cell division of strain DCB-1. *Applied and Environmental Microbiology* 56:1206–1211. (#6)

Mohn, W.W. and Tiedje, J.M. 1990. Catabolic thiosulfate disproportionation and carbon dioxide reduction in strain DCB-1, a reductively dechlorinating anaerobe. *Journal of Bacteriology* 172:2065–2070. (#2)

Möller, D., Schauder, R., Fuchs, G., and Thauer, R.K. 1987. Acetate oxidation to CO_2 via a citric acid cycle involving an ATP-citrate lyase: a mechanism for the synthesis of ATP via substrate level phosphorylation in *Desulfobacter postgatei* growing on acetate and sulfate. *Archives of Microbiology* 148:202–207. (#2)

Möller-Zinkhan, D. and Thauer, R.K. 1988. Membrane-bound NADPH dehydrogenase- and ferredoxin:NADP oxidoreductase activity involved in electron transport during acetate oxidation to CO_2 in *Desulfobacter postgatei*. *Archives of Microbiology* 150:145–154. (#2)

Möller-Zinkhan, D. and Thauer, R.K. 1990. Anaerobic lactate oxidation to 3 CO_2 by *Archaeoglobus fulgidus* via the carbon monoxide dehydrogenase pathway: demonstration of the acetyl-CoA carbon-carbon cleavage reaction in cell extracts. *Archives of Microbiology* 153:215–218. (#2)

Möller-Zinkhan, D., Börner, G., and Thauer, R.K. 1989. Function of methanofuran, tetrahydromethanopterin, and coenzyme F_{420} in *Archaeoglobus fulgidus*. *Archives of Microbiology* 152:362–368. (#2)

Montagna, P.A. 1982. Sampling design and enumeration statistics for bacteria extracted from marine sediments. *Applied and Environmental Microbiology* 43:1366–1372. (#7)

Moore, W.E.C., Johnson, J.L., and Holdeman, L.V. 1976. Emendation of *Bacteroidaceae* and *Butyrivibrio* and description of *Desulfomonas* gen. nov. and ten new species in the genera *Desulfomonas, Butyrivibrio, Eubacterium, Clostridium*, and *Ruminococcus*. *International Journal of Systematic Bacteriology* 26:238–252. (#6)

Morris, J.G. 1986. Anaerobiosis and energy-yielding metabolism, pp. 1–21. In Barnes, E.M and Mead, G.C. (editors) *Anaerobic Bacteria in Habitats Other Than Man*. Blackwell Scientific Publications, Oxford. (#1)

Mountfort, D.O., R.A. Sher, E.L. Mays, and J.M. Tiedje. 1980. Carbon and electron flow in mud and sand flat intertidal sediments at Delaware inlet, Nelson, New Zealand. *Applied and Environmental Microbiology* 39:686–694. (#7)

Moura, I., Lino, A.R., Moura, J.J.G., Xavier, A.V., Fauque, G., Peck, H.D. Jr., and LeGall, J. 1986. Low-spin sulfite-reductases: a new homologous group of non-heme iron-siroheme proteins in anaerobic bacteria. *Biochemical and Biophysical Research Communications* 141:1032–1041. (#3)

Moura, I., Fauque, G., LeGall, J., Xavier, A.V., and Moura, J.J.G. 1987a. Characterization of the cytochrome system of a nitrogen-fixing strain of a sulfate reducing bacterium: *Desulfovibrio desulfuricans* strain Berre-Eau. *European Journal of Biochemistry* 162:547–554. (#5)

Moura, I., LeGall, J., Lino, A.R., Peck, H.D. Jr., Xavier, A.V., DerVartanian, D.V., Moura, J.J.G., and Huynh, B.H. 1987b. Characterization of two dissimilatory sulfite reductases (desulfoviridin and desulforubidin) from the sulfate reducing bacteria: Mössbauer and EPR studies. *Journal of the American Chemical Society* 110:1075–1082. (#3)

Moura, I., Teixeira, M., Huynh, B.H., LeGall, J., and Moura, J.J.G. 1988. Assignment of individual heme EPR signals of *Desulfovibrio baculatus* (strain 9974) tetraheme cytochrome c_3. *European Journal of Biochemistry* 176:365–369. (#6)

Moura, I., Tavares, P., Moura, J.J.G., Ravi, N., Huynh, B.H., Liu, M.Y., and LeGall, J. 1990. Purification and characterization of desulfoferredoxin. *Journal of Biological Chemistry*. 265:21596–21602. (#5)

Moura, J.J.G., Xavier, A.V., Bruschi, M., LeGall, J., Hall, D.O., and Cammack, R. 1976. A molybdenum-containing iron-sulfur protein from *Desulfovibrio gigas*. *Biochemical and Biophysical Research Communications* 72:782–789. (#2)

Moura, J.J.G., LeGall, J., and Xavier, A.V. 1984. Interconversion from 3Fe into 4Fe clusters in the presence of *Desulfovibrio gigas* cell extracts. *European Journal of Biochemistry* 141:319–322. (#3)

Moura, J.J.G., Costa, C., Liu, M.-Y., Moura, I., and LeGall, J. 1991. Structural and functional approach toward a classification of the complex cytochrome *c* system found in sulfate-reducing bacteria. *Biochimica Biophysica Acta* 1058:61–66. (#5)

Mullenbach, G.T., Tabrizi, A, Irvine, B.D., Bell, G.I., Trainer, J.A., and Hallewell, R.A. 1988. Selenocysteine's mechanism of incorporation and evolution revealed in cDNAs of three glutathione peroxidases. *Protein Engineering* 2:239–246. (#3)

Nakagawa, A., Nagashima, E., Higuchi, Y., Kusunoki, M., Matsuura, Y., Yasuoka, N., Katsube, Y., Chichara, H., and Yagi, T. 1986. Crystallographic study of cytochrome c_{553} from *Desulfovibrio vulgaris*. *Journal of Biochemistry* 99:605–606. (#5)

Nakagawa, A., Higuchi, Y., Yasuoka, N., Katsube, Y., and Yagi, T. 1990. S-Class cytochromes *c* have a variety of folding patterns: structure of

cytochrome c_{553} from *Desulfovibrio vulgaris* determined by the anomalous dispersion method. *Journal of Biochemistry* 108:701–703. (#5)
Nakano, K., Kikumoto, Y., and Yagi, T. 1983. Amino acid sequence of cytochrome c_{553} from *Desulfovibrio vulgaris* Miyazaki. *Journal of Biological Chemistry* 258:12409–12412. (#5)
Nanninga, H.J. and Gottschal, J.C. 1987. Properties of *Desulfovibrio carbinolicus* sp. nov. and other sulfate-reducing bacteria isolated from an anaerobic-purification plant. *Applied and Environmental Microbiology* 53:802–809. (#2)
Nannipierei, P. 1984. Microbial biomass and activity measurements in soil: ecological significance, pp. 515–521. In Klug, M.J. and Reddy, C.A. (editors) *Current Perspectives in Microbial Ecology*. American Society for Microbiology, Washington, D.C. (#7)
Nazina, I.N. and Pivovarova, T.A. 1979. Submicroscopic organization and sporulation in *Desulfotomaculum nigrificans*. *Microbiologiya* (English translation) 48:302–306. (#1)
NBS 1980. *Economic Effects of Metallic Corrosion in the U.S.* National Bureau of Standards Special Publication 511-1. Superintendent of Documents, U.S. Govt. Printing Office (#8)
Nethe-Jaenchen, R. and Thauer, R.K. 1984. Growth yields and saturation constant for *Desulfovibrio vulgaris* in chemostat culture. *Archives of Microbiology* 137:236–240. (#1, 3)
Niviere, V., Hatchikian, C., Cambilleau, C., and Frey, M. 1987. Crystallization, preliminary X-ray study and crystal activity of hydrogenase from *Desulfovibrio gigas*. *Journal of Molecular Biology* 195:969–971. (#5)
Norqvist, A. and Roffey, R. 1985. Biochemical and immunological study of cell envelope proteins in sulfate-reducing bacteria. *Applied and Environmental Microbiology* 50:31–37. (#1, 8)
Novelli, P.C., Scranton, M.I., and Michener, R.H. 1987. Hydrogen distributions in marine sediments. *Limnology and Oceanography* 32:565–576. (#7)
Odom, J.M. 1990. Industrial and environmental concerns with sulfate-reducing bacteria. *ASM News* 56:473–476. (#1)
Odom, J.M. and Peck, H.D. Jr. 1981a. Hydrogen cycling as a general mechanism for energy coupling in the sulfate reducing bacteria, *Desulfovibrio* sp. *FEMS Microbiology Letters* 12:47–50. (#3, 5)
Odom, J.M. and Peck, H.D. Jr. 1981b. Localization of dehydrogenases, reductases and electron transfer components in the sulfate reducing bacterium, *Desulfovibrio gigas*. *Journal of Bacteriology* 147:161–169. (#3, 8)
Odom, J.M. and Peck, H.D. Jr. 1984. Hydrogenase, electron transfer proteins and energy coupling in the sulfate reducing bacteria *Desulfovibrio*. *Annual Review of Microbiology* 38:551–592. (#3, 5)
Odom, J.M. and Wall, J.D. 1987. Properties of a hydrogen-inhibited mutant of *Desulfovibrio desulfuricans* ATCC 27774. *Journal of Bacteriology* 169:1335–1337. (#4)
Odom, J.M., Jesse, K., Knodel, E., and Emptage, M. 1991. Immunological cross-reactivities of adenosine-5′-phosphosulfate reductases from sulfate-reducing and sulfide-oxidizing bacteria. *Applied and Environmental Microbiology* 57:727–733. (#8)
Ogata, M., Arihara, K., and Yagi, T. 1981. D-Lactate dehydrogenase of *Desulfovibrio vulgaris*. *Journal of Biochemistry* 89:1423–1431. (#2)
Ollivier, B., Cord-Ruwisch, R., Hatchikian, E.C., and Garcia, J.L. 1988. Characterization of *Desulfovibrio fructosovorans* sp. nov. *Archives of Microbiology* 149:447–450. (#2, 5)
Olmstead, F.H. and Hamlin, H. 1900. Converting portions of the Los Angeles outfall sewer into a septic tank. *Engineering News* 44:317–318. (#8)

Olsen, G.J. 1985. Earliest phylogenetic branchings: comparing rRNA-based evolutionary trees inferred with various techniques. *Cold Spring Harbor Symposia on Quantitative Biology*, Volume LII, pp. 825–837. (#6)
Olsen, G.J., Lane, D.J., Giovannoni, S.J., Pace, N.R., and Stahl, D.A. 1986. Microbial ecology and evolution: a ribosomal RNA approach. *Annual Review of Microbiology* 40:337–365. (#6)
Oppenberg, B. and Schink, B. 1990. Anaerobic degradation of 1,3-propanediol by sulfate-reducing and by fermenting bacteria. *Antonie van Leeuwenhoek* 57:205–213. (#2)
Oremland, R.S. 1988. The biogeochemistry of methanogenic bacteria, pp. 405–447. In Zehnder, A. (editor) *The Biology of Anaerobic Microorganisms*. John Wiley, New York. (#7)
Oremland, R.S. and Capone, D.G. 1988. Use of "specific" inhibitors in biogeochemistry and microbial ecology. *Advances in Microbial Ecology* 10:285–383. (#7)
Oremland, R.S. and Policin, S. 1982. Methanogenesis and sulfate reduction: competitive and non-competitive substrates in estuarine sediments. *Applied and Environmental Microbiology* 4:1270–1276. (#7)
Oremland, R.S., Culbertson, C.W., and Winfrey, M.R. 1991. Methylmercury decomposition in sediments and bacterial cultures: involvement of methanogens and sulfate reducers in oxidative demethylation. *Applied and Environmental Microbiology* 57:130–137. (#7,8)
Orr, W.L. 1978. Geological and geochemical controls on the distribution of H_2S in nature. *Advances in Organic Geochemistry, Proceedings of the International Meeting* 7:571. (#8)
Ostrowski, J., Wu, J.-Y., Rueger, D.C., Miller, B.E., Siegel, L.M., and Kredich, N.M. 1989. Characterization of the cysJIH regions of *Salmonella typhimurium* and *Escherichia coli* B:DNA sequences of cysI and cysH and a model for the siroheme-Fe_4S_4 active center of sulfite reductase hemoprotein based on amino acid homology with spinach nitrite reductase. *Journal of Biological Chemistry* 264:15726–15737. (#3)
Oyaizu, H. and Woese, C.R. 1985. Phylogenetic relationships among the sulfate respiring bacteria, myxobacteria and purple bacteria. *Systematic and Applied Microbiology* 6:257–263. (#6)
Pace, B. and Campbell, L.L. 1971. Homology of the ribosomal ribonucleic acid of *Desulfovibrio* species with *Desulfovibrio vulgaris*. *Journal of Bacteriology* 106:717–719. (#6)
Pace, N.R., Stahl, D.A., Lane, D.J., and Olsen, G.J. 1986. The analysis of microbial populations by ribosomal RNA sequences. *Advances in Microbial Ecology* 9:1–55. (#6)
Pankhania, I.P., Spormann, A.M., Hamilton, W.A., and Thauer, R.K. 1988. Lactate conversion to acetate, CO_2, and H_2 in cell suspensions of *Desulfovibrio vulgaris* (Marburg): indications for the involvement of an energy driven reaction. *Archives of Microbiology* 150:26–31. (#2, 3, 8)
Pankhurst, E.S. 1971. The isolation and enumeration of sulfate-reducing bacteria, pp. 223–240. In Shapton, D.A. and Board, R.G. (editors) *Isolation of Anaerobes*. Academic Press, New York. (#1)
Parker, C.N.J., Seal, K.J., and Robinson, M.J. 1988. Hydrogen absorption during the microbial corrosion of steel. *Biodeterioration* 7:391–397. (#8)
Paul, E.A. and Voroney, R.P. 1984. Field interpretation of microbial biomass activity measurements, pp. 509–514. In Klug, M.J. and Reddy, C.A. (editors) *Current Perspectives in Microbial Ecology*. American Society for Microbiology, Washington, D.C. (#7)

Peakman, T., Crouzet, J., Mayaux, J.F., Busby, S., Mohan, S., Harborne, N., Wootton, J., Nicolson, R., and Cole, J. 1990. Nucleotide sequence, organization and structural analysis of the products of genes in the nirB-cysG region of the *Escherichia coli* K12 chromosome. *European Journal of Biochemistry* 191:315–323. (#3)

Peck, H.D., Jr. 1959. The ATP-dependent reduction of sulfate with hydrogen in extracts of *Desulfovibrio desulfuricans. Proceedings of the National Academy of Sciences of the United States of America* 45:701–708. (#3)

Peck, H.D., Jr. 1960. Evidence for oxidative phosphorylation during the reduction of sulfate with hydrogen by *Desulfovibrio desulfuricans. Journal of Biological Chemistry* 235:2734–2738. (Dedication, #1, 3)

Peck, H.D., Jr. 1962. Symposium on metabolism of inorganic compounds. V. Comparative metabolism of inorganic sulfur compounds in microorganisms. *Bacteriological Reviews* 26:67–94. (#3, 4)

Peck, H.D., Jr. 1966. Phosphorylation coupled with electron transfer in extracts of the sulfate reducing bacterium, *Desulfovibrio gigas. Biochemical and Biophysical Research Communications* 22:112–118. (#1, 3)

Peck, H.D., Jr. 1974. Sulfation linked to ATP cleavage, pp. 651–669. In Boyer, P.D. (editor) *The Enzymes*, 3rd edition, vol. 10. Academic Press, New York. (#3)

Peck, H.D., Jr. 1984. Physiological diversity of the sulfate-reducing bacteria, pp. 309–335. In Strohl, W.R. and Tuovinen, O.H. (editors) *Microbial Chemoautotrophy*. Ohio State University Press, Columbus. (#1, 4, 6)

Peck, H.D., Jr. and LeGall, J. 1982. Biochemistry of dissimilatory sulphate reduction. *Philosophical Transactions of the Royal Society (London)* B298:443–466. (#1, 2, 3)

Peck, H.D. Jr. and Lissolo, T. 1988. Assimilatory and dissimilatory sulfate reduction, pp. 99–132. In Cole, J. and Ferguson, S. (editors) *The Nitrogen and Sulfur Cycles*. Cambridge University Press, Cambridge, UK. (#3)

Peck, H.D. Jr. and Odom, J.M. 1984. Hydrogen cycling in *Desulfovibrio*: a new mechanism for energy coupling in anaerobic microorganisms, pp. 215–243. In Cohen, Y., Castenholz, R.W., and Halvorson, H.O. (editors) *Microbial Mats: Stromatolites*. A.R. Liss, New York. (#1)

Peck, H.D., Jr., Le Gall, J., Lespinat, P.A., Berlier, Y., and Fauque, G. 1987. A direct demonstration of hydrogen cycling employing membrane-inlet mass spectrometry. *FEMS Microbiology Letters* 40:295–299. (#3)

Peterson, B.J., Howarth, R.W., Lipschultz, F., and Ashendorf, D. 1980. Salt marsh detritus: an alternative interpretation of stable carbon isotope ratios and the fate of *Spartina alterniflora. Oikos* 34:173–177. (#7)

Pfennig, N. and Biebl, H. 1976. *Desulfuromonas acetoxidans* gen. nov. and sp. nov., a new anaerobic, sulfur-reducing, acetate-oxidizing bacterium. *Archives of Microbiology* 110:3–12. (#1, 8)

Pfennig, N. and Biebl, H. 1981. The dissimilatory sulfur-reducing bacteria, pp. 941–947. In Starr, M.P., Stolp, H., Trüper, H., Balows, A., and Schlegel, H.G. (editors) *The Prokaryotes*, Vol. 1. Springer-Verlag, New York. (#1, 8)

Pfennig, N. and Widdel, F. 1981. Ecology and physiology of some anaerobic bacteria from the microbial sulfur cycle, pp. 169–177. In Bothe, H. and Trebst, A. (editors) *Biology of Inorganic Nitrogen and Sulfur*. Springer-Verlag, Berlin. (#1)

Pfennig, N. and Widdel, F. 1982. The bacteria of the sulphur cycle. *Philosophical Transactions of the Royal Society (London)* B298:433–441. (#1)

Pfennig, N., Widdel, F., and Trüper, H.G. 1981. The dissimilatory sulfate-reducing bacteria, pp. 926–940. In Starr, M.P., Stolp, H., Trüper, H.G.,

Balows, A., and Schlegel, H.G. (editors) *The Prokaryotes*, Vol. 1. Springer-Verlag, Heidelberg. (#1, 6, 8)
Pierrot, M., Haser, R., Frey, M., Payan, F., and Astier, J.P. 1982. Crystal structure and electron transfer properties of cytochrome c_3. *Journal of Biological Chemistry* 257:14341–14348. (#5)
Platen, H., Temmes, A., and Schink, B. 1990. Anaerobic degradation of acetone by *Desulfococcus biacutus*. *Archives of Microbiology* 154:355–361. (#2)
Plauchud, M. 1877. Recherches sur la formation des eaux sulfereuses naturelles. *Comptes Rendus* Vol LXXXIV:235–238. (#8)
Pohl, M., Bock, E., Rinken, M., Aydin, M., and Konig, W.A. 1984. Volatile sulfur compounds produced by methionine degrading bacteria and the relationship to concrete corrosion. *Zeitschrift fuer Naturforschung* 39:240–243. (#8)
Pollock, W.B.R., Chemerika, P.J., Forrest, M.E., Beatty, J.T., and Voordouw, G. 1989. Expression of the gene encoding cytochrome c_3 from *Desulfovibrio vulgaris* (Hildenborough) in *Escherichia coli*: export and processing of the apoprotein. *Journal of General Microbiology* 135:2319–2328. (#5)
Pollock, W.B.R., Loutfi, M., Bruschi, M., Rapp-Giles, B.J., Wall, J.D., and Voordouw, G. 1991. Cloning, sequencing and expression of the gene encoding the high-molecular weight cytochrome *c* from *Desulfovibrio vulgaris* Hildenborough. *Journal of Bacteriology* 173:220–228. (#3, 5)
Porterfield, W.W. 1984. *Inorganic Chemistry: A Unified Approach*, 688 pp. Addison-Wesley, Reading, Mass. (#1)
Postgate, J.R. 1951. The reduction of sulphur compounds by *Desulphovibrio desulfuricans*. *Journal of General Microbiology* 5:725–738. (#3)
Postgate, J.R. 1952. Growth of sulphate reducing bacteria in sulphate-free media. *Research* (London) 5:189–190. (#3)
Postgate, J.R. 1954. Presence of cytochrome in an obligate anaerobe. *Biochemical Journal* 56:xi–xii. (Dedication, #4, 5)
Postgate, J.R. 1956. Cytochrome c_3 and desulphoviridin: pigments of the anaerobe *Desulphovibrio desulphuricans*. *Journal of General Microbiology* 14:545–572. (Dedication, #3)
Postgate, J.R. 1963. A strain of *Desulfovibrio* able to use oxamate. *Archiv für Mikrobiologie* 46:287–295. (#2)
Postgate, J.R. 1965. Recent advances in the study of the sulphate-reducing bacteria. *Bacteriology Reviews* 29:425–441. (#5)
Postgate, J.R. 1973. *A Plain Man's Guide to Jazz*, 146 pp. Hanover Books, London. (Dedication)
Postgate, J.R. 1979. *The Sulphate-Reducing Bacteria*, 151 pp. Cambridge University Press, Cambridge. 151 pp. (Dedication, Introduction, #7)
Postgate, J.R. 1982a. *The Fundamentals of Nitrogen Fixation*, 252 pp. Cambridge University Press, Cambridge. UK. (Dedication)
Postgate, J.R. 1982b. Economic importance of sulphur bacteria. *Philosophical Transactions of the Royal Society* (London) B298:583–600. (#1)
Postgate, J.R. 1984a. *The Sulphate-Reducing Bacteria*, 2nd edition, 208 pp. Cambridge University Press, Cambridge, UK. (Dedication, Introduction, #1, 2, 3, 4, 5, 6, 7, 8)
Postgate, J.R. 1984b. Genus *Desulfovibrio* Kluyver and van Niel 1936,397[AL], pp. 666–672. In Krieg, N.R. and Holt, J.G. (editors) *Bergey's Manual of Systematic Bacteriology*, vol. 1. Williams and Wilkins, Baltimore. (#2)
Postgate, J.R. 1986. This week's citation classic. *Current Contents* 29:16. (Dedication)

Postgate, J.R. and Campbell, L.L. 1966. Classification of *Desulfovibrio* species, the nonsporeforming sulfate-reducing bacteria. *Bacteriology Reviews* 30:732–738. (#1, 6, 7)

Postgate, J.R. and Kent, H.M. 1984. Derepression of nitrogen fixation in *Desulfovibrio gigas* and its stability to ammonia or oxygen stress in vivo. *Journal of General Microbiology* 131:2119–2122. (#5)

Postgate, J.R. and Kent, H.M. 1985. Diazotrophy within *Desulfovibrio*. *Journal of General Microbiology* 131:2119–2122. (#5)

Postgate, J.R., Kent, H.M., Robson, R.L., and Chesshyre, J.A. 1984. The genomes of *Desulfovibrio gigas* and *Dv. vulgaris*. *Journal of General Microbiology* 130:1597–1601. (#4, 5)

Postgate, J.R., Kent, H.M., and Robson, R.L. 1986. DNA from diazotrophic *Desulfovibrio* strains is homologous to *Klebsiella pneumoniae* structural nif DNA and can be chromosomal or plasmid-borne. *FEMS Microbiology Letters* 33:159–163. (#4, 5)

Postgate, J.R., Kent, H.M., and Robson, R.L. 1987. Nitrogen fixation by *Desulfovibrio*, pp. 457–471. In Cole, J.A. and Ferguson, S.J. (editors) *The Nitrogen and Sulphur Cycles*. Cambridge University Press, Cambridge. (#5)

Postgate, J.R., Kent, H.M., and Robson, R.L. 1988. Nitrogen fixation by *Desulfovibrio*, pp. 457–471. In Cole, J.A. and Ferguson, S.J. (editors) *The Nitrogen and Sulphur Cycles*. Cambridge University Press, Cambridge, U.K. (#4, 5)

Powell, B., Mergeay, M., and Christofi, N. 1989. Transfer of broad host-range plasmids to sulphate-reducing bacteria. *FEMS Microbiology Letters* 59:269–274. (#4, 5)

Prickril, B.C., Czechowski, M.H., Przybyla, A.E., Peck Jr., H.D., and LeGall, J. 1986. Putative signal peptide on the small subunit of the periplasmic hydrogenase from *Desulfovibrio vulgaris*. *Journal of Bacteriology* 167:722–725. (#5)

Prickril, B.C., He, S.-H., Li, C., Menon, N., Choi, E.S., Przybyla, A.E., Der-Vartanian, D.V, Peck Jr., H.D., Fauque, G., LeGall, J., Texeira, M., Moura, I., Moura, J.J.G., Patil, D., and Huyn, B.J. 1987. Identification of three distinct classes of hydrogenase in the genus *Desulfovibrio*. *Biochemical and Biophysical Research Communications* 149:369–377. (#5)

Prickril, B.C., Kurtz, D.M., LeGall, J., and Voordouw, G. 1991. Cloning and sequencing of the gene for rubrerythrin from *Desulfovibrio vulgaris* (Hildenborough). *Biochemistry* 30:11118–11123. (#5)

Priefer, U.B., R. Simon, and A. Puhler. 1985. Extension of the host range of *Escherichia coli* vectors by incorporation of RSF1010 replication and mobilization functions. *Journal of Bacteriology* 163:324–330. (#4)

Qatibi, A.-I. 1990. Fermentation du lactate, du glycérol et des diols par les bactéries sulfato-réductrices du genre *Desulfovibrio*. Ph. D. thesis, Université de Provence Aix-Marseille 1. (#2)

Qatibi, A.I., Cayol, J.L., and Garcia, J.L. 1991a. Glycerol and propanediols degradation by *Desulfovibrio alcoholovorans* in pure culture in the presence of sulfate or in syntrophic association with *Methanospirillum hungatei*. *FEMS Microbiology Ecology* 85:233–240. (#2)

Qatibi, A.I., Nivière, V., and Garcia, J.L. 1991b. *Desulfovibrio alcoholovorans* sp. nov., a sulfate-reducing bacterium able to grow on glycerol, 1,2- and 1,3-propanediol. *Archives of Microbiology* 155:143–148. (#2)

Quesnel, L.E., Al-Najjar, A.R., and Buddhavudhikrai, P. 1978. Synergism between chlorhexidine and sulfadiazine. *Journal of Applied Bacteriology* 45:397–405. (#8)

Ragsdale, S.W., Clark, J.E., Ljungdahl, L.G., Lundie, L.L., and Drake, H.L. 1983. Properties of purified carbon monoxide dehydrogenase from *Clostri-*

dium thermoaceticum, a nickel, iron-sulfur protein. *Journal of Biological Chemistry* 258:1984–2364. (#3)

Rapp, B.J. and Wall, J.D. 1987. Genetic transfer in *Desulfovibrio desulfuricans*. *Proceedings of the National Academy of Sciences of the United States of America* 84:9128–9131. (#4)

Rapp, B.J. and Wall, J.D. 1989. Plasmid transfer by conjugation in *Desulfovibrio desulfuricans*. *Abstracts of the Annual Meeting of the American Society for Microbiology* H-266. (#4)

Reeburgh, W.S. 1980. Anaerobic methane oxidation: rate depth distributions in Skan Bay sediments. *Earth and Planetary Science Letters* 47:345–352. (#7)

Reeburgh, W.S. and Heggie, D.T. 1977. Microbial methane consumption reactions and their effect on methane distributions in freshwater and marine environmental. *Limnology and Oceanography* 22:1–9. (#7)

Regensburger, A., Ludwig, W., and Schleifer, K.H. 1988. DNA probes with different specificities from a cloned 23S rRNA gene of *Micrococcus luteus*. *Journal of General Microbiology* 134:1197–1204. (#6)

Reysenbach, A. and Deming, J.W. 1991. Effects of hydrostatic pressure on growth of hyperthermophilic archaebacteria from the Juan de Fuca Ridge. *Applied and Environmental Microbiology* 57:1271–1274. (#7)

Richaud, P., Vignais, P.M., Colbeau, A., Uffen, R.L., and Cauvin, B. 1990. Molecular biology studies of the uptake hydrogenase of *Rhodobacter capsulatus* and *Rhodocyclus gelatinosus*. *FEMS Microbiology Reviews*, in press. (#5)

Ricklefs, R.E. 1990. *Ecology*, 3rd edition, 896 pp. W.H. Freeman, New York. (#7)

Rieder, R., Cammack, R., and Hall, D.O. 1984. Purification and properties of the soluble hydrogenase from *Desulfovibrio desulfuricans* (strain Norway 4). *European Journal of Biochemistry* 145:637–643.

Riederer-Henderson, M.A. and Peck, H.D., Jr. 1986. Properties of formate dehydrogenase from *Desulfovibrio gigas*. *Canadian Journal of Microbiology* 32:430–435. (#2)

Riederer-Henderson, M. and Wilson, P.W. 1970. Nitrogen fixation by sulphate-reducing bacteria. *Journal of General Microbiology* 61:27–31. (#5)

Robinson, J.A. and Tiedje, J.M. 1984. Competition between sulfate-reducing and methanogenic bacteria for H_2 under resting and growing conditions. *Archives of Microbiology* 137:26–32. (#7)

Robinson, J.B. and Tuovinen, O.H. 1984. Mechanisms of microbial resistance and detoxification of mercury and organo-mercury compounds: physiological, biochemical and genetic analyses. *Microbiological Reviews* 48:95–124. (#8)

Rogers, G.S. 1919. The Sunset-Midway oil field, California. *United States Geological Survey* Paper 117, pt. 2, pp. 26–29. (#8)

Rohde, M., Furstenau, V., Meyer, F., Przybyla, A.E., Peck, H.D. Jr., LeGall, J., Choi, E.S., and Menon, N.K. 1990. Localization of membrane-associated (NiFe) and (NiFeSe) hydrogenases of *Desulfovibrio vulgaris* using immunoelectron microscopic procedures. *European Journal of Biochemistry* 191:389–396. (#3, 5)

Rousset, M., Dermoun, Z., Hatchikian, C.E., and Belaich, J.P. 1990. Cloning and sequencing of the locus encoding the large and small subunit genes of the periplasmic [NiFe] hydrogenase from *Desulfovibrio fructosovorans*. *Gene* 94:95–101. (#5)

Rousset, M., Dermoun, Z., Chippaux, M., and Belaich, J.P. 1991. Marker exchange mutagenesis of the *hydN* genes in *Desulfovibrio fructosovorans*. *Molecular Microbiology* 5:1735–1740. (#4)

Rozanova, E.P. and Pivovarova,T.A. 1988. Reclassification of *Desulfovibrio thermophilus* (Rozanova, Khudyakova, 1974). *Microbiology* 57:85–90. (#5)

Rozanova, E.P., Nazina, T.N., and Galushko, A.S. 1988. A new genus of sulfate-reducing bacteria and the description of its new species, *Desulfomicrobium apsheron* gen. nov., sp. nov. *Mikrobiologiya* 57:634–641. (#6)
Sadana, J.C. 1954. Pyruvate oxidation in *Desulphovibrio desulphuricans*. *Journal of Bacteriology* 67:547–553. (#1)
Sadana, J.C. and Jagganathan, V. 1954. Purification of hydrogenase from *Desulfovibrio desulfuricans*. *Biochimica et Biophysica Acta* 14:287–288. (#3)
Saiki, R., Gelfand, D.H., Stoffel, S., Scharf, S.J., Higuchi, R., Horn, G.T., Mullis, K.B., and Erlich, H.A. 1988. Primer-directed enzymatic amplification of DNA with a thermostable DNA polymerase. *Science* 239:487–491. (#6)
Samain, E., Dubourgier, H.C., and Albagnac, G. 1984. Isolation and characterization of *Desulfobulbus elongatus* sp. nov. from a mesophilic industrial digestor. *Systematic and Applied Microbiology* 5:391–401. (#2)
Samain, E., Patil, D.S., DerVartanian, D.V., Albagnac, G., and LeGall, J. 1987. Isolation of succinate dehydrogenase from *Desulfobulbus elongatus*, a propionate oxidizing, sulfate reducing bacterium. *FEBS Letters* 216:140–144. (#2)
Sand, W. and Bock, E. 1984. Concrete corrosion in the Hamburg sewer system. *Environmental Technology Letters* 5:517–528. (#8)
Sayavedra-Soto, L.A., Powell, G.K., Evans, H.J., and Morris, R.O. 1988. Nucleotide sequence of the genetic loci encoding subunits of *Bradyrhizobium japonicum* uptake hydrogenase. *Proceedings of the National Academy of Sciences of the United States of America* 85:8395–8399. (#5)
Schaschl, E. 1980. Elemental sulfur as a corrodent in deaerated neutral aqueous environments. *Materials Performance* 19:9–12. (#8)
Schauder, R., Eikmanns, B., Thauer, R.K., Widdel, F., and Fuchs, G. 1986. Acetate oxidation to CO_2 in anaerobic bacteria via a novel pathway not involving reactions of the citric acid cycle. *Archives of Microbiology* 145:162–172. (#2, 6)
Schauder, R., Widdel, F., and Fuchs, G. 1987. Carbon assimilation pathways in sulfate-reducing bacteria. II. Enzymes of a reductive citric acid cycle in the autotrophic *Desulfobacter hydrogenophilus*. *Archives of Microbiology* 148:218–225. (#2, 6, 7)
Schauder, R., Preuss, A., Jetten, M., and Fuchs, G. 1989. Oxidative and reductive acetyl CoA/carbon monoxide dehydrogenase pathway in *Desulfobacterium autotrophicum*. 2. Demonstration of the enzymes of the pathway and comparison of carbon monoxide dehydrogenase. *Archives of Microbiology* 151:84–89. (#2)
Schelel, M. and Trüper, H.G. 1979. Purification of *Thiobacillus denitrificans* siroheme sulfite reductase and investigation of some molecular and catalytic properties. *Biochimica et Biophysica Acta* 568:454–467. (#3)
Schidlowski, M. 1979. Antiquity and evolutionary status of bacterial sulfate reduction: sulfur isotope evidence. *Origins of Life* 9:299–311. (#1)
Schidlowski, M., Hayes, J.M., and Kaplan, I.R. 1983. Isotopic inferences of ancient biochemistries: carbon, sulfur, hydrogen and nitrogen, pp. 149–186. In Schopf, J.W. (editor) *Earth's Earliest Biosphere*. Princeton University Press, Princeton, N.J. (#3)
Schlegel, H. G. 1986. *General Microbiology*, 6th edition, 587 pp. Cambridge University Press, Cambridge, UK. (#1)
Schnell, S. and Schink, B. 1991. Anaerobic aniline degradation via reductive deamination of 4-aminobenzoyl-CoA in *Desulfobacterium anilini*. *Archives of Microbiology* 155:183–190. (#2)
Schnell, S., Bak, F., and Pfennig, N. 1989. Anaerobic degradation of aniline and dihydroxybenzenes by newly isolated sulfate-reducing bacteria and

description of *Desulfobacterium anilinii*. *Archives of Microbiology* 152:556–563. (#2, 8)

Schopf, J.W. and Walter, M.R. 1982. Origin and early evolution of cyanobacteria: the geological evidence, pp. 543–564. In Carr, N.G. and Whitton, B.A. (editors) *The Biology of Cyanobacteria. Botanical Monographs no. 19.* Blackwell, Oxford, UK. (#3)

Schopf, J.W., Hayes, J.M., and Walter, M.R. 1983. Evolution of earth's earliest ecosystems: recent progress and unsolved problems, pp. 361–384. In Schopf, J.W. (editor) *Earth's Earliest Biosphere.* Princeton University Press, Princeton, N.J. (#1)

Schubauer, J.P. and Hopkinson, C.S. 1984. Above- and belowground emergent macrophyte production and turnover in a coastal marsh ecosystem, Georgia. *Limnology and Oceanography* 29:1052–1065. (#7)

Scranton, M.L., Novelli, P.C., and Loud, P.A. 1984. The distribution and cycling of hydrogen gas in the waters of two anoxic marine environments. *Limnology and Oceanography* 29:993–1003. (#7)

Seitz, H. and Cypionka, H. 1986. Chemolithotrophic growth of *Desulfovibrio desulfuricans* with hydrogen coupled to ammonification of nitrate or nitrite. *Archives of Microbiology* 146:63–67. (#7)

Senez, J.C. and Leroux-Gilleron, J. 1954. Preliminary note on the anaerobic degradation of cysteine and cystine by sulfate-reducing bacteria. *Bulletin de la Société de Chimie Biologique* 36:553–559. (#2)

Senior, E., Lindstrom, E.B., Banat, I.M., and Nedwell, D.B. 1982. Sulfate reduction and methanogenesis in the sediment of a salt marsh on the east coast of the United Kingdom. *Applied and Environmental Microbiology* 43:987–996. (#7)

Shariat, M., Anderson, A.C., and Mason, J.W. 1979. Screening of common bacteria capable of demethylation of methylmercuric chloride. *Bulletin of Environmental Contamination and Toxicology* 221:255. (#8)

Sharpe, G.S. 1984. Broad-host-range cloning vectors for Gram-negative bacteria. *Gene* 29:93–102. (#4)

Shelton, D.R. and Tiedje, J.M. 1984. Isolation and partial characterization of bacteria in an anaerobic consortium that mineralizes 3-chlorobenzoic acid. *Applied and Environmental Microbiology* 48:840–848. (#6)

Shinkai, W., Hase, T., Yagi, T., and Matsubara, H. 1980. Cytochrome c_3 from *Desulfovibrio vulgaris. Journal of Biochemistry* (Tokyo) 87:1747–1756. (#5)

Sieker, L.C., Stenkamp, R.E., Jensen, L.H., Prickril, B., and LeGall, J. 1986. Structure of rubredoxin from the bacterium *Desulfovibrio desulfuricans. FEBS Letters* 208:73–76. (#5)

Singleton, R. Jr., Denis, J., and Campbell, L.L. 1982. Cytochrome c_3 from the sulfate reducing anaerobe, *Desulfovibrio africanus*: antigenic properties. *Journal of Bacteriology* 152:527–529. (#1)

Singleton, R., Jr., Denis, J., and Campbell, L.L. 1984. Antigenic diversity of cytochromes c_3 from the anaerobic, sulfate-reducing bacteria, *Desulfovibrio. Archives of Microbiology* 139:91–95. (#1, 6, 8)

Singleton, R., Jr., Denis, J., and Campbell, L.L. 1985. Whole-cell antigens of members of the sulfate-reducing genus *Desulfovibrio. Archives of Microbiology* 141:195–197. (#1, 6)

Singleton, R., Jr., Ketchum, R.B., and Campbell, L.L. 1988. Effect of calcium cation on plating efficiency of the sulfate-reducing bacterium *Desulfovibrio vulgaris. Applied and Environmental Microbiology* 54:2318–2319. (#4)

Sivela, S. and Sundman, V. 1972. Demonstration of the Thiobacillus type bacteria which utilize methyl sulfides. *Archives of Microbiology* 103:303–304. (#8)

Skyring, G.W. and Donnelly, T.H. 1982. Precambrian sulfur isotopes and a possible role for sulfite in the evolution of biological sulfate reduction. *Precambrian Research* 17:41–61. (#3)

Skyring, G.W., Oshrain, R.L., and Wiebe, W.J. 1979. Sulfate reduction rates in Georgia marshland soils. *Geomicrobiology Journal* 1:389–400. (#7)

Sleytr, U., Adam, H., and Klaushofer, H. 1969. Die Feinstruktur der Zellwand und Cytoplasmamembran von *Clostridium nigrificans*, dargestellt mit Hilfe der Gefrierätz- und Ultradünnschnittechnik. Archiv für Mikrobiologie 66:40–58. (#1)

Smets, B., Rittmann, B.E., and D.A. Stahl. 1990. The role of genes in biological processes. *Environmental Science and Technology* 24:162–169. (#6)

Smith, A.D. 1982. Immunofluorescence of sulphate-reducing bacteria. *Archives of Microbiology* 133:118–121. (#1, 8)

Smith, D.W. 1980. An evaluation of marsh nitrogen fixation, pp. 135–142. In Kennedy, V.S. (editor) *Estuarine Perspectives*. Academic Press, New York. (#7)

Smith, D.W. and Strohl, W.R. 1991. Sulfur-oxidizing bacteria, pp. 121–146. In Shively, J.M. and Barton, L.L. (editors) *Variations in Autotrophic Life*. Academic Press, London. (#7)

Smith, R.L. and Oremland, R.S. 1982. Big soda lake (Nevada). 2. Pelagic sulfate reduction. *Limnology and Oceanography* 32:794–803. (#7)

Sneddon, R. 1951. Casing failures traced to bacterial action. *Petroleum Engineering*, August, pp. 7–12. (#8)

Sorensen, J., Christensen, D., and Jorgensen, B.B. 1987. Volatile fatty acids and hydrogen as substrates for sulfate reducing bacteria in anaerobic marine sediments. *Applied and Environmental Microbiology* 42:5–11. (#7)

Sorokin, Y. 1966a. Role of carbon dioxide and acetate in biosynthesis of sulphate-reducing bacteria. *Nature* 210:551. (#1, 2)

Sorokin, Y. 1966b. Sources of energy and carbon for biosynthesis in sulfate-reducing bacteria. *Microbiology* 35:643–647. (#2)

Speich, N. and Trüper, H.G. 1988. Adenylylsulphate reductase in a dissimilatory sulphate-reducing archaebacterium. *Journal of General Microbiology* 134:1419–1425. (#6)

Spormann, A.M. and Thauer, R.K. 1988. Anaerobic acetate oxidation to CO_2 by *Desulfotomaculum acetoxidans*. Demonstration of enzymes required for the operation of an oxidative acetyl-CoA/carbon monoxide dehydrogenase pathway. *Archives of Microbiology* 150:374–380. (#2)

Spormann, A.M. and Thauer, R.K. 1989. Anaerobic acetate oxidation by *Desulfotomaculum acetoxidans*. Isotopic exchange between CO_2 and the carbonyl group of acetyl-CoA and topology of enzymes involved. *Archives of Microbiology* 152:189–195. (#2)

Stackebrandt E., Murray, R.G.E., and Trüper, H.G. 1988. *Proteobacteria* classis nov., a name for the phylogenetic taxon that includes the "purple bacteria and their relatives". *International Journal of Systematic Bacteriology* 38:321–325. (#6)

Stahl, D.A. 1986. Evolution, ecology, and diagnosis: unity in variety. *Bio/Technology* 4:623–628. (#6)

Stahl, D.A. and Amann, R.I. 1991. Development and application of nucleic acid probes, pp. 205–248. In Stackebrandt, E. and Goodfellow, M. (editors) *Nucleic Acids Techniques in Bacterial Systematics*. John Wiley & Sons, Inc., Chichester. (#6)

Stahl, D.A., Lane, D.J., Olsen, G.J., and Pace, N.R. 1984. Analysis of hydrothermal vent-associated symbionts by ribosomal RNA sequences. *Science* 224:409–411. (#6)

Stahl, D.A., Lane, D.J., Olsen, G.J., and Pace, N.R. 1985. Characterization of a

Yellowstone hot spring microbial community by 5S ribosomal RNA sequences. *Applied and Environmental Microbiology* 49:1379–1384. (#6)
Stahl, D.A., Flesher, B., Mansfield, H.R., and Montgomery, L. 1988. Use of phylogenetically-based hybridization probes for studies of ruminal microbial ecology. *Applied and Environmental Microbiology* 54:1079–1084. (#6)
Staley, J.T. and Konopka, A. 1985. Measurement of in situ activities of nonphotosynthetic microorganisms in aquatic and terrestrial habitats. *Annual Review of Microbiology* 39:321–346. (#7)
Stams, A.J.M. and Hansen, T.A. 1982. Oxygen-labile L(+) lactate dehydrogenase activity in *Desulfovibrio desulfuricans. FEMS Microbiology Letters* 13:389–394. (#2)
Stams, A.J.M. and Hansen, T.A. 1986. Metabolism of L-alanine in *Desulfotomaculum ruminis* and two marine *Desulfovibrio* strains. *Archives of Microbiology* 145:277–279. (#2)
Stams, A.J.M., Veenhuis, M., Weenk, G.H., and Hansen, T.A. 1983. Occurrence of polyglucose as a storage polymer in *Desulfovibrio* species and *Desulfobulbus propionicus. Archives of Microbiology* 136:54–59. (#2)
Stams, A.J.M., Kremer, D.R., Nicolay, K., Weenk, G.H., and Hansen, T.A. 1984. Pathway of propionate formation in *Desulfobulbus propionicus. Archives of Microbiology* 139:167–173. (#2)
Stams, A.J.M., Hansen, T.A., and Skyring, G.W. 1985. Utilization of amino acids as energy substrates by two marine *Desulfovibrio* strains. *FEMS Microbiology Ecology* 31:11–15. (#2)
Stams, A.J.M., Hoekstra, L.G., and Hansen, T.A. 1986. Utilization of L-alanine as carbon and nitrogen source by *Desulfovibrio* HL21. *Archives of Microbiology* 145:272–276. (#2)
Stanier, R.Y., Doudoroff, M., and Adelberg, E.A. 1970. *The Microbial World*, 3rd edition, 873 pp. Prentice-Hall, Englewood Cliffs, NJ. (#7)
Starkey, R.L. 1961. Sulfate-reducing bacteria, their production of sulfide and their economic importance. *Tappi* 44:493–496. (#1)
Steenkamp, D.J. and Peck, H.D. Jr. 1981. On the proton translocation association with nitrite respiration in *Desulfovibrio desulfuricans. Journal of Biological Chemistry* 256:5450–5458. (#1, 3)
Steiner, P. 1983. Organics lead the way in corrosion inhibitors. *Chemical Marketing Reporter*, Feb. 7, pp 38–39. (#8)
Stephenson, M. and Stickland, L.G. 1931. Hydrogenase: II. The reduction of sulphate to sulphide by molecular hydrogen. *Biochemical Journal* (London) 25:215–220. (#3)
Stetter, K.O. 1988. *Archaeoglobus fulgidus* gen. nov., sp. nov.: a new taxon of extremely thermophilic archaebacteria. *Systematic and Applied Microbiology* 10:172–173. (#6)
Stetter, K.O. and Gaag, G. 1983. Reduction of molecular sulphur by methanogenic bacteria. *Nature* 305:309–311. (#3)
Stetter, K.O., Lauerer, G., Thomm, M., and Neuner, A. 1987. Isolation of extremely thermophilic sulfate reducers: evidence for a novel branch of archaebacteria. *Science* 236:822–824. (#6)
Stewart, D.E., LeGall, J., Moura, I., Moura, J.J.G., Peck, H.D. Jr., Xavier, A.V., Weiner, P.K., and Wampler, J.E. 1987. Electron transport in sulfate reducing bacteria: a hypothetical model of the flavodoxin-tetraheme cytochrome c_3 complex. *Biochemistry* 27:2444–2450. (#3)
Stewart, D.E., LeGall, J., Moura, I., Moura, J.J.G., Peck, H.D. Jr., Xavier, A.V., Weiner, P.K., and Wampler, J.E. 1989. Electron transport in sulfate reducing bacteria: a hypothetical model of the rubredoxin-tetraheme cytochrome c_3 complex. *European Journal of Biochemistry* 185:695–700. (#3)

Stieb, M. and Schink, B. 1989. Anaerobic degradation of isobutyrate by methanogenic enrichment cultures and by a *Desulfococcus multivorans* strain. *Archives of Microbiology* 151:126–132. (#2)

Stokkermans, J., van Dongen, W., Kaan, A., van den Berg, W. and Veeger, C. 1989. *hydγ*, a gene from *Desulfovibrio vulgaris* (Hildenborough) encodes a polypeptide homologous to the periplasmic hydrogenase. *FEMS Microbiology Letters* 58:217–222. (#3, 5)

Stolp, H. 1988. *Microbial Ecology: Organisms, Habitats, Activities*, 308 pp. Cambridge University Press, Cambridge, UK. (#7)

Stott, J.F.D. and Herbert, B.N. 1986. The effect of pressure and temperature on sulfate-reducing bacteria and the action of biocides in oil field water injection systems. *Journal of Applied Bacteriology* 60:57–66. (#8)

Stouthamer, A. 1988. Dissimilatory reduction of oxidized nitrogen compounds, pp. 245–303. In Zehnder, A.J.B. (editor) *Biology of Anaerobic Organisms*. John Wiley, New York. (#3)

Sukenaga, Y., Ishida, K., Takeda, T., and Takagi, K. 1987. cDNA sequence coding for human glutathione peroxidase. *Nucleic Acids Research* 15:7178. (#5)

Szewzyk, R. and Pfennig, N. 1987. Complete oxidation of catechol by the strictly anaerobic sulfate-reducing *Desulfobacterium catecholicum* sp. nov. *Archives of Microbiology* 147:163–168. (#2, 8)

Sznyter, L.A., Slatko, B., Moran, L., O'Donnell, K.H., and Brooks, J.E. 1987. Nucleotide sequence of the *DdeI* restriction-modification system and characterization of the methylase protein. *Nucleic Acids Research* 15:8249–8266. (#4, 5)

Szostak, J.W., Stiles, J.I., Tye, B.-K., Chiu, P., Sherman, F., and Wu. R. 1979. Hybridization with synthetic oligonucleotides. *Methods in Enzymology* 68:419–428. (#6)

Tabushi, I., Nishiya, T., Yagi, T., and Inokuchi, H. 1981. Efficient electron channel through self-aggregation of cytochrome c_3 on an artificial membrane. *Journal of the American Chemical Society* 103:6963–6965. (#3)

Tabushi, I., Nishiya, T., Shimomura, M., Kunitake, T., Inokuchi, H., and Yagi, T. 1984. Cytochrome c_3 modified artificial liposome. Structure, electron transport and pH gradient generation. *Journal of the American Chemical Society* 106:219–226. (#3)

Tamura, A., Kawate, T., Ogata, M., and Yagi, T. 1988. Interaction of cellular hydrogenase, cytochrome c_3 and desulfoviridin in *Desulfovibrio vulgaris* Miyazaki with their antibodies. *Journal of Biochemistry* (Tokyo) 104:722–726. (#3)

Tan, J., Helms, L.R., Swenson, R.P., Cowan, J.A. (1991). Primary structure of the assimilatory sulfite reductase from *Desulfovibrio vulgaris*, Hildenbourough. *Biochemistry* 30:9900–9902. (#3)

Tan, J., Helms, L.R., Swenson, R.P., and Cowan, J.A. 1991. Primary structure of the assimilatory-type sulfite reductase from *Desulfovibrio vulgaris* (Hildenborough): Cloning and nucleotide sequence of the reductase gene. *Biochemistry*, 30:9900–9907. (#5)

Tanaka, K. 1990. Several new substrates for *Desulfovibrio vulgaris* strain Marburg and a spontaneous mutant of it. *Archives of Microbiology* 155:18–21. (#2)

Taylor, B.F. and Oremland, R.S. 1979. Depletion of adenosine triphosphate in *Desulfovibrio* by oxyanions of Group VI elements. *Current Microbiology* 3:101–103. (#8)

Taylor, J. and Parkes, R.J. 1985. Identifying different populations of sulfate-reducing bacteria within marine sediment systems using fatty acid biomarkers. *Journal of General Microbiology* 131:631–642. (#8)

Tempest, D.W. 1978. The biochemical significance of microbial growth yields: a reassessment. *Trends in Biochemical Sciences* 3:180–184. (#3)

Texeira, M., Fauque, G., Moura, I., Lespinat, A.V., Berlier, Y., Prickril, B., Peck Jr., H.D., Xavier, A.V., LeGall, J., and Moura., J.J.G. 1987. Nickel-[iron-sulfur]-selenium containing hydrogenases from *Desulfovibrio baculatus* (DSM 1743). Redox centers and catalytic properties. *European Journal of Biochemistry* 167:47–58. (#5)

Thauer, R.K. 1988. Citric-acid cycle, 50 years on. Modifications and alternative pathway in anaerobic bacteria. *European Journal of Biochemistry* 176:497–508. (#2)

Thauer, R.K. 1989. Energy metabolism of sulfate-reducing bacteria, pp. 397–413. In Schlegel, H.G. and Bowien, B. (editors) *Autotrophic Bacteria*. Science Tech Publishers, Madison, Wis. (#1, 2)

Thauer, R.K. and Badziong, W. 1980. Respiration with sulfate as electron acceptor, pp. 65–85. In Knowles, G.J. (editor) *Diversity of Bacterial Respiratory Systems*, Vol. 2. CRC Press Inc., Boca Raton, Fla. (#1, 3)

Thauer, R.K. and Morris, J.G. 1984. Metabolism of chemotrophic anaerobes: old views and new aspects, pp. 123–168. In Kelly, D.P. and Carr, N.G. (editors) *The Microbe 1984, Part II, Prokaryotes and Eukaryotes*. Cambridge University Press, Cambridge, UK. (#3)

Thauer, R.K., Jungerman, K., and Decker, K. 1977. Energy conservation in chemotrophic anaerobes. *Bacteriological Reviews* 41:100–180. (#1)

Thauer, R.K., Moller-Zinkham, D., and Spormann, A.M. 1989. Biochemistry of acetate catabolism in anaerobic chemotrophic bacteria. *Annual Review of Microbiology* 43:43–67. (#7)

Thebrath, B., Dilling, W., and Cypionka, H. 1989. Sulfate activation in *Desulfotomaculum*. *Archives of Microbiology* 152:296–301. (#1)

Thiele, J.H. and Zeikus, J.G. 1988. Control of interspecies electron flow during anaerobic digestion: significance of formate transfer versus hydrogen transfer during syntrophic methanogenesis in flocs. *Applied and Environmental Microbiology* 54:20–29. (#2)

Tiller, A.K. 1982. Aspects of microbial corrosion, pp. 115–159. In Parkins, R.N. (editor) *Corrosion Processes*. Applied Science Publishers, New York. (#8)

Toerien, D.F., Thiel, P.G., and Hattingh, M.M. 1968. Enumeration, isolation and identification of sulfate-reducing bacteria of anaerobic digestion. *Water Research* 2:505. (#8)

Traore, A.S., Hatchikian, E.C., Belaich, J.P., and LeGall, J. 1981. Microcalorimetric studies of the growth of sulfate-reducing bacteria: energetics of *Desulfovibrio vulgaris* growth. *Journal of Bacteriology* 145:191–199. (#3)

Trinkerl, M., Breunig, A., Schauder, R., and König, H. 1990. *Desulfovibrio termitidis* sp. nov., a carbohydrate-degrading sulfate-reducing bacterium from the hindgut of a termite. *Systematic and Applied Microbiology* 13:372–377. (#2)

Trousil, E.B. and Campbell, L.L. 1974. The amino acid sequence of cytochrome c_3 from *Desulfovibrio vulgaris*. *Journal of Biological Chemistry* 249:386–393. (#5)

Tsien, H.C., Bratina, B.J., Tsuji, K., and Hanson, R.S. 1990. Use of oligodeoxynucleotide signature probes for identification of physiological groups of methylotrophic bacteria. *Applied and Environmental Microbiology* 56:2858–2865. (#6)

Tsuji, K. and Yagi, T. 1980. Significance of hydrogen burst from growing cultures of *Desulfovibrio Vulgaris* Miyazaki and the role of hydrogenase and cytochrome c_3 in energy producing system. *Archives of Microbiology* 125:35–42. (#3)

Turner, N., Barata, B., Bray, R.C., Deistung, J., LeGall, J., and Moura, J.J.G. 1987. The molybdenum iron-sulphur protein from *Desulfovibrio gigas* as a

form of aldehyde oxidase. *Biochemical Journal* 243:755–761. (#2)
Twigg, R.S. 1945. Oxidation-reduction aspects of resazurin. *Nature* 145:401. (#1)
Tyagi, R.D., Tran, F.T., and Polprasert, C. 1988. Bioconversion of lignosulfonate into lignin and H_2S by mutualistic bacterial system. *Journal of Microbial Biotechnology* 5:90–98. (#8)
Uffen, R.L. 1973. Anaerobic growth of a Rhodopseudomonas species in the dark with carbon monoxide as sole carbon and energy substrate. *Proceedings of the National Academy of Sciences of the United States of America* 73:3298–3302. (#3)
Uffen, R.L., Colbeau, A., Richaud, P., and Vignais, P.M. 1990. Cloning and sequencing of the genes encoding uptake-hydrogenase subunits of *Rhodocyclus gelatinosus*. *Molecular and General Genetics* 221:49–58. (#5)
Uyeda, K. and Rabinowitz, J.G. 1971. Pyruvate-ferredoxin oxidoreductase III. Purification and properties of the enzyme. *Journal of Biological Chemistry* 246:3111–3119. (#3)
Valiella, I., Teal, J.M., and Persson, N.Y. 1976. Production and dynamics of experimentally enriched salt marsh vegetation: below-ground biomass. *Limnology and Oceanography* 23:798–812. (#7)
Vamos, R. 1958. H_2S the cause of "bruzone" (Akiochi) disease of rice. *Soil Plants and Food* 4:37–40. (#8)
van Delden, A. 1903. Breitrag zur Kenntnis der Sulfatreduktion durch Bakterien. Zentralblatt für Bakteriologie (Abt. 2) Vol II, pp. 81–94. (#8)
Van den Berg, W.A.M., Stokkermans, J.P.W.G., and van Dongen, W.M.A.M. 1989. Development of a plasmid transfer system for the anaerobic sulphate reducer, *Desulfovibrio vulgaris*. *Journal of Biotechnology* 12:173–184. (#4, 5)
Van der Westen, H.M., Mayhew, S.G., and Veeger, C. 1978. Separation of hydrogenase from intact cells of *Desulfovibrio vulgaris*. *FEBS Letters* 86:122–126. (#5)
Van Dongen, W., Hagen, W., van den Berg, W., and Veeger, C. 1988. Evidence for an unusual mechanism of membrane translocation of the periplasmic hydrogenase of *Desulfovibrio vulgaris* (Hildenborough), as derived from expression in *Escherichia coli*. *FEMS Microbiology Letters* 50:5–9. (#5)
Van Rooijen, G.J.H., Bruschi, M., and Voordouw, G. 1989. Cloning and sequencing of the gene encoding cytochrome c_{553} from *Desulfovibrio vulgaris* Hildenborough. *Journal of Bacteriology* 171:3573–3578. (#5)
Van den Berg, W.A.M., van Dongen, W.M.A.M., and Veeger, C. 1991. Reduction of the amount of periplasmic hydrogenase in *Desulfovibrio vulgaris* (Hildenborough) with antisense RNA: Direct evidence for an important role of this hydrogenase in lactate metabolism. *Journal of Bacteriology*, 173:3688–3694. (#4, 5)
Vega, J.M. and Kamin, H. 1977. Spinach nitrite reductase. Purification and properties of a siroheme-containing iron-sulfur enzyme. *Journal of Biological Chemistry* 252:896–909. (#3)
Vogels, G.D., Keltjieus, J.T., and Van Der Drift, C. 1988. Biochemistry of methane production, pp. 707–770. In Zehnder, A.J.B. (editor) *Biology of Anaerobic Organisms*. John Wiley, New York. (#3)
von Wolzogen Kuhr, C.A.H. 1922. On the occurrence of sulfate-reduction in the deeper layers of the earth. *Proceedings of the Koninklijke Nederlandse Akademie van Wetenschappen* 25:188–198. (#8)
von Wolzogen Kuhr, C.A.H. and van der Vlugt, L.S. 1934. De grafiteering gietijzer als electrobiochemische proces in anaerobe gronden. *Water* (Netherlands) 18:147–165. (#8)
Voordouw, G. 1987. Molecular biology of redox proteins in sulphate reduction,

pp. 147–160. In Cole J.A. and Ferguson S. (editors) *The Nitrogen and Sulphur Cycles* Society for General Microbiology Symposium 42. (#5)

Voordouw, G. 1988. Cloning of genes encoding redox proteins of known amino acid sequence from a library of the *Desulfovibrio vulgaris* (Hildenborough) genome. *Gene* 69:75–83. (#5)

Voordouw, G. 1990. Hydrogenase genes in Desulfovibrio, pp. 37–51. In Belaich, J.P., Bruschi, M., and Garcia, J.L. (editors) *Microbiology and Biochemistry of Strict Anaerobes Involved in Interspecies Hydrogen Transfer*. Plenum Press, New York. (#5)

Voordouw, G. and Brenner, S. 1985. Nucleotide sequence of the gene encoding the hydrogenase from *Desulfovibrio vulgaris* (Hildenborough). *European Journal of Biochemistry* 148:515–520. (#3, 5)

Voordouw, G. and Brenner, S. 1986. Cloning and sequencing of the gene encoding cytochrome c_3 from *Desulfovibrio vulgaris* (Hildenborough). *European Journal of Biochemistry* 159:347–351. (#3, 5)

Voordouw, G. and Wall, J.D. 1992. Genetics and molecular biology of sulfate-reducing bacteria, pp. 451–468. In Sebald M. (editor) *Genetics and Molecular Biology of Anaerobic Bacteria*. Springer-Verlag, New York. (#5)

Voordouw, G., Walker, J.E., and Brenner, S.. 1985. Cloning of the gene encoding the hydrogenase from *Desulfovibrio vulgaris* (Hildenborough) and determination of the NH_2-terminal sequence. *European Journal of Biochemistry* 148:509–514. (#5)

Voordouw, G., Hagen, W.R., Kruse-Wolters, M., van Berkel-Arts, A., and Veeger, C. 1987a. Purification and characterization of *Desulfovibrio vulgaris* (Hildenborough) hydrogenase expressed in *Escherichia coli*. *European Journal of Biochemistry* 162:31–36. (#5)

Voordouw, G., Kent, H.M., and Postgate, J.R. 1987b. Identification of the genes for hydrogenase and cytochrome c_3 in Desulfovibrio. *Canadian Journal of Microbiology* 33:1006–1010. (#5)

Voordouw, G., Menon, N.K., LeGall, J., Choi, E.-S., Peck Jr, H.D., and Przybyla, A.E. 1989a. Analysis and comparison of nucleotide sequences encoding the genes for [NiFe] and [NiFeSe] hydrogenase from *Desulfovibrio gigas* and *Desulfovibrio baculatus*. *Journal of Bacteriology* 171:2894–2899. (#5)

Voordouw, G., Strang, J.D., and Wilson, F.R. 1989b. Organization of the genes encoding [Fe] hydrogenase in *Desulfovibrio vulgaris* subsp. *oxamicus* Monticello. *Journal of Bacteriology* 171:3881–3889. (#5)

Voordouw, G., Niviere, V., Ferris, F.G., Fedorak, P.M., and Westlake, D.W.S. 1990a. Distribution of hydrogenase genes in *Desulfovibrio* spp. and their use in identification of species from the oil field environment. *Applied and Environmental Microbiology* 56:3748–3754. (#3, 5, 8)

Voordouw, G., Pollock, W.B.R., Bruschi, M., Guerlesquin, F., Rapp-Giles, B.J., and Wall, J.D. 1990b. Functional expression of *Desulfovibrio vulgaris* Hildenborough cytochrome c_3 in *Desulfovibrio desulfuricans* following conjugational gene transfer from *Escherichia coli*. *Journal of Bacteriology* 172:6122–6126. (#4, 5)

Wakao, N. and Furusaka, C. 1976. Presence of microaggregates containing sulfate-reducing bacteria in a paddy field soil. *Soil Biology and Biochemistry* 8:157–159. (#8)

Walch, M., Ford, T.E., and Mitchell R. 1989. Influence of hydrogen-producing bacteria on hydrogen uptake by steel. *Corrosion* 45:705–709. (#8)

Wake, L.V., Christopher, R.K., Rickard, P.A.D., Andersen, J.E., and Ralph, B.J. 1977. A thermodynamic assessment of possible substrates for sulphate-reducing bacteria. *Australian Journal of Biological Science* 30:155–172. (#1)

Wall, J.D., Rapp-Giles, B.J., and Concannon, S.P. 1990. Identification of a small cryptic plasmid in *Desulfovibrio desulfuricans*. *Abstracts of the Annual*

Meeting of the American Society for Microbiology H-158. (#4)
Wallace, R.B., Shaffer, J., Murphy, R.F., Bonner, J., Hirose, T., and Itakura, K. 1979. Hybridization of synthetic oligodeoxyribonucleotides to Φ 174 DNA: the effect of single base pair mismatch. *Nucleic Acids Research* 6:3543–3557. (#6)
Ward, D.M., Weller, R., and Bateson, M.M. 1990. 16S rRNA sequences reveal numerous uncultured microorganisms in a natural community. *Nature* 345:63–65. (#6)
Warthmann, R. and Cypionka, H. 1990. Sulfate transport in *Desulfobulbus propionicus* and *Desulfococcus multivorans*. *Archives of Microbiology* 154:144–149. (#3)
Watenpaugh, K.D., Sieker, L.C., Jensen, L.H., LeGall, J., and Dubourdieu, M. 1972. Structure of the oxidized form of a flavodoxin at 2.5Å resolution: resolution of the phase ambiguity by anomalous scattering. *Proceedings of the National Academy of Sciences of the United States of America* 69:3185–3188. (#5)
Wayne, L.G., Brenner, D.J., Colwell, R.R., Grimont, P.A.D., Kandler, O., Krichevsky, M.I., Moore, L.H., Moore, W.E.C., Murray, R.G.E., Stackebrandt, E., Starr, M.P., and Trüper, H.G. 1987. Report of the Ad Hoc Committee on Reconciliation of Approaches to Bacterial Systematics. *International Journal of Systematic Bacteriology* 37:463–464. (#6)
Weimer, P.J., van Kavelaar, M.J., Michel, C.B., and Ng, T.K. 1988. Effect of phosphate on the corrosion of carbon steel and on the composition of the corrosion products in two-stage chemostat cultures of *Desulfovibrio desulfuricans* G100A. *Applied and Environmental Microbiology* 54:386–396. (#4, 8)
Weisburg, W.G., Barns, S.M., Pelleitier, D.A., and D.J. Lane. 1991. 16S ribosomal DNA amplification for phylogenetic study. *Journal of Bacteriology* 173:697–703. (#6)
Weller, R. and Ward, D.M. 1989. Selective recovery of 16S rRNA sequences from natural microbial communities in the form of cDNA. *Applied and Environmental Microbiology* 55:1818–1822. (#6)
Westrich, J.T. and Berner, R.A. 1984. The role of sedimentary organic matter in bacterial sulfate reduction: The G model tested. *Limnology and Oceanography* 29:236–249. (#7)
Widdel, F. 1980. Anaerober Abbau von Fettsäure und Benzoesäure durch neu isolierte Arten Sulfate-reduzierender Bakterien. Dissertation, Univ. of Göttingen, Göttingen, Germany. (#1, 2)
Widdel, F. 1986. Sulfate-reducing bacteria and their ecological niches, pp 157–184. In: Barnes, E.M. and Mead, G.C. (editors) *Anaerobic Bacteria in Habitats Other than Man*. Blackwell Scientific Publications, Oxford. (#8)
Widdel, F. 1987. New types of acetate-oxidizing, sulfate-reducing *Desulfobacter* species, *D. hydrogenophilus* sp. nov., *D. latus* sp. nov., and *D. curvatus* sp. nov. *Archives of Microbiology* 148:286–291. (#1, 6, 7)
Widdel, F. 1988. Microbiology and ecology of sulfate- and sulfur-reducing bacteria, pp. 469–585. In Zehnder, A.J.B. (editor) *Biology of Anaerobic Organisms*. John Wiley, New York. (Dedication, #1, 2, 3, 4, 6, 7)
Widdel, F., and Bak, F. 1992. Gram-negative mesophilic sulfate-reducing bacteria., pp. 3353–3378. In: Balows, A., Trüper, H.G., Dworkin, M., Harder, W., and Schleifer, K.-H. (editors) The Prokaryotes: A handbook on the biology of bacteria: Ecophysiology, isolation, identification, applications. 2nd ed. Springer-Verlag, New York. (#6)
Widdel, F. and Hansen, T.A. 1991. The dissimilatory sulfate- and sulfur-reducing bacteria, pp. 583–624. In Balows, A., Trüper, H.G., Dworkin, M., Harder, W., and Schleifer, K.-H. (editors) *The Prokaryotes*, 2nd edition, vol. I. Springer-Verlag, New York. (#1)
Widdel, F. and Pfennig, N. 1977. A new anaerobic, sporing, acetate-oxidizing,

sulfate-reducing bacterium, *Desulfotomaculum (emend.) acetoxidans*. *Archives of Microbiology* 112:119–122. (#2, 8)

Widdel, F. and Pfennig, N. 1981a. Studies on dissimilatory sulfate-reducing bacteria that decompose fatty acids. I. Isolation of new sulfate-reducing bacteria enriched with acetate from saline environments. Description of *Desulfobacter postgateii* gen. nov., sp. nov. *Archives of Microbiology* 129:395–400. (#6, 8)

Widdel, F. and Pfennig, N. 1981b. Sporulation and further nutritional characteristics of *Desulfotomaculum acetoxidans*. *Archives of Microbiology* 129:401–402. (#8)

Widdel, F. and Pfennig, N. 1982. Studies on dissimilatory sulfate-reducing bacteria that decompose fatty acids. II. Incomplete oxidation of propionate by *Desulfobulbus propionicus* gen. nov., sp. nov. *Archives of Microbiology* 131:360–365. (#6, 8)

Widdel, F. and Pfennig, N. 1984. Dissimilatory sulfate- or sulfur-reducing bacteria, pp. 663–679. In Krieg, N.R. and Holt, J.G. (editors) *Bergey's Manual of Systematic Bacteriology*, Vol. 1. The Williams & Wilkins Co., Baltimore. (#6)

Wiebe, W.J. 1984. Some potentials for the use of microorganisms in ecological theory, pp. 17–21. In Klug, M.J. and C.A. Reddy (editors) *Current Perspectives in Microbial Ecology*. American Society for Microbiology, Washington, D.C. (#7)

Wieringa, K.T. 1940. The formation of acetic acid from carbon dioxide and hydrogen by anaerobic spore-forming bacteria. *Antonie van Leeuwenhoek* 6: 251–262. (#3)

Winfrey, M.R. and Ward, D.M. 1983. Substrates for sulfate reduction and methane production in intertidal sediments. *Applied and Environmental Microbiology* 45:193–199. (#7)

Winfrey, M.R. and Zeikus, J.G. 1977. Effect of sulfate on carbon and electron flow during microbial methanogenesis in freshwater sediments. *Applied and Environmental Microbiology* 33:275–281. (#7)

Wisotzkey, J.D., Jurtshuk, P. Jr., and Fox, G.E. 1990. PCR amplification of 16S rDNA from lyophilized cell cultures facilitates studies in molecular systematics. *Current Microbiology* 21:325–327. (#6)

Woese, C.R. 1987. Bacterial evolution. *Microbiological Reviews* 51:221–271. (#3, 6)

Woese, C.R. and Fox, G.E. 1977. Phylogenetic structure of the prokaryotic domain: the three primary kingdoms. *Proceedings of the National Academy of Sciences of the United States of America* 74:5088–5090. (#6)

Woese, C.R. and Olsen, G.J. 1986. Archaebacterial phylogeny: perspectives on the urkingdoms. *Systematic and Applied Microbiology* 7:161–177. (#6)

Woese, C.R., Stackebrandt, E., Macke, T.J., and Fox, G.E. 1985. A phylogenetic definition of the major eubacterial taxa. *Systematic and Applied Microbiology* 6:143–151. (#6)

Woese, C.R., Kander, O., and Wheelis, M.L. 1990. Towards a natural system of organisms: proposal for the domains Archaea, Bacteria, and Eucarya. *Proceedings of the National Academy of Sciences of thee United States of America* 87:4576–4579. (#6)

Wolin, M.J,, Wolin, E.A., and Jacobs, N.I. (1961) Cytochrome-producing anaerobic vibrio *Vibrio succinogenes* sp.n. *Journal of Bacteriology* 81:911–917. (#3)

Wood, P.M. 1978. A chemiosmotic model for sulfate respiration. *FEBS Biochemistry Letters* 95:12–18. (#3)

Wright, R.T. and Coffin, R.B. 1984. Ecological significance of biomass and activity measurements, pp. 485–494. In Klug, M.J. and Reddy, C.A. (editors) *Current Perspectives in Microbial Ecology*. American Society for Microbiology, Washington, D.C. (#7)

Yagi, T. 1958. Enzymatic oxidation of carbon monoxide. *Biochimica et Biophysica Acta* 30:194–195. (#3)

Yagi, T. 1970. Solubilization, purification and properties of particulate hydrogenase from *Desulfovibrio vulgaris*. *Journal of Biochemistry* (Tokyo) 68:649–657. (#3)

Yagi, T. 1979. Purification and properties of cytochrome *c*-553, an electron acceptor for formate dehydrogenase of *Desulfovibrio vulgaris*, Miyazaki. *Biochimica et Biophysica Acta* 548:96–105. (#2)

Yagi, T. and Ogata, M. 1990. Electron carrier proteins in *Desulfovibrio vulgaris* Miyazaki, pp. 237–248. In Belaich, J.P., Bruschi, M., and Garcia, J.L. (editors) *Microbiology and Biochemistry of Strict Anaerobes Involved in Interspecies Hydrogen Transfer*. Plenum Press, New York. (#5)

Yagi, T., Kimura, K., Daidoji, H., Sakai, F., Tamura, S., and Inikuchi, H. 1976. Properties of purified hydrogenase from the particulate fraction of *Desulfovibrio vulgaris* Miyazaki. *Journal of Biochemistry* 79:661–671. (#5)

Yagi, T., Kimura, K., and Inokuchi, H. 1985. Analysis of the active center of hydrogenase from *Desulfovibrio vulgaris* Miyazaki by magnetic measurements. *Journal of Biochemistry* 97:181–187. (#5)

Yang, D., Kaine, B.P., and C.R. Woese. 1985. The phylogeny of archaebacteria. *Systematic and Applied Microbiology* 6:251–256. (#6)

Yates, M.G. 1967. Stimulation of the phosphoroclastic system by nucleotide triphosphates. *Biochemical Journal* (London) 103:32c–34c. (#3)

Yavitt, J.B. and Lang, G.E. 1990. Methane production in contrasting wetland sites: response to organic-chemical components of peat and to sulfate reduction. *Geomicrobiology Journal* 8:27–46. (#7)

Zehnder, A.J.B. (editor) 1988 *Biology of Anaerobic Organisms*, 872 pp. John Wiley, New York. (#1)

Zeikus, J.G., Dawson, M.A., Thompson, T.E., Ingvorsen, K., and Hatchikian, E.C. 1983. Microbial ecology of microbial sulphidogenesis: isolation and characterization of *Thermodesulfobacterium commune* gen. nov. sp. nov. *Journal of General Microbiology* 129:1159–1169. (#6)

Zellner, G., Messner, P., Kneifel, H., and Winter, J. 1989a. *Desulfovibrio simplex* spec. nov., a new sulfate-reducing bacterium from a sour whey digester. *Archives of Microbiology* 152:329–334. (#2)

Zellner, G., Stackebrandt, E., Kneifel, H., Messner, P., Sleytr, U.B., De Macario, E.C., Zabel, H., Stetter, K.O., and Winter, J. 1989b. Isolation and characterization of a thermophilic, sulfate reducing archaebacterium, *Archaeoglobus fulgidus* strain Z. *Systematic and Applied Microbiology* 11:151–160. (#6)

Zellner, G., Kneifel, H., and Winter, J. 1990. Oxidation of benzaldehydes to benzoic acid derivatives by three *Desulfovibrio* strains. *Applied and Environmental Microbiology* 56:2228–2233. (#2, 6, 8)

Zinoni, F., Birkman, A., Stadtman, T.C., and Bock, A. 1986. Nucleotide sequence and expression of the selenocysteine containing polypeptide of formate dehydrogenase (formate-hydrogen-lyase linked) from *Escherichia coli*. *Proceedings of the National Academy of Sciences of the United States of America* 83:4650–4654. (#5)

Ziomek, E. and Williams, R.E. 1989. Modification of lignins by growing cells of the sulfate-reducing anaerobe *Desulfovibrio desulfuricans*. *Applied and Environmental Microbiology* 55:2262–226*x*. (#8)

ZoBell, C.E. 1958. Ecology of sulfate-reducing bacteria. *Producers Monthly* 22:12–29. (#8)

Zuckerkandl, E. and Pauling, L. 1965. Molecules as documents of evolutionary history. *Journal of Theoretical Biology* 8:357–366. (#6)

Index

Brock/Springer Series in Contemporary Bioscience

(Continued from page ii)

John H. Andrews
COMPARATIVE ECOLOGY OF MICROORGANISMS AND MACROORGANISMS

Ryszard J. Chróst (Editor)
MICROBIAL ENZYMES IN AQUATIC ENVIRONMENTS

Alan G. Williams and Geoffrey S. Coleman
THE RUMEN PROTOZOA

John H. Andrews and Susan S. Hirano (Editors)
MICROBIAL ECOLOGY OF LEAVES

Satoshi Ōmura (Editor)
THE SEARCH FOR BIOACTIVE COMPOUNDS FROM MICROORGANISMS

Madeleine Sebald (Editor)
GENETICS AND MOLECULAR BIOLOGY OF ANAEROBIC BACTERIA

J.M. Odom and Rivers Singleton, Jr.
THE SULFATE-REDUCING BACTERIA: CONTEMPORARY PERSPECTIVES

MIX
Papier aus verantwortungsvollen Quellen
Paper from responsible sources
FSC® C105338

If you have any concerns about our products,
you can contact us on
ProductSafety@springernature.com

In case Publisher is established outside the EU,
the EU authorized representative is:
Springer Nature Customer Service Center GmbH
Europaplatz 3, 69115 Heidelberg, Germany

Printed by Libri Plureos GmbH
in Hamburg, Germany